ORDER IN LIVING ORGANISMS

Order in Living Organisms

A Systems Analysis of Evolution

by

RUPERT RIEDL
Chairman of the Zoological Institute
University of Vienna, Austria

translated by

R. P. S. JEFFERIES
Department of Palaeontology
British Museum (Natural History)

A Wiley-Interscience Publication

JOHN WILEY & SONS

Chichester · New York · Brisbane · Toronto

The original edition was published by Verlag Paul Parey, Hamburg and
Berlin, under the title *Die Ordnung des Lebendigen* © 1975 Verlag Paul
Parey, Hamburg and Berlin, who retain copyright of all illustrations in
this edition.
Copyright © 1978 by John Wiley & Sons Ltd.

Library of Congress Cataloging in Publication Data:

Riedl, Rupert.
  Order in living organisms.

  Translation of Die Ordnung des Lebendigen. Includes
  bibliographical references and indexes. 1. Evolution.
  2. Life (Biology). I. Title.

QH366.2.R5313   575   77-28245
ISBN 0 471 99635 1

Printed in Great Britain by The Pitman Press, Bath.

CONTENTS

PREFACE TO THE GERMAN EDITION

Man has long sought to understand his own origin. The *Book of Genesis* foresaw an answer which has gradually been perfected by the study of evolution. Indeed evolutionary theory has long been one of the most inclusive constructions of human thought. Moreover, since the theory of descent was propounded, each new structure discovered helps us to reconstruct the course of evolution.

The mechanism of the process, however, still causes discussion. Without doubt the mechanism of selection (Darwinism), of mutation (Neodarwinism) and of population dynamics, which together make up the synthetic theory of evolution, have a fundamental explanatory value. But it is very doubtful whether these hitherto proven mechanisms will by themselves explain the regularities of evolution, or its gross course (i.e. transpecific evolution), or the fundamental order in living organisms which these imply. This order must surely exist for thousands of books have been written about it.

The particular concept of phylogeny implied by the synthetic theory is based on the opportunistic and short-sighted selection of occasional chance mistakes in the transmission of building instructions. It seems to fail when the nascent regularity takes on a powerful, everlasting form. Our ignorance of the ordering mechanism represents a gap in the concept. Darwin himself was conscious of this, but since his time the endeavour to fill this gap has led not to solutions but to battle fronts. Formerly it was Neodarwinism against Neolamarckism, Weismannism against vitalism; now it is superempiricism and reductionism against systems theory and holism. Indeed, many assert that only the experimental sciences can answer questions about causes. They feel justified in excluding the study of form from science, although it is the basis of morphology, systematics and broad phylogeny.

I shall try to elucidate the nature of this ordering mechanism. It is the causal connection which produces the regularities of macro-evolution as an inevitable result, as well as the ordered, predictable diversity of organic forms. Indeed it is the reason why organic nature is not an indescribable confusion, but a describable order which corresponds to our thought patterns and to their orderly results (i.e. to our civilization). Given such a universal effect there is a danger of confusing real with subjective order. I must, therefore, start by explaining the pattern of real order. The mechanism will then become almost self-evident. The key to it is the insight that the innumerable riddles which still arise from the directionality of evolution and the predictability of form and of development are all consequences of the same general ordering principle. They represent instances of a law. But so as not to be deceived about the reality and extent of this orderliness we need to be able to measure regularity objectively. This is where we shall begin.

xii

My theory starts from a more widely applicable causal concept. That is to say, it begins with the insight that the effects of the evolutionary mechanism react on what we call its causes. I shall show that the prospects of success for a mutational change are different for the different levels of organization. Consequently the prospect of success for a change in a feature (i.e. in a phene or event) governs that for a change in a gene (i.e. in a genetic 'decision') just as the prospect of success for a change in a gene governs that for a feature. Decisions are connected with the resulting events by way of a feedback mechanism, so as to form a total system of effects. Essentially this is a selective mechanism which extracts the increasing improbability of living organization from the laws of chance probability. The possible modes of interdependence produce four known switching patterns of genetic decisions and also four corresponding morphological ordering patterns for phenomena. I shall call these the standard part, hierarchy, interdependence, and traditive inheritance. From the dynamic viewpoint all this signifies 'self-design'. This is a self-steering quality in evolution which explains many unsolved problems such as trend and orthogenesis in phylogeny (to mention but two), homoeostasis and the law of recapitulation in ontogeny, the operon and regulator systems in genetics, homology and the morphotype in morphology and the reality of systematic groupings and of natural classification in systematics.

The consequences of the theory are paths of evolution which regulate, govern, and design themselves. We ourselves are neither the product of blind accident, nor in some way preordained; we are neither meaningless nor with an *a priori* meaning. Instead we are, so to speak, the product of a strategy of nascent systems of law and order — a strategy opposed to entropy and decay. Our meaning, so far as we have one, we have earned ourselves. We are neither stuck in a cul-de-sac nor have we found the road to perfection. Rather this road can be encountered from time to time so long as the profits extracted from chance can be paid back to the laws of probability. The mechanisms which work to our benefit have indeed become canalized, but so has our prospect of freeing ourselves from deep-rooted evils. Freedom is not a question of throwing our burden away, but the gradual perfecting of mass laws into individual laws. Our environment is not the hunting ground of opportunism, nor the fountain of youth of the reformists, but a reflection and caricature of its own creatures. The prospects of our road to humanity lie in the creation of a humane environment.

I wish to explain my theory because many people are filled with disquiet or doubt. It is hard to say whether it suits the spirit of the times, for the old controversies no longer attract much attention. Nevertheless I hope that it may do, for I can see the liberating and reconciling qualities which an understanding of the orderliness of genesis will give. Otherwise I should never have started this work. Indeed much in our present situation does not favour acceptance of the theory: verification is a thing of the future; the molecular genetics of higher organisms is only just beginning; the measurement of living order is full of contradictions; the problem of form is unsolved; its methodology is in decay; and the literature is almost unsurveyable for a single individual. On the other hand many things do favour its acceptance: verification seems to be approaching; the problem is being narrowed down so that its solution has been anticipated in almost all its parts by scientists of both camps; and we now have too much insight to see ourselves as the product of pure accident, or even as legitimate despoilers or manipulators of our environment.

I gained the courage to write this text partly from information theory and thermodynamics. They teach us that we have overlooked our own orderliness. I gained it also from those few remaining libraries in which superficial thinking has locked morphology up, despite the fact that comparative anatomy and high-level systematics are based upon it. These two subjects contain one of the most profound of human perceptions, and perhaps the most liberating — the knowledge of man's own origin.

I am deeply grateful to my wife who made possible a very difficult task in these working years on either side of the Atlantic. I would also like to acknowledge the understanding help proferred by the publishers Paul Parey. My thanks also go out to my coworker Daniela Auer, who drafted the illustrations, to An Painter and Hermi Troglauer who always took care of my text (on either side of the Ocean), and to my friend Harald Rohracher who had to witness the whole operation.

Vienna and North Carolina
December 1973   Rupert Riedl

PREFACE TO THE ENGLISH EDITION

This book seeks to solve a whole group of evolutionary problems that challenge contemporary biology. The reaction to its German version has confirmed Ernst Mayr's prediction that it will be an eye-opener for many, while many others will be taken aback. And for good reasons the same reaction may await this English translation.

The *material* presented here has partly escaped the attention of the English-speaking scientific community. There are probably two main reasons for this. Firstly, in central Europe it has been believed that structural patterns could not be explained entirely in terms of immediate function; and the search for a deeper explanation, which began with Goethe and continued throughout the nineteenth century, became confused with German idealistic philosophy — a fact which made it both difficult and suspect for English-speaking scientists. Secondly, the key literature on the epistemological background of morphology, such as Remane's *Die Grundlagen des natürlichen Systems der vergleichenden Anatomie und der Phylogenetik*, has never been translated into English.

As a result the word 'morphology' became disreputable in English. The study of structure, so as to show itself as a respectable science of good family, avoided using the term. In central Europe, on the other hand, morphology saw itself as the epistemological base on which comparative anatomy, systematics, and taxonomy could be built.

The *problem* of macro- or transspecific evolution was the main point of contention in these European battlefields. The English-speaking world, on the other hand, concentrated on the micro- or intraspecific phenomena, probably because experimentation was thought to be superior to description. Also progress in 'New Systematics' caused its background in 'Old Systematics' to be forgotten, and 'Numerical Taxonomy' erroneously accused old taxonomy of circular reasoning.

As a consequence, most of the unexplained phenomena in macro-evolution were first minimized, then swept under the carpet and finally forgotten. This happened although such phenomena are incompatible with, and even contradict, the current Neodarwinian explanation of evolutionary mechanisms. Instead of feeling excited when contradictions and traps appeared in our basic theory, people have tended to be disappointed. They have treated these difficulties as gothic or baroque ornaments, unfitted to the plain architecture of modern biology.

The *solution* which this book offers includes the previously unexplained cause for the existence of homologues, the reason for Haeckel's law and, the reason for the existence of taxa. The book explains why the 'natural system' is in hierarchic order, why the system of gene-interactions shows the same pattern and how the genome acquires this rational-seeming organization. The solutions to these problems require nothing more than simple systems theory. In addition to the currently accepted transmission of information from genome to phenome, the solution given here presents evidence and rationale for feedback loops from phenome to genome. This feedback information causes to develop, by trial and error, those gene interactions which improve their own adaptive speed or success. The book proposed an additional type of selection which precedes Darwinian selection, just as industrial testing of products cautiously precedes the selection which the market is expected to apply.

Consequently the material on which this theory is based may seem more of a challenge than the solution does. This is because the mechanism visualized is current in present-day scientific thought. The theory explains the cause of living order by the same systems conditions as the order in our industrialized civilization. And this is now understood to be a feedback of information, looping between individuals and society as well as between market and industry. But the theory can do this only by taking into account a whole group of puzzling phenomena, both in morphology and transspecific evolution, which are nowadays either obscured or put aside. This is its challenge.

The *translation* was done with much care and scholarly experience by R. P. S. Jefferies of the British Museum (Natural History) in London. The text presented great difficulties for it had to be fairly compact to squeeze all the necessary evidence into a single volume. Moreover, it involved many new viewpoints and perspectives, all of them approached by systems analysis, in a way that was fresh to this branch of evolutionary study. I am very much obliged for his cooperation. I also wish heartily to thank Mrs R. Smolker, of the publishers Paul Parey, and Dr Janet Boullin and the other personnel of the John Wiley Co., who accepted the challenge of presenting a heterodox way of thought to the English-speaking scientific community,.

Vienna   Rupert Riedl
September 1978

INTRODUCTION

I must warn the reader to expect some hard labour, caused entirely by my lack of skill. Not to warn him would be a discourtesy and I can avoid that at least. Before starting into the unknown I shall describe certain handholds that I shall use repeatedly. They will be understood more readily if the reader and I remember that we are both systems of a large number of molecules sufficiently highly organized to consider even molecular probabilities.

Even within the scope of this investigation it is necessary to establish what we can actually know and how we can know it. I shall therefore sketch out those ideas which in my view contain the key to our problem, and indeed the mechanism for solving it.

a. Accident and necessity

Experience has shown that everything which we observe in this world can be ascribed either to accident or necessity. This distinction depends on the possibility of prediction. Predictability (i.e. the possibility of explaining in terms of necessity) increases with increasing insight into a mechanism, and it increases at the expense of uncertainty (i.e. explanation in terms of accident). But necessity — the number of necessary consequences, can also increase objectively as when we improve a machine, or when an organism acquires an additional regular feature. Like the subjective increase in necessity, this happens because additional features lose their accidental distribution, becoming subject to predictable arrangement. Thus bits of metal from a junk box can be used to make the seconds hand of a watch, the biological molecules of a tissue can form an eye, or the behavioural features of a group of people can be used in extending the rules of a society.

The subjective limit between accident and necessity is determined by the possibilities of our perceptive apparatus. It is situated where the predictable passes into the unpredictable. In daily practice in the macroscopic world this is the point where a phenomenon escapes from investigation. Thus we ascribe a heads-or-tails decision to pure accident although we may feel certain that the thrown coin follows exclusively physical laws.

The aim of science is to widen the subjective limits of predictability and to discover the objective ones. Up to now, however, many scientists will only acknowledge one single objective limit. This is in the atomic realm (in the micro-world of physics) where the instant when an atomic change will occur seems to be unpredictable in principle.

b. Decision and event

Our perceptive apparatus also helps us in a second respect by distinguishing between decisions and resultant events. It does this so consistently that we seem to be distinguishing between cause and effect.

The distinction between decisions and resultant events is purely of practical significance. This is true even in those simple cases where we are convinced that the event is only the simplified expression of a great number of decisions arranged in particular patterns. Examples are the events that follow the pressing of a button on a desk calculator, or after the sowing of grass seed or sending a telegram. Indeed the desk calculator, somewhat like the seed, is built with the intention of including all these decisions. These can be taken for granted by its user, together with the logic of the wiring.

So far as we know, all events are made up, in the last analysis, from decisions of molecular type. This makes it necessary in practice to distinguish between decisions and events. That is to say, insight into the mode of action of the decisions is either entirely prevented, or only attainable by complicated methodological analysis. For although the best sense perceptions penetrate almost to the atomic realm, we can see neither a quantum of light nor the oscillation of a molecule. Moreover, even the smallest sense or nerve cell is constituted by billions (10^{12}) of atomic decisions. It can therefore be understood why we do not describe the observable world in terms of decisions but in terms of the complexes of decisions that we call events — a crystal and its growth, or an organism and its embryo.

c. Mutation and selection

One of the strangest features of evolution is the fact that heritable change occurs only in the realm of accident and the micro-world while selection happens in the realm of necessity and the macro-world. Between the two there yawns a gap in degree of complexity which can reach trillions (10^{18}) or quadrillions (10^{24}) in order of magnitude.

This must be the reason for imagining a unidirectional mechanism for evolution — a theory which involves the necessary selection of changed events that were caused by accidentally changed decisions. If decision and event are interpreted as cause and effect, then the possibility of an influence exerted by the events in the decisions would seem to be excluded by definition. This crucial element in Neodarwinism is decreed by present-day dogma as it was by the Weismann doctrine of early genetics. In this dogma all evolutionary change would be the result of pure accident.

The requirements of selection would in this case be defined purely by the environment. However, the clash between the newly formed events and the newly exploitable environment is equally unpredictable. Consequently all the products of evolution would themselves be the result of pure accident.

These two consequences — the accidental nature of evolutionary change and of evolution's products — would be inescapable if the issuing of evolutionary commands could be seen as a series of decisions of equal value — rather like the playing instructions of a giant puzzle written down in Morse code on a long strip of paper. In actual fact, however, we are not dealing with a mere enumeration of proteins but with a stratified system of decisions and events which stretches over a whole trillions-wide range of complexity.

d. The system of phenomena

According to the events they produce, decisions range from the formation of the simplest biological molecules to that of whole organisms. Among all these many levels of complexity, including even groups of organs and body regions, not one is known which cannot be controlled by a single decision of lower or higher level. And experience shows that accidental changes can likewise occur with equal probability at every level of these decisions.

Decisions are therefore dependent on each other. Among the types of interdependence which are built into the system all the geometrically possible forms seem to be realized. These can be sorted into simultaneous or successive, unilateral or reciprocal interdependences between equal or unequal phenomena — applying the word phenomenon both to the decisions and to the events produced by them. This implies a causal interconnection of decisions with one another which extends beyond the unidirectional concept of Neodarwinism.

This causal connection expresses itself structurally in a definite number of patterns of interdependence. In the molecular realm these appear as four switching or wiring patterns. In the morphological realm of events they are the predictable patterns of morphological order. These patterns of interdependence are the reason why the world of organisms is describable.

All in all, a reciprocal interdependence prevails of decisions with one another. This interdependence of decisions happens indirectly by the selection of the interdependent events that result from the decisions. Decisions are selected by events which are themselves selected. There exists a 'strategy of the accidental'.

e. Cause and effect

The mechanism of evolution is thus multidirectional. Its causality is not unidirectional but involves feedback. Effects influence their causes. The following playing instructions will give an idea of the simplest mechanical model.

Every player (i.e. genotype) follows the rule (identical molecular rules) that a coin shall first be thrown four times (four accidental gene decisions, each with a chance of success of 0.5). The object of the game is to reach a winning pattern which is unknown to the player and determined by the bank (external conditions). Let us suppose that the bank will honour the pattern of: (1) heads, (2) heads, (3) heads, and (4) tails (as giving a selective advantage in the environment). Each player is allowed, however, to change his strategy by random experiment, say after every 20 turns (the chance of ranking genes by mutation) by omitting a decision through retaining or 'remembering' an earlier one instead. In practice, the chance of success will double with every accidentally chosen correct retention, as for example with (1) for both (1) and (2), it will rise from 0.5^4 to 0.5^3; with every incorrect retention, however, it will fall to nil, as for example (3) for (3) and (4). If continual losers are uniformly excluded from the game and the winners, together with their strategy, multiply by identical replication, then the strategy of accident will soon have replicated the secret winning pattern determined by the bank.

If decisions thus come together to form systems, it will happen that, under conditions of selection, the pattern of decisions will copy the pattern of events demanded by the selective conditions. This is especially true when the functional dependence of events (i.e. what I later call the 'burden' of features) no longer permits events to change separate from each other. Events work backwards on their producers. Cause and effect acquire new dimensions. The system of effects and reverse effects reaches colossal dimensions since this rule of play in living organisms operates with mutation rates of one in ten thousand, and with very high replication rates and selection rates over millions of individual instances. The required harmony of effects is imposed by force upon the accident of decisions.

Accident, of course, cannot be swindled. The advantages reaped only hold for the currently applicable winning pattern. But since that can change, the advantage gained must be paid for by a narrowing of potentialities. The phenomena, i.e. the patterns of decisions as well as those of events, which means the possible patterns of cause and effect, become canalized. The result is an eternally constant order — the order of living organisms.

f. Material and methods

I have just asserted that accident and necessity, acting on decisions and events, combine by mutation and selection to form self-ordering systems of reciprocal relationships. If this is true, two consequences would be expected which are of interest when discussing the living world.

From the statical point of view a degree of order of quite unimaginable dimensions must completely penetrate all levels of the living world. Regularity must reign and predictions be possible where formerly we only reckoned on accident. This will be true whether we speak of the structure of the molecular code or of the form of whole phyla of organisms, of the transmission of data in the epigenetic system or that in our own thinking apparatus. Every gap in the orderliness will contradict the theory. That is why this book has so many pages.

From the dynamic viewpoint, the concept of evolution moves away from the meaninglessness of blind accident. It passes into the levels of necessity, of self-planning, of self-target-setting (for which there is no word), of fixated hopes and evils. Its courses and prospects become predictable. To assert this, however, is a heavy responsibility. It is encouraging if true but a deception if false. This is why I shall proceed with all caution.

I must mention one thing further before beginning in earnest. I shall try, so far as at all possible, to avoid technical jargon — the 'Double Dutch' of the specialisms. I do this so as to be accessible to the individual disciplines of biology and also to be understood by other scientists and educated laymen. So far as possible I shall therefore try to explain the important ideas of each subject in such a way as neither to bore the specialist nor overload the non-specialist. If either should happen, however, the reader may cheerfully skip the passage in question and continue at the points which interest him. For I shall adhere to the structure of the argument. Chapters I and II describe general and biological order; Chapter III describes molecular order; and Chapters IV to VII deal with morphological order, each with sections: (A) Definitions, (B) Evidence, and (C) Mechanism. Chapter VIII summarizes the whole with sections: (A) Mechanism and (B) Consequences.

CHAPTER I

WHAT IS ORDER?

Order is an expression of conformity to law. The *Brockhaus Encyclopaedia* defines it as: 'A meaningful connection between independent quantities according to internal laws.' I shall later define it as: 'Law times the number of instances where it applies.'

Order is a universal concept. Clearly there is no field of thought that can do without it. The specialisms that make use of it extend from art to thermodynamics and from religious ethics to traffic management. Our relationship to order is likewise universal. It is the basis for a baby's earliest concept formation, on the one hand, and for scientific method on the other. Ethics, metaphysics, and epistemology all equally demand it:

'A world without order would have no meaning.'

'The order of the world, if it did not exist, would have to be demanded.'

'A world without order would be neither recognizable nor conceivable.'

A. THREE APPROACHES TO THE PROBLEMS OF ORDER

By contrast with this unanimous expectation the forms of the expected order make a far more discordant picture. Indeed it seems to be a general characteristic of order that there will always be arguments about its forms, its limits, and even its existence.

1. For and against order

Under such circumstances a search for the cause of order would seem completely hopeless. This is particularly true if, as I propose, scientific method is adhered to. In point of fact the problem of order, even until very recently, has been studied in philosophy, law, religious ethics, and the social sciences only by use of historical and humanistic methods.

This ought to be warning enough for a scientist, except that a quantitative concept of order has arisen from the theory of probability and chance. This is the concept of negative entropy so ingeniously proposed by Schrödinger.[1] By inverting entropy, which is a measure for chaos, for the freedom of accident or for unpredictability, he showed us the right direction to look.

Nevertheless to help the reader I shall briefly describe the obstacles which block the scientific path to the recognition of order.

a. *Order as presupposition*

We cannot think without order. The first obstacle therefore consists in the difficulty of distinguishing between real and subjective order. Indeed we must admit that we tend

1

to assume the reign of a pre-existing order whenever we lack insight into the cause of any regular phenomenon. In such cases order is a substitute for insight.

Order seems intuitively to be an inseparable constituent of the world. Order is its regulated predictable component, directed reassuringly along particular routes, in a sea of uncertainty and lurking confusion. This necessity of thought is not merely prescientific; it is probably as old as thought itself. It began in the prehistory of early cultures.

Man has clearly always been convinced of the reign of order. For as prehistory gradually discloses 'primitive' world-views the imagined gods and creators become gradually more palpable. The dictates of these beings were intended to explain the otherwise inexplicable part of an obviously eternal order. Even today we only need to ask ourselves the questions that lie beyond scientific method, such as the purpose of creation or the goal of evolution, or the aim of birth and death, to find once again that only a belief in order can offer hope. It does not matter whether we call this belief by its true name or use some philosophical or nature-philosophical periphrasis[2] like entelechy or vitalism[3]. *Deus lex mundi.* From this we learn that the most valuable and humane of these conceptions of order are the basis of our modern culture. We also realize that the sought-for cause, which this culture is always striving after, has repeatedly been shown to exist, even though in unexpected guise (Fig. 1).

A biologist will be excused from pursuing this subject further, for experts have discussed it exhaustively. It should only be emphasized how deeply anchored the concept of the reign of order seems to be.

Fig.1 The supposition of order beyond the perceivable world. Beyond the celestial sphere the artist expects to find that standard parts, symmetries, and even orderly mechanisms are once again dominant. After a German woodcut of the fifteenth or sixteenth century. From Zinner (1931).

b. *Order and reality*

Ever since some degree of understanding was gained in the nineteenth century of what matter, life, and evolution were about, a belief in the reality of such order has become uncertain. In my opinion there are three reasons for this doubt, which underlies the modern extreme opinion that expects no order.

In the first place it is often taken as proven that the concept of order will disappear either when the particular orderly phenomenon is found not really to exist and the hypothesis of order is falsified, or else, as is commoner, when the hypothesis is confirmed. In the latter case the phenomena can then more appropriately be called 'the instance of a law'. Order would be a transitional condition of our insight, prior to the recognition of causes.

In the second place there are the two universal evolutionary theories of science, those of physics and biology. These have consequences which, as seems at present, in no way support a concept of order. In physics there is the second law of thermodynamics, or law of entropy, which states that every phenomenon in this Universe leads finally to an increase in disorder. (However, this tells us nothing about how order arose, though obviously there must have been enough of it originally for it still to be decreasing now.) In biology there is the synthetic theory, synthesizing Darwinism and genetics. This asserts that the evolution of organisms can be explained by environmental selection of random mistakes which occur now and then during the replication of genetically stipulated decisions. (Up till now none of the many theories requiring the supplementary action of an orderly principle has been verified.) In a broad context it would therefore seem unnecessary to presuppose order at all.

In the third place, the supposed orderly pattern of the outer world agrees strikingly with our own thought patterns. How does this conjunction arise? Is it not plausible to suppose that what we take for real order is, in truth, only the projection of the fact that our thoughts require order. It would be an artefact, so to speak, of the limitations of our thinking apparatus. For the only alternative is that our thought patterns are the selection product of the pattern of reality, which seems far-fetched.

In this introduction I cannot deal further with the concept of order. I shall return to all three questions as appropriate. I shall then show, first, that instance, law, and order are all connected together; second, that our theories of evolution seem to be incomplete; and third, that the hypothesis that thought patterns are a selection product of the pattern of reality is more probable than the alternative.

2. Entropy, negentropy, order and chaos

The question of what order is has been given new impetus by the researches of physicists and I now wish to explain why they feel hopeful of finding an answer. Obviously I must leave out the purely physical arguments for their optimism. I only mention that these arguments are independent of anything mentioned above, and arise from the dimensions of expectation and uncertainty.

Physics has long taught us that all isolated systems change towards greater equilibrium, which is the same as loss of energy. Perpetual motion is impossible. Thermodynamics also proves that this decrease corresponds to an increase in atomic disorder (**D**) as given by the Boltzmann constant ($k = 1.38 \times 10^{-16}$ erg/°C). This decrease in energy, or degree of increase in atomic disorder, is called entropy (S):

$$S = k \log \mathbf{D} \tag{1}$$

This measure for disorder was used by Schrödinger (1944) as the basis for a numerical expression of atomic order or negentropy (see also Section I B3*a*). It is agreed today that negentropy (N) corresponds to a function of the reciprocal of atomic disorder ($1/\mathbf{D}$) :

$$N = k \log 1/\mathbf{D} \tag{2}$$

In this way the physicists have set up a measure for order. Moreover, they developed it especially for biologists since they worked it out in connection with the phenomenon of life and with Schrödinger's question: 'What is life?'. For order is at its most obvious in living phenomena and thus a belief in the reality of order begins to return. 'Life seems to be orderly and lawful behaviour of matter, not based exclusively on its tendency to go over from order to disorder. . . .'[4] On the contrary: 'It springs to the eye that the tendency of living organisms is to organize their surroundings, that is to produce order where formerly there was disorder. Life then appears in some way to oppose the otherwise universal drive to disorder. Does it mean that living organisms do or may violate the second law of thermodynamics?'[5] No, that is not the case. The biosphere, including its input and output, obeys the law of entropy, but its open systems, the organisms, are able to evade it. 'The entire process is exentropic owing to the flow of energy from the sun to outer space, but the local processes may lead to order such as a rotifer, a sonnet, or the smile on the face of Mona Lisa'.[6] I share the optimism of the physicists.

The details of how organisms evade the law of entropy we shall also have to learn from the physicists. However, I shall myself try to elucidate what mechanisms lead to the orderly patterns of living structures, and what those patterns are. In the first instance, however, it is important that negentropy, as a measure of material order, can prove the reality of the latter and provide a starting point for its study.

3. Accident and necessity, certainty and uncertainty

Before I continue with the numerical concept of material order it will be useful to clarify the concepts that are closely linked with it.

All events which can be studied by scientific method can be regarded as either accidental or necessary. This world of accident and necessity seems to contain no third alternative. Of course this distinction does not mean much at first. It is clear, however, that with respect to events we can take up one of two positions. In one group of instances we possess a particular expectation and expect that it will be confirmed in repetitions of the event. In the other group of instances we have no such expectation and have to take notice of the unexpected, with varying degrees of uncertainty. However, the border region between accident and necessity is large. There is scarcely an experience in which some surprise is not mixed with fulfilled expectation, or conversely. This will concern us later. But what are accident and necessity?

We begin to be convinced of the necessity of an occurrence if our expectation is fulfilled so reliably that the last traces of uncertainty disappear — as we can say for simplicity. In such cases we tend to assume a cause for such regularity in the reign of order or of conformity to law, as we usually say. We do this because of our experience that the same thing does not repeat itself for no reason. Whether the supposed cause will be confirmed, or whether it will eventually give place to another, is asked only later — think for example of our changing conceptions of the cause of the sun's movement.

The word accidental, on the other hand, is used of those events for which the methods available do not allow us to form expectations. This may be because the event simply is not repeated, as with historical events, or else because the time when it will be repeated is not determinable; again this indeterminacy may be because details — such as relate to the causes of motions for example — either are left out of consideration (as with the heads-tails decision), are too complicated to be followed (as with the accidental meeting of friends), or for fundamental reasons simply escape examination (as with the breakdown of atoms). In this way, each part of the event corresponds to pure accident, concerning which no appropriate prediction can be made. The greater the play given to accident the greater is the uncertainty.

B. ORDER AS PROBABILITY

Where are we then? What do 'certain' and 'uncertain' mean? In answering these questions we must make the strange attempt to judge certainty starting from uncertainty. In connection with all our further questions we need to know what we think about 'conformity to law' and 'law content'. I must therefore ask the reader to follow me into a region lying between epistemology and probability theory which is assuredly as simple as it looks difficult. The key to a solution is indeed very straightforward. It lies in the double nature of what we call probability — in the strange complementary interpretation of 'information' in present-day science.

1. Indeterminacy and determinacy

Uncertainty and predictability have an inverse relationship. We wish to measure degree of surprise or accident, so I now want to define the degree of predictability or of necessity. I shall begin with what is known.

a. *Information content*

Information theory has developed a measure for specifying the degree of surprise. The so-called information content (I) of an accidental event corresponds to the inverse of its probability (P). I increases with the number of unpredictable accidental possibilities and thus with the degree of uncertainty. In the simplest case, that of tossing a coin, the probability (P) of the next event being 'tails' is (x), then $P_x = \frac{1}{2}$. The reciprocal $1/P_x$ is therefore a measure of uncertainty and equals 2.

In using the idea of 'information'[7] in this way we must remember that it differs most surprisingly from the colloquial idea of information, including that of genetic information. In its colloquial use factual situations are the contents of a piece of information with qualities like 'important', 'correct', or 'understandable'. But in its technical use the information content depends exclusively on the degree of probability of the occurrence. The two ideas only agree assuming an uninformed receiver.[8]

The unit of measurement for I is usually the *bit* — the digital yes-no decision as used in electronics. The binary choice between 2, 4, 8, and 16 events requires respectively 1, 2, 3, and 4 *bits*. This implies a relationship involving a logarithm to the base 2. Thus the information content (I) of an event (x) in *bits* is the logarithm to the base 2 of the reciprocal of its probability:

$$I_x = \log_2 1/P_x \tag{3}$$

Suppose, for example, that out of a range of 32 possible and equally probable events (e.g. a roulette wheel with 32 positions), there is a series of six individual events such as

15, 2, 12, 9, 12, 20. Then, for each event $P = 1/32$ and $I = 5$ *bits*. One could also say that accident must make five equal decisions in order to select one individual event out of 32. The whole series of six events would contain $I = 6 \times 5 = 30$ *bits*.

Now, as a rare case, 'chance could decree' that in the six throws of the roulette the series 1, 2, 3, 4, 5, 6 appeared — an apparently meaningful series. If this really was accidental, then the series would contain the same 30 *bits* of information.

If however the 'meaningful series' was produced intentionally, as for example by the mechanism of the machinery, then, as soon as we were convinced of this, the information content would disappear entirely. For as soon as the occurrence of an event can be predicted with certainty, then all surprise disappears, as also is evident from equation 3. The probability (P) is then precisely 1, its reciprocal is 1 and $\log_2 1 = 0$.

This conclusion will be important for our definition of determinacy content and must therefore be examined with care. This is especially true because an opposite idea of information has been developed in the natural sciences. I wish to avoid tangling the threads together here, but will come back to this matter after clarifying the determinacy concept (Sections I B3b and c).

b. Predictability

Predictability, therefore, decides whether the events 1, 2, 3, 4, 5, 6 (out of 32 possibilities for each) should count as 30 *bits* of information or as none. The ability to predict requires five preconditions. Two of these must exist in the observer, and three in the systems that produce the events. To examine these preconditions we need an objective standpoint. I shall therefore consider the observer as the receiver and the producer of the events as the source.

In this connection I shall follow the objective procedure of communication technology and information theory in supposing that the receiver initially knows nothing about the structure of the source and will learn no more than what he deduces from its transmissions. Indeed I shall not go beyond these assumptions at this point, but shall merely seek to define what must be taken as given from the epistemological viewpoint. The preconditions are as follows:

1. The source must repeat its transmissions, for only in this way can the receiver recognize those regularities from which law or meaning can be deduced. For the repetition of an event in the same fashion will only happen by accident over a long period of time.

Thus the transmission 'S & 5' does not allow a meaning to be recognized any more than '& ¼ ≠ S & 5' etc. The transmission 'S & 5 S & 5 S & 5' on the other hand does show a meaning i.e. the repetition of the group of events S & 5. This phenomenon of repetition is of such importance that we must examine it later in detail.

2. The receiver must have a memory.

3. The receiver must be able to compare, for otherwise he could neither recognize the repetition of an event in a series of events, nor know the number of events which might possibly be transmitted.

4. The programmes of a large number of sources must be so organized that the receiver can learn to distinguish between individual events, on the one hand, and series of events, on the other. Thus if the individual events of the series '1 2 3 4 5' always occurred as 1 2 3 4 5 1 2 3 4 5 etc. then the receiver would no more experience a regularity than if he received S S S S. He needs to become acquainted with the events 1 to 5 predominantly in other combinations so as to become convinced of their individual existence. For only after many comparisons does it become improbable that a transmission (i.e. a series of natural events) is differentiated in a repeated regular pattern by mere accident.

5. The programme of a source must remain within the same limits long enough for the receiver to appreciate these limits (as for example the range of a set of symbols). For only a large number of comparisons make it improbable that the limits in a series of natural events have remained the same by accident alone.

All this shows that the preconditions in the receiver are also the minimal preconditions for thought. And I shall anticipate my argument by saying that the minimal preconditions in the source represent those of order. Naturally they are fulfilled wherever we recognize laws.

c. Measuring the improbability of accident

I now return to the important question: What criteria show that the occurrence of an event is dominated by accident rather than necessity?

This can be illustrated by an example. In a coin-tossing contest how often must my opponent throw tails (on which he has bet) before I doubt the reign of pure accident? It does not require many throws. The probability of his first success is still ½, but with his second, third, fifth, tenth or hundredth success it falls to 1/4, 1/8, 1/32, 1/1024, and $1/1.3 \times 10^{30}$. My faith in accident will fall still faster if, out of 32 cards in a game of skat, he exclusively draws the Jack. For the probabilities at the first, second and tenth occasion are 1/32, 1/1024 and $1/1.1 \times 10^{15}$.

This decrease of probability can also be expressed as an increase in improbability (the reciprocals of the just-quoted reciprocals). Thus we can say that the improbability of drawing the Jack ten times in succession is 1.1×10^{15} or of throwing tails one hundred times in succession is 1.3×10^{30}. However, if one explanatory hypothesis becomes impossible — namely the supposition that we are dealing with accidental events — then we are forced to seek another.

Naturally, in any given condition of investigation there will always remain a minute degree of probability that the result is due to accident. The degree of improbability at which we become convinced of intent or trickery is a matter of taste or of faith. Eventually, however, it will assuredly be reached. We only need to continue the game long enough.

When the Jack is drawn for the hundredth time the degree of improbability is already 3.3×10^{150}. And when tails are thrown for the thousandth time it is 1.07×10^{301}. Numbers like these are already beyond all physical possibilities. If the whole of humanity (2×10^9) experimented every second of every day of every year since the origin of the Universe (3×10^7) it would scarcely have completed 10^{27} experiments. And between 10^{123} and 10^{274} times as many experiments would be needed to achieve such a result accidentally on one single occasion.

d. The probability of accident versus the probability of determinacy

If an explanation in terms of accident is impossible then experience shows that we must suppose its opposite, variously called intent, stipulation or conformity to law. In this world of accident and necessity we must then assume the reign of necessity. In future I shall refer to the reign of necessity as determinacy.

This is the next important step in defining order — that of realizing that events (indeed one and the same set of events) can be seen from opposite viewpoints, assuming either the reign of accident, or of necessity, in the source.

There are now two probabilities (P) to be considered: the probability of an accidental or indeterminate event (P_I) and the probability of a determinative event (P_D). The probability of indeterminacy (P_I) measures the extent to which an observer can expect an event, supposing that no regularity affects its occurrence. The probability of determinacy (P_D), on the other hand, measures the extent to which the observer can expect the event assuming the action of a regularity D.

This consideration, as well as the splitting of P into P_I and P_D, is beyond the usual scope of information theory. It can be seen as an extension called 'determinacy theory' which is indeed based on the corresponding probability theorem but which would have no meaning if information is defined only as a measure of lack of predictability.

e. The probability of determinacy

The probability that an event or series of events is to be seen as determinative or indeterminate must depend on the ratio of P_I to P_D. This is because the probabilities of the reign of determinative or indeterminate processes in an event would be expected to behave reciprocally to each other. We can express this ratio (the probability of law) as the degree to which we expect determinacy — this will be the probability with which we are constrained to suppose conformity to a law (P_l). At the limits, complete certainty of the reign of law will be $P_l = 1$ and the greatest improbability of the reign of law will be $P_l = 0$. We then have the quotient

$$P_l = P_D/(P_D + P_I) \tag{4}$$

This ratio will also give the degree of expectation of determinacy. At the beginning of research into any natural event (the programme of a source unknown to us) we shall have no knowledge concerning its background. This situation will correspond to intermediate values between 0 and 1. With increasing experience, however, the certainty will increase either that we are dealing with determinative or with indeterminate events and the ratio will closely approach either 0 or 1.

Suppose, for example, that my suspicion of determinacy comes to be confirmed — tails will be thrown because my opponent is cheating as I suspected. With a single throw I shall still be in great doubt because the probability of accident is still ½.

$$P_l = P_D/(P_D + P_I) = 0.5/(0.5 + 0.5) = 0.5/1 = 0.5$$

Only the continued success of my opponent will justifiably increase my distrust.

Thus my experience will increase with the repetition of an event. I have already mentioned the importance of such repetition for our ability to come to a conclusion (Section I B1c). But I can now go further by showing that, when our accidental or chance expectation is not confirmed, then the number of occurrences (i.e. the number of disappointed expectations) enters the equations as a power.

Thus the chance probability (P_I) that tails will be thrown two, three, five or ten times in succession decreases as 1/2, 1/8, 1/32 and 1/1024 i.e. as $(1/2)^2$, $(1/2)^3$, $(1/2)^5$, and $(1/2)^{10}$. Let the number of occurrences of the same state of an event be a, the expectation of regularity be (P_{la}) and the expectation of regularity with respect to the number of occurrences be P_{la}. Using equation 4 we can then write:[9]

$$P_{la} = P_D^a /(P_D^a + P_I^a) \tag{5}$$

Suppose on the basis of uninterrupted fulfilment of our prediction in a series of events we can maintain our supposition of determinacy $(P_{la} = 1)$ then the equation will simplify to $P_{la} = 1/(1 + P_I^a)$.

Thus if we predict that tails will always fall, we remain very uncertain after the first throw for: $P_{la} = 0.5/(0.5 + 0.5) = 0.5$. But if it is confirmed at the second, fifth, and tenth throw then the probability that our prediction is right increases as $0.5/(0.5 + 0.5^2)$ to $0.5/(0.5 + 0.5^5)$ to $0.5/(0.5 + 0.5^{10})$ i.e. $P_{la} = 0.66$, 0.94, and 0.998. With the hundredth occurrence the probability that we are dealing with a determinative process has reached virtual certainty: $P_{la} = 1/(1 + 0.5^{100})$ $= 1/(1 + 7.9 \times 10^{-31})$. This corresponds to a number near to unity with more than 30 nines following the point.

Of course it is a question of taste at what approximation to unity we assume the reign of determinacy. Since the experiment can be continued as long as we like, however, it is certain that such a value will eventually be reached.

The same is true in the opposite sense. If our expectation of regularity is repeatedly disappointed (on average at every second throw) then the improbability of our assumption of determinacy also increases as the power. On the contrary the expectation of an event under accidental conditions $(P_I = 0.5)$ will be repeatedly confirmed, so the formula that applies will still be $P_{la} = P_D/(P_D + P_I)$.

One can say therefore that, if expectation is confirmed, then the repetition does not change the probability of an occurrence, but repeatedly confirms it. Thus from the fifth to the tenth throw P_{la} will sink from $0.5^5 / (0.5^5 + 0.5) = 0.0588$ to $0.5^{10} / (0.5^{10} + 0.5) = 0.0019$. We shall certainly be convinced of the reign of pure accident after the hundredth throw for then: $P_{la} = 0.5^{100} / (0.5^{100} + 0.5)$ $= 7.9 \times 10^{-31} /(7.9 \times 10^{-31} + 0.5) \sim 1.6 \times 10^{-30}$. This is a number with 30 zeros after the point

But if our expectation that tails will be thrown $(P_D = 1/2)$ is sometimes not confirmed then P_D would sink, with each disappointment, to a half of its value. In the inverse example P_I would sink in the same fashion but we should nevertheless expect (if the probability of tails failing $P_I = \frac{1}{2}$) to be disappointed on average at every second event (cf. example in Sections II B2a and I B1f).

f. Specifying the determinacy content

As soon as the probability seems sufficiently high that a series of events is ruled by determinative decisions rather than accidental ones, we are justified in trying to calculate the determinacy content (D). We can specify this for a series of natural events, in the same way as the indeterminacy content (I) is computed by current information theory.

1. *The simplified solution.* As already mentioned, the information content (I) of a single event, which is the logarithm to the base 2 of the reciprocal of its chance probability P_I, corresponds to the least number of accidental decisions necessary in the system in order to produce it once. Consequently the maximal determinacy content (D_{max}) of an event (E), which is the $\log_2$ of the various possible individual events $(\log_2 E)$, corresponds to the least number of determinative decisions which must already have been established in the structure of the source. For the whole set of numbers (again E), therefore:

$$D_{max} = E \cdot \log_2 E \tag{6}$$

It is easy to see that, if a machine has a range of numbers of 32 symbols, at least five digital decisions must be built into it, in order to be able to select any one of the 32 $(\log_2 32 = 5)$. Thus the assumed accidental decisions reappear in the form of determinative decisions.

2. *The general solution.* The maximal determinacy content (D_{max}), however, is an extreme case. It only applies when there is complete certainty of the exclusively determinative character of a source – as for example when we have made the source ourselves. In analysing natural phenomena, however, we always have to reckon with the possible effect of both types of decision – accidental $(bits_I)$ and determinative $(bits_D)$. This corresponds to the learning process.

For example, so long as we are able to consider the message 1 2 3 4 5 6 (out of a range of 32 equally probable numbers) as a product of accident, then we can specify its information content as 30 $bits_I$. If, however, with repetition P_{la} becomes very high, then we specify 30 $bits_D$. But if, again, the message continues so as to give 1 2 3 4 5 6 12 32 15 8 8 3, then these 30 $bits$ seem to return to the realm of accident.

Again the accidental series 16 2 8 30 4 12 4 28 26 etc. might seem to provide 5 $bits_I$ per event, until we discover that only even numbers occur. The range that we thought to contain 32 symbols has been reduced to 16. We now find only 4 $bits_I$ per event. The lacking *bit* reappears as a command, as the decision 'no odd numbers', and thus as 1 bit_D.

This process of learning the regularities in behaviour corresponds to a decrease in uncertainty, which is a decrease in the maximal possible information content of a message. Thus the general determinacy content will consist of the difference between maximal information content (I_I) and factual information content (I_D). We could also express it as the information content according to the accidental theory *minus* the information content according to the determinative theory.

$$D = I_I - I_D \tag{7}$$

This means that, in every chain of events, the information content I reaches a maximum I_I, when all events are set by accident. If it becomes possible, however, to predict an event more precisely than can be done by chance probability, then I decreases to the factual information content I_D. The difference must correspond to the determinacy content as recognized above.

Applying $I_I = \log_2 1/P_I$ and $I_D = \log_2 1/P_D$ (cf. equation 3) we obtain the special determinacy content of an event (or chain of events) as $D = \log_2 1/P_I - \log_2 1/P_D$, or

$$D = \log_2 (P_D/P_I) \tag{8}$$

Taking the example of an initial range of 32 symbols, but missing odd numbers out, then, for each event, we can reckon

$$D = \log_2 (P_D/P_I) = \log_2 [(1/16)/(1/32)] = \log_2 (32/16) = \log_2 2 = 1 \; bit.$$

This is the *bit D* whose existence we predicted on the basis of the general determinative command 'no odd numbers'.

For example, if we had no insight into the phenomenon of gravitation, then I, at the first dropping experiment, would be I_I (maximal). As insight increased, by the verification of suppositions implying determinacy, the remaining quantity of uncertainty falls to I_D. The difference lies in D (the prognoses now possible). As knowledge increases D continually approaches I_I until the formulation of the law of gravitation. This formulation is D_{max} (= $I_I = E \cdot \log_2 E$; cf. equation 6). Its law content is a transmutation of that maximal information which was to be extracted from the phenomenon of gravitation starting from complete uncertainty or ignorance.

In the case of mixed indeterminate and determinative events the uncertainty remaining when the determinacy content is completely known will be $I_D = I_I - D$; cf. equation 7. This is a measure of the remaining freedom of the system.

g. *The limits of systems and methods*

The concepts of information and of determinacy both demand definite limits of consideration so that the probability of events can be specified. These could be the limits of a range of symbols or of the structure of the source. One of the basic learning processes is to make the limits of the methods or of thinking approach those of the system.

Particularly interesting in this connection are those systemic limits which Nature draws between accidental and determinative events. We have long been accustomed to investigate accidental phenomena where they pass into determinative phenomena, and vice versa.

For example, in investigating the determinacy content of the law of gravity our interest ceases at accidental features such as the time of day, the number of observers, or languages used during the experiments. Likewise, in investigating the information content of a game of dice, the dice players' interest ceases at the determinants specifying the colour of the dice, the age of the dice-cup or of the experimenter.

It is useful, however, to define these limits and it is necessary to consider every increase in the range of symbols, even when it passes over the limit between accidental and determinative phenomena.

For example, if, after numerous observations, we have established a range of numbers from 1 to 32, whose occurrence is specified by pure accident, then at the same time we have established that: '33 does not occur'. This determinant, like many others in the system, is already established, for example by the manufacturer of a roulette wheel. (In the same way the seventh surface of a die is excluded, or the occurrence of an edge on a sphere.) In investigating a roulette wheel $I_I = D + I_D$ still applies, for D is equal to the determinacy content of the manufacturing process, and I_D equals the information content given by the behaviour of the ball.

The same is required for changes in the course of time, when a system moves from the world of accidental events into that of determinative laws, or leaves it again. Take for example such a simple chain of events as the tenfold transmission of the sequence 1 to 16 out of 32 numbers. We should be compelled to recognize that this conformed to law since its P_{la} affords complete certainty. This would remain true even if, both before and afterwards, the source produced completely unpredictable events. Concerning our certainty of the orderliness of living organisms, it matters even less that their orderliness arises from the world of pure accident and returns to complete chaos after death.

At this point we begin to see how to define order content quantitatively. But the present state of theory demands that we should now relate order content to a closely allied concept — that of law content.

2. Redundancy content and law content

As we have seen, there is little agreement between the statistical and colloquial concepts of information. Similarly it is strange to find that the key to recognizing and quantifying conformity to law lies in the phenomenon of redundancy.

We usually use the word 'redundant' for that part of a message which can be left out without decreasing the information content, (as we shall say cautiously at first).[10] Thus for example, the telegram 'boy arrived' gives no less information than a doubling of its individual events such as 'boy boy arrived arrived' or 'bbooyy aarriivveedd'.

For the recognition of redundancy we must presuppose a receiver, of the type already described, with a memory and the ability to make comparisons. The recognition of redundancy implies the re-recognition of a message already received, and has the same meaning as the prediction that, for example, a message in the condition 'boy boy arrived arriv . . .' will be followed by 'ed'. The probability of determinacy (P_D) will thus be $P_D = 1$.

This shows the first important characteristic of redundancy: It has meaning only in the context of determinative events. (The next occurrence of the number 32 in a game of roulette, for example, is not predictable.) Instead of 'information content', therefore, we can say, more precisely, 'determinacy content' (D).

An apparent limiting case exists when we believe that we have acquired 'insight' into a 'law of chance'. Thus after many observations of the results from a roulette wheel with 32 equal-chance positions we can predict that $P_I = 1/32$ and that the number 33, fractions, letters, and so forth do not occur. In actual fact, however, we have not attained the paradox of 'insight into accident'. We have only achieved insight into the determinative decisions which the manufacturer of the game has set as limits to the working of accident. Insight, for example, into how the ball 'chooses' position 32 is prevented by the lay-out of the apparatus. If we were to study this behaviour of the ball we should immediately have to consider the laws of motion and would gain D at the expense of I_D.

a. Redundancy content

The redundancy content (R) of a message, that is to say of a determinacy content in $bits_D$, can again be recognized by the methods which we have used previously. We can specify it by the minimal number of determinative decisions necessary in the mechanism of the source in order to transmit the redundant determinacy content $(bits_R)$. That is to say, by the number of supernumerary decisions (a special case of $bits_D$).

As already mentioned, if we receive the message 1 2 3 4 5 6 out of a known range of 32 numbers, then at first it will contain $I_I = 6 \times 5 = 30\ bits_I$ of information. After 10 repetitions, however, i.e. 11 transmissions, we have to abandon the accident hypothesis because of high P_{la}. We have then received $D = 11 \times 6 \times 5 = 330\ bits_D$ of which $10 \times 6 \times 5 = 300$ are $bits_R$.

The point when we discover determinacy in the course of the message depends only on P_{la}. Possibly it will only be reached after many repetitions. As soon as the discovery has been made, however, we can quantify the redundancy retrospectively, back to the first repetition.

In defining redundancy content in this manner we have made the simplifying assumption that, in the transmission and reception of the message, mistakes either do not occur or will not be noticed. If mistakes do occur within a determinative sequence, however, they will usually be noticeable and then we can distinguish two forms of redundancy. But the difference between these forms will again disappear after quantitative analysis, as shown below.

b. Useful and empty redundancy

We call redundancy content 'useful' if, by correct repetition, it cancels out mistakes in the original message or removes misunderstandings. Thus, in a system that makes a mistake in every second event, the message 'boy arrived' will perhaps become **r o z u r n i i e a** and will be totally incomprehensible (unchanged events are printed in bold type). However, the doubled version **b b o o y y a a r r i i v v e e d d** becomes **r b z o u y n a i r a r h i z v a e n d** which is still decipherable.

In analysing redundancy content, however, the following peculiarity must be recognized. Assume that an observer, such as the designer of a source-receiver system, receives the message '1 2 14 1 2 3 1 2 3' (out of a range of 32 numbers) and recognizes 14 as a permutation of 3. At the moment of recognition the 5 $bits_I$ that corresponded to 14, assuming no previous knowledge, will transform into 5 $bits_D$ and the 5 $bits_D$ of the first '3' in the repetition will transform into 5 $bits_R$. This is taken as given in linguistics and communication theory. Usually, however, the research worker is himself the receiver, in that he seeks to reconstruct the mechanism of an unknown source only on the basis of its message. The first thing that such a primary receiver must do is to attain certainty concerning the determinative character and the mistakes in the message, by way of a high expectation of regularity (the probability P_{la}). At the moment when he ceases to be surprised by mistakes in determinacy, because he can correct them, useful redundancy turns into empty redundancy.

Analysis shows, therefore, that the distinction between useful and empty redundancy always disappears. We can therefore continue with a single redundancy concept. Only in the design of source-receiver systems by a third party does useful redundancy again acquire a meaning – as a design element, so to speak. It appears in the determinacy of the genetic code, in the evolution of languages, and the development of communications equipment.

In the same curious way, therefore, that 'information' surprises and determinacy informs, we find, after analysis, that useful redundancy is as empty as empty redundancy is useful.

c. *Redundancy content and law content*

We have already met two of the parameters applicable to redundancy (a and R). The number of identical occurrences of a message (a) is crucial for the recognition of a determinative occurrence (Section I B1e). I have used the redundancy content R, on the other hand (Section I B2a) to indicate the number of redundant decisions in such a determinative occurrence, assuming provisionally that all these decisions occur (for with systemization, as later shown, it is possible to eliminate many of them).

Having defined R we can subtract these recurring decisions from the total quantity of information, i.e. from the determinacy content D. The remainder L corresponds to the content of the original communication or statement, i.e. to the law content of a determinative occurrence.

$$L = D - R \tag{9}$$

This important entity L, which determines all the repetitions of a determinative event, corresponds to the idea of law, conformity to law or regularity in ordinary speech. Thus determinacy content of an occurrence, up till now measured in $bits_D$, can be stated in a more differentiated form as law content plus redundancy content, i.e. $bits_L + bits_R$.

In the same way L can be defined from the quotient of the determinacy content D and the length of the series or relative redundancy r (where $r = D/L$), that is $L = D/r$. It follows that

$$D = L \cdot r \tag{10}$$

We can therefore describe determinacy as regularity times the repeated occurrence of decisions. But, again, such instances conforming to a law are what we understand by order (see also equation 18).

If we describe order as the product of law times the number of instances where the law applies we satisfy the widespread assumption that a law, if not applied, does not lead to order. Such a description also answers to our feeling that the quantity of order does not depend only on the complexity of the regularity (see also p.22).

We know from experience that a law, promulgated as a string of subordinate paragraphs in a mass of complicated text (law content) and hedged about with 'ifs' and 'buts', so that it can scarcely be applied and always ambiguously, will often be superseded a few years later. The law of gravitation, on the other hand, can be formulated very simply and the material world seems to have obeyed it since creation.

In this way the order content (D) can be derived from probability considerations. A further indication that it is in fact the product of law content times the number of instances will be found when we consider the real dimensions of law and order (Section I B3). First, however, I shall finish discussing redundancy by explaining what features can be expected in its production.

Determinacy, whether in the form of regularity or of redundancy, requires in the source, as in the receiver, at least a very simple form of memory. It also requires a constant decoding mechanism, as explained in detail later. Otherwise we cannot explain how the decisions needed for the selection or determination of a possible event, within the range of possible events, always follow each other in correct sequence.

The simplest conceivable form of such a memory, in my opinion, would consist of two different states linked together to form a chain. The various decisions could be differentiated by inequalities in the constitution, surface, electric charge or, as in punch tape, in position. Thus the chain or tape, as soon as the reading direction was fixed, would by its material form determine the sequence of individual decisions.

d. Visible and hidden redundancy

When we visualize in material form the chain of decisions needed for a determinative occurrence, it is obvious that not every redundant decision shows itself as a redundant event. Thus we can distinguish visible from hidden redundancy. By visible redundancy I mean repetitions of features or events; and by hidden redundancy I mean unnecessary complexity in decisions of a sort which is not immediately reflected by events. At first sight, hidden redundancy seems to be a phenomenon of subordinate importance but its influence on the pattern of order will soon become clear. I shall therefore consider its arithmetical basis.

To explain why I am considering this decision redundancy, which cannot in itself directly influence the form of a message, I shall refer both to an earlier and a later phase in the argument. Looking back, it must be remembered that a quantitative approach to the phenomenon of redundancy is only possible by way of the decisions that bring it about. Looking ahead, I mention that the pattern of decision redundancy, because it necessarily comes to be dismantled, will produce the basic forms of orderly pattern. This will be true whether the redundancy of the individual event is immediately visible or not. This is a crucial point for my theory and I shall consider it in detail later.

1. *Visible redundancy (R')* must depend on a repetition of those decisions which define the law content (L) of the message. It is required that the whole chain must contain in material form as many replicas (a of them) as are repeated in the series of events (E). The maximal content of visible redundancy will then be:

$$R'_{max} = E \cdot \log_2 E \, (a-1) - x \tag{11}$$

The identicality of the determinants of visible redundancy therefore depends on an identical sequence, or total replication, of all yes-no (or A-B) decisions. The whole pattern repeats itself.

As already mentioned, the transmission of E events (for example 16 events, each with a possible range of 16) demands that $E \cdot \log_2 E = 16 \times 4 = 64$ $bits_L$; or, for a transmissions, $E \cdot \log_2 E \cdot a$ $bits_D$. As a consequence R_{max} (the maximal redundancy content) increases with 1, 100, and 10 000 repetitions ($a = 2$, 101, and 10 001) from 64 to 6400 to 640 000 $bits_R$. From this total we subtract only x decisions as not redundant. For example assuming provisionally that each transmission requires the command 'go' then $x = a - 1$.

Thus in the message 'La mia bella amica' the redundancy of three of the '*a*s' is visible because the first '*a*' already defines the gender.

Hidden redundancy, on the other hand, must be a question of long-windedness in the determinative decisions. We cannot therefore use letters as an example though the long-windedness would have its analogy in 'thought decisions'. However, the principle corresponds exactly. Identicality of the determination of hidden redundancy depends on identical position (i.e. ranking) of particular agreeing yes-no (A-B) preliminary decisions.

This dependency is not immediately obvious. I shall therefore illustrate it in its two basic forms — individual and special ranking.

2. *Hidden individual redundancy (R")* exists when individual decisions of identical position (i.e. rank) become redundant. This can be illustrated by the fixing of the sequence of particular decisions. Thus with reference to the accompanying table, for the transmission of the events (E) I-VIII we require $E \cdot \log_2 E = 8 \times 3 = 24$ $bits_D$. These can be visualized as the holes 1 to 24 in a punch tape. It is then evident that, first, the 3 $bits_D$ per event consist of a first and second preliminary decision and a final decision, and second that some decisions of the same rank (three ranks here)[11] will always be identical. For example, the decisions at 1 4 7 10 or 14 17 etc.

Number of decision	1	2	3	4	5	6	7	8	9	10	11	12	13	14	15	16	17	18	19	20	21	22	23	24
1st predecision	a			*a*			*a*			*a*			b			b			b			b		
2nd predecision		a			*a*			b			b			a			a			b			b	
Final decision			a			b			a			b			a			b			a			b
EVENT			I			II			III			IV			V			VI			VII			VIII

If we assume a decoding mechanism that remembers or 'retains' a preliminary decision until it is reversed by an opposing decision (e.g. from no. 1 until 13) then all those decisions (e.g. 4, 7, 10) are redundant which have been printed in *italics* in the punched tape shown. For they can be left out without decreasing the content of the message. The number of such redundant decisions corresponds to the difference:

$$R''_{max} = E \cdot \log_2 E - \sum_{i=1}^{\log_2 E} 2^i \tag{12}$$

This is the difference between the maximal required number of decisions ($E \cdot \log_2 E$) and the minimal required number ($2^1 + 2^2 + 2^3$). This is $8 \times 3 - (2 + 4 + 8) = 24 - 14 = 10$ $bits_R$.

It is obvious that a decoding mechanism with at least a minimal memory is definitely presupposed if hidden redundancy is to be reduced. It is easy to foresee that communications technology will use such a mechanism. More important, however, is the fact that the molecular genetic system also includes it (cf. Section IIIC). I shall show this in discussing the cause of the pattern of biological order. All these assumptions will be fully confirmed later. Indeed the whole phenomenon of decoding will need to be considered, for without decoding no code reveals its 'meaning'.

3. *Hidden serial redundancy (R''')* exists when whole series of preliminary decisions, differing from each other in relative rank, become redundant to the same extent. This occurs when, out of a possible range of numbers of a source, only a few are transmitted (E) while other alternatives (e) are completely excluded. In this case (R''') is additional to (R'').

$$R'''_{max} = (E - 1) \cdot [\log_2 (E + e) - \log_2 E] \tag{13}$$

For example if only the events I to VIII out of a possible range of 1024 numbers are transmitted ($E = 8$, $e = 1016$) then we obtain the following picture with the adjoining calculation.

Number of decision	1	2	3	4	5	6	7	8	9	10	11	12	13	14	15	16	17	18	19	20	21	22
1st predecision	a										*a*										*a*	
2nd predecision		a										*a*										*a*
3rd predecision			a										*a*								and so on	
4th predecision				a										*a*							according to	
5th predecision					a										*a*						equation 13	
6th predecision						a										*a*						
7th predecision							a										*a*					
8th predecision								a										*a*			and so on	
9th predecision									a										*a*		according to	
Final decision										a										b	equation 12	
Event										I										II		

From equation 13 it follows that $R''' = (8 - 1) \cdot [\log_2 (8 + 1016) - \log_2 8] = 7 \times (10 - 3) = 49\,bits_R$. This means that seven decisions to exclude *b* will repeat themselves seven times identically (for the events II to VIII). Given a memory for commands, they are redundant.

In summary, the interesting result is reached that the maximal total redundancy (consisting of *R'*, *R''*, *and R'''*) becomes very high even for very simple types of composite transmission or determinative events. I shall discuss the significance of the pattern of redundancy later. The composite maximal redundancy content, using equations 11, 12, and 13, will be:

$$R_{max} = E \cdot \log_2(E + e) \cdot a - \sum_{i=1}^{\log_2 E} 2^i - [\log_2(E + e) - \log_2 E] - x \qquad (14)$$

Thus if our source, with a range of $(E + e) = 1024$ different individual events, only transmits those (E) from I to VIII, but sends these out 10 000 times, then $E = 8$, $e = 1016$ and $a = 10^4$. For x there are two limiting cases. In the maximal case the decoding mechanism will need the command 'go' for every replication of L $(x = a - 1)$. In the minimal case it will operate with the commands 'on' or 'off' $(x = 2)$.

The R_{max} of such a message can be calculated from the maximal possible decisions, $E \cdot \log_2 (E + e) \cdot a$ (cf. equation 11) minus the minimal required decisions which consist of the terms:

$$\sum_{i=1}^{\log_2 E} 2^i \text{ and } \log_2 (E + e) - \log_2 E \text{ (cf. equations 12 and 13)}$$

as well as the value for x. For the second limiting case we obtain:

$8 \times 10 \times 10\,000 - (2 + 4 + 8) - (10 - 3) - 2 = 799\,977 bits_R$, or with 800,000 $bits_D$ only 23 $bits_L$.

In such systems r, i.e. $D/L = 8 \times 10^5/23 = 3.5 \times 10^4$, will already reach values between 10^4 and 10^5. It can be shown that, given the complexity of organisms, orders of magnitude between 10^5 and 10^{20} can occur.

Such *conserved redundancy* specifies not only the probability of law but also the number of instances when the law content is applied. When we appreciate its dimensions we can imagine the extraordinarily high statistical probability with which the laws governing living order can be recognized. We can also foresee the almost unimaginable dimensions which living order, seen as law times the number of instances when the law applies, will reach.

e. The origin and fate of redundancy

A chain of events which never repeats itself therefore contains the pure expression of a law, as already shown, but its events can in no way be predicted. Every phenomenon and every process which we can foresee in this Universe as conforming to law, possesses within its determinacy content a large, or even extraordinarily large, quantity of redundancy. A world without redundancy, if such existed, would entirely escape our powers of concept formation.

1. The origin of redundancy represents a problem whose biological aspect can be solved (Section III B2b). Whether the general problem is soluble, in so far as it applies to law and the application of law in the inorganic world, does not need to be decided here. In my opinion it is chiefly a philosophical problem to decide whether our world of events, which is redundant in high degree, may have arisen from a world with less redundancy or none — from pure unrepeated conformity to law. Or whether the redundancy of events is always accompanied by a redundancy of decisions. Or whether it arose from almost pure redundancy by the splitting-up and elaboration of an original or minimal law. For the moment I shall leave all this to one side, but I shall come back to it in Section VIII B7c.

In the living world we shall find that regularity and redundancy of decisions and events form a system for the accumulation of determinacy, and thus of order. To appreciate this we need only think of the increase in conformity to law involved in the differentiation of evolutionary ground plans or of the visible redundancy that goes with the mass reproduction of the individuals of a species.

2. The accumulation of redundant decisions seems, however, to be a general necessity. Only a teleological final cause in Nature would be able to avoid it completely. Such a final cause, however, cannot be demonstrated even for the construction of the determinacy code of living organisms. New decisions will be inserted or rejected in the first place on the basis of immediate selective advantage. This will happen without regard to whether, seen from outside, a decision is redundant or not, and oblivious to whether it could 'by pondering' be avoided without decreasing the information content.

3. Accumulation of redundant decisions, however, takes on a completely new aspect if a principle of economy is introduced. The introduction of such a principle is permitted, for example, when the insertion, conservation, and decoding of determinative events costs energy, as in the design of biological or of source-receiver systems.

Thus even with such a simple message as the transmission 10 000 times of the numbers 1 to 8 (out of total of 1024 possibilities) there will only be 23 unavoidable decisions as opposed to 799 977 which could be avoided by better decoding.

When avoidable decisions begin to outnumber unavoidable decisions by several orders of magnitude then avoiding them will to the same degree become important, and consequently will happen, as it does in the design of machinery.

4. This dismantling of redundant determinative decisions will happen, in designing machinery, when the attainable profit becomes larger than the labour of rethinking, or in the evolution of organisms, greater than the selectional cost of trial and error.

This connection will prove to be the key to the cause of the biological patterns of order and will be dealt with in detail below. Here I shall only mention that it is a general principle and that the accumulation of redundant decisions does not tend towards any maximum.

Redundancy of decisions and redundancy of events (i.e. r and a) behave differently to each other. In systems governed by a principle of economy, the increase of redundant decisions is counteracted by a regulator, but we cannot discern this by inspecting the events, or at least not by inspecting events in the same plane. The spread and application of a regularity only comes up against a limit much later. This happens at the very limits of range of determinative systems, where their conditions for stability vanish. In technology this occurs when the market for a machine is exhausted. With organisms and their communities it happens when niches are filled or ranges totally occupied.

3. Order, determinacy, and negentropy

In the preceding sections I have tried to present what is generally understood by order and have developed a method which permitted a quantitative description. I proposed that order could be specified as resulting from determinative decisions which define its law content. If this is accepted, then we can take the next important step towards describing biological order.

Here, however, we meet a problem which ought to be mentioned, although my later argument does not presuppose its solution. This is the problem of the connection of entropy, negentropy, and information. I mention this problem because what I have already said contributes to its solution, and because this solution further supports my theory.

I shall deal with this matter in three stages. In the first I shall recapitulate the concepts. In the second I shall describe the problem. And in the third I shall propose a solution.

a. Entropy and negentropy

1. *What is entropy?* The matter was excellently dealt with by Schrödinger, writing for the educated layman (1969, p.76). He said: 'Let me first emphasize that it is not a hazy concept or idea, but a measurable physical quantity just like the length of a rod, the temperature at any point of a body . . . To give an example, when you melt a solid, its entropy increases by the amount of the heat of fusion divided by the temperature at the melting point. You see from this, that the unit in which entropy is measured is cal/°C.' Schrödinger's classic account continues (1969, p.77): 'Much more important for us here is the bearing on the statistical concept of order and disorder, a connection that was revealed by the investigations of Boltzmann and Gibbs in statistical physics. This too is an exact quantitative connection, and is expressed by

$$\text{entropy} = k \log \mathbf{D} \qquad \text{(cf. 1)}$$

where k is the so-called Boltzmann constant (= 3.2983×10^{-24} cal/°C) [nowadays more often written 1.38×10^{-16} erg/°C] and $\mathbf{D}$ is a quantitative measure of the atomic disorder of the body in question. To give an exact explanation of this quantity $\mathbf{D}$ in brief non-technical terms is well nigh impossible. The disorder it indicates is partly that of heat motion, partly that which consists in different kinds of atoms or molecules being mixed at random, instead of being neatly separated . . .' 'An isolated system. . . increases its entropy and more or less rapidly approaches the inert state of maximum entropy. We now recognize this fundamental law of physics to be just the natural tendency of things to approach the chaotic state (the same tendency that the books of a library or the piles of papers and manuscripts on a writing desk display) unless we obviate it. (The analogue of irregular heat motion, in this case, is our handling those objects now and again without troubling to put them back in their proper places.)'

2. *What is negentropy?* Schrödinger then makes an important statement (p. 79):'If $\mathbf{D}$ is a measure of disorder, its reciprocal, $1/\mathbf{D}$ can be regarded as a direct measure of order. Since the logarithm of $1/\mathbf{D}$ is just minus the logarithm of $\mathbf{D}$ we can write Boltzmann's equation thus:

$$-(\text{entropy}) = k \log (1/\mathbf{D}) \qquad \text{(cf. 2)}$$

Hence the awkward expression 'negative entropy' can be replaced by a better one: entropy, taken with the negative sign, is itself a measure of order.'

After the criticism which he at first encountered Schrödinger added: 'Besides, "negative entropy" is in no way my own discovery. It is the very idea round which Boltzmann's independent discussion revolves.'

b. Certainty and uncertainty

A second part of the argument illuminates the connection between chaos and probability. As long ago as 1894, Boltzmann considered entropy as a measure for lack of information.[12] Entropy (S), being a measure for disorder, is thus also connected with probability.

1. *Information as entropy.* Every closed physical system changes, as its entropy increases, from a less probable to a more probable total condition. This relationship between entropy (S) and probability (P) is given by the Boltzmann-Planck equation:

$$S = k \log P \qquad\qquad\qquad\qquad\qquad (15)$$

in which k is again the Boltzmann constant (cf. equation 1). On the other hand $\mathbf{D}$ (atomic disorder) is replaced by the probability P, which indicates the number of 'elementary complexes' in the system. These are the individual distinguishable configurations which atomic systems can take on by discontinuous changes from one metastable structure to another (Planck, cf. Brillouin 1956, p.120).

It seems inadvisable to go further into this subject, both because of the author's specialism and the reader's patience. Also my basic theme does not require it. By way of illustration, however, I add that P increases with the number of possibilities in the system, i.e. with the general confusion. In an ideal single crystal at absolute zero temperature it reaches a minimum value of 1, i.e. each atom now has a single possible and predictable position ($P = 1$). In this case $S = k \log 1 = 0$ corresponding to the least possible disorder.

Thus Boltzmann already saw entropy as a measure for lack of information. A quantitative interpretation was foreshadowed by Smoluchowski (1914), discovered by Szilard (1929), but was not understood, was forgotten and then largely rediscovered by Wiener, Shannon and Weaver[13] at the end of the 1940s. They define information (I) (as in our equation 3) as the logarithm of the reciprocal of the probability of a particular occurrence $I = K \log 1/P$ (with K as a constant) or the logarithm of the number of possibilities.

$$I = K \log P_0 \tag{16}$$

The correspondence of equations 15 and 16 is obvious. If, instead of the constant, K, we write the Boltzmann constant k then we measure information in the energy units of the law of entropy.

The inversion of $1/P$ to give P_0 can be illustrated by our roulette wheel with 32 individual equally probable events. Each individual event has a probability of $P = 1/32$ so $1/P = 32$ and 32 is equal to the number of possibilities.

We can therefore follow Shannon and Weaver[14] in stating: 'Information turns out to be exactly that which is known in thermodynamics as entropy. For in both cases it is a question of number of possibilities and freedom of choice. Entropy, chaos, mixing together, freedom of choice, and information are thus identical as is accepted in information theory and in physics.'[15]

2. *Information as negentropy.* We could equally well say the converse: 'If we obtain more information about the problem, we may be able to specify that only one out of the P_0 outcomes is actually realized. The greater the uncertainty in the initial problem is, the greater P_0 will be, and the larger will be the amount of information required to make the selection.' This theorem, which in principle goes back to Brillouin (1956, p.1), implies the opposite conclusion to that reached in the previous paragraph, i.e. it implies that negentropy, order, organization, the separation of mixtures, design, and information are identical. So many biophysicists and cyberneticists have accepted this position[16] that the growing literature gives the impression that the matter is closed.

Which position is adopted certainly depends on the points of view favoured by the study of natural laws on the one hand and the study of living organization on the other. Shannon and Weaver sought to measure disorder or unpredictability while Schrödinger and Brillouin sought to measure order or predicatability. Schrödinger already compared order with negentropy, while Boltzmann compared chaos with lack of information.

3. *Entropy or negentropy.* However, I think it unlikely that only one of the two apparently contradictory theorems is correct, although the discussion about who could have confused entropy with negentropy has still not broken off.[17] Equally improbable is Brillouin's suspicion that Shannon and Weaver had already confused the two. I wish to suggest another possibility. It may seem unlikely that an anatomist could teach the information theorists about information. Nevertheless, I suspect that both interpretations are correct and indeed that each presupposes the other.

When we speak of information we must remember the strange difference between the everyday and the quantitative concepts, and also the double relationship of the latter to determinacy and indeterminacy, to order and chaos.

c. Information as entropy and negentropy

In this position I need to remove yet another obstacle in order to fit our elaborated formulation of order into the structure of the relevant theorems. This obstacle is the obvious contradiction that information in the first place can increase with the degree of uncertainty and disorder, and, in the second place, with that of predictability or order. For, although neither theorem in itself appears to be contradictory, nevertheless it cannot be expected that entropy could be the same as negentropy.

Indeed we have already worked out the answer in that our concept of order implies the distinction between accidental and determinative phenomena. Let us take the proven statement that both entropy and negentropy start by specifying probability. To make the two points of view concordant with each other, we then merely need to ask: 'Probability of what?'

This is perhaps an unorthodox question, but we must remember that in nature 'probability in itself' does not exist. This world contains both accident and necessity. In the first place it contains only these, because, apart from accidental and non-accidental, no third alternative is possible. In the second place, it contains both these, because without experience they cannot be separated. Thus, when we ask about the probability of an event, then we can always mean either the probability of explaining the event by accidental decisions ($bits_I$) or by determinative decisions ($bits_D$).

Technically it is agreed that information is a measure for lack of predictability or of knowledge, a measure of the rarity of events, for the surprising, the new, or the unexpected. It is identical with the number of decisions which are required to explain a phenomenon, or to describe it, or to establish it. It is a measure for the improbability that these decision should coincide in large numbers. But what decisions are we talking about?

1. *Information as indeterminacy.* In the first place let us consider indeterminate events. These depend on accidental decisions which by definition cannot be predicted by the observer. It is therefore a question of the information that I should have, if only I could get it — information about the roulette wheel, about the history or the movement of molecules which, apart from Maxwell's Demon, nobody possesses. Such information increases in a consistent fashion with the range of numbers, the number of symbols or of possibilities in the source, and thus with lack of limits, lack of selection, disorder, lack of meaning, or in consequence entropy.

The Demon imagined by Maxwell was able to open a little door between two equal gas-filled spaces. As molecules moved here and there by thermal motion he would open the door, for example, only when a molecule was moving from the right-hand space towards the left-hand one. For molecules passing in the reverse direction he would keep the door shut. The pressure gradient, the free energy, or perpetual motion, which he could thus build up would correspond to the advantage in information which he had over us concerning the movement of individual molecules.

In designing apparatus for generating randomness or for games of chance we increase the information in this sense. We do this by widening the limits of the possibilities that we allow to accident, by decreasing determinative rules.

The information content of accidental happenings corresponds to the number of decisions which I must suppose, or concede, when I cannot gain insight into the time and nature of the decisions. I have given it the dimensions of $bits_I$, described it as I_D and now name it *indeterminacy content.*

2. *Information as determinacy.* In the second place we consider non-accidental or determinative events as I have called them. These depend on determinative decisions whose mode of happening has been fundamentally explained (i.e. causally understood) and indeed can be arranged by us. It is thus a question of information which I have as soon as it is gained; it is information about the design of the apparatus, about causal regularity,

about organization. Such information increases in correspondence with the distance from thermodynamic equilibrium and from the most probable condition of mixture and lack of selection. It increases with the exclusion of accident and thus with organization, specialization, cost of construction, previously laid-down conditions, with meaning, with the degree of order, and consequently with negentropy.

In designing apparatuses, whether machines or organizations, we increase this information by increasing the limits within which exclusively determinative decisions operate, i.e. by pushing back the accidental events.

The information content of determinative events therefore corresponds to the number of those totally distinct decisions whose position and type, as we must suppose, are accessible to us. That is to say, they are understandable, describable and in the last analysis, predictable, and they were established by necessity, by purpose or by causality. To this type of information content I have given the dimensions of $bits_D$, described it as D and named it *determinacy content*.

Another example will illustrate this necessary antagonism of probabilities. If I rattle a prearranged puzzle about in its box, then I can calculate probability assuming the reign of accident, and the unselected mixing and entropy of the pieces becomes more and more probable. But the chance of my finding any particular piece in its proper place according to the rules of play becomes more and more improbable.

On the other hand, if I arrange the haphazard mixture according to the rules of play, then I can calculate excluding accident, and then for each piece the probability increases of finding it in a particular position. But with order, with the negentropy of the game, the product viewed as an accidental result becomes more and more improbable.

3. The synthesis is therefore very simple. Probability in itself has no meaning, for the improbability of order can be understood from the probability of chaos. And the improbability of chaos can be understood only from the probability of order (cf. Fig. 2a-d).

Content of:

Information, I_I (in *bits*)	Indeterminacy (accident) I_D (in *bits*$_I$)	Determinacy (necessity), D (in *bits*$_D$)	Law L (in *bits*$_L$)	Redundancy R (in *bits*$_R$)
200	200	0	0	0
200	100	100	20	80
200	5	195	100	95
200 (4.2 x 10⁶)	0	200 (4.2 x 10⁶)	200 (4.2 x 10⁶)	0

Grid (a):
```
E E R D V F T Z D S
F K A I U U C L P R
Q O J A R B R M F Q
N D A W G Q R Y Q Z
```

Grid (b):
```
N R A O J P X F N H
A A A A A B B B B B
C C C C C D D D D D
R O L I G Y A M Q Q
```

Grid (c):
```
A B C D E F G H I J
K L M N O P Q R S T
A B C D E F M H I J
K L M N O P Q R S T
```

Grid (d):
```
W H A N - T H A T -
A P R I L L - W I T
H - H I S - S H O U
R E S - S O O T E -  ...
```

Fig. 2 a-d. Games with their contents of accident and necessity, with messages with the same number of events (40) and same number of symbols (32 = 5 bits). (a) Assuming the reign of pure accident. (b) With the message half determined. (c) Determined with one mistake. (d) Determined (in parentheses are the values for those who know the *Canterbury Tales* beyond this message).

Nevertheless it is true that science needed first to recognize the agreement between information $(I = K \log P)$ and chaos $(S = K \log P)$ for only chaos was convincingly defined. Only later did it emphasize the relationship between information and order, since the study of dominant order is a main scientific aim. The fact that interested voices raised the possibility of a contradiction shows how meticulous science is.

All this, however, has already been pointed out in the great literature on this subject. Wiener is reported to have said: 'Order is essentially a lack of accidentalness.'[18] In practice I go only a small step further when I conclude that the sum of the indeterminacy content (I_D) and of the determinacy content (D) of a defined system, i.e. the general information content, will remain the same:

$$I_D + D = \text{constant} \tag{17}$$

This is because in this world, just as we can only choose between the alternatives of accident and necessity, so also we can only choose between understanding causally and not understanding causally. This is true however often we lose our way on the journey towards knowledge. This too has almost been said already. Even Democritus said that: 'Everything that exists in the Universe is the fruit of accident and necessity.' Indeed this has neither been successfully refuted, nor forgotten. Monod (1971) put this sentence as a motto at the beginning of his book.

A final example may be helpful. I possess a source with a range of 32 symbols which can communicate at least 21 000 lines, each with about 40 apparently meaningless individual events. Since $\log_2 32 = 5$ we calculate $21\,000 \times 40 \times 5 = 4.2 \times 10^6$ bits$_I$ and thus more than 4 million apparently random accidental events. It may be sufficient to quote a single line.

Event number	. . . . 5 . . . . 10 . . . . 15 20
Type of event	23 08 01 14 27 20 08 01 20 27 01 16 18 09 12 12 27 23 09 20

	. . . . 25 . . . . 30 . . . . 35 40
	08 27 08 09 19 27 19 08 15 21 18 05 19 27 19 15 15 20 05 27

But now I merely reveal that the symbols 1 to 26 represent the letters of the alphabet while 27 to 32 represent : space . , ; ! ? The first line of the Chaucer's prologue then becomes obvious for: 'Here bygynneth the book of the tales of Caunterbury' (cf. Fig. 2d).

Event number	. . . . 5 . . . . 10 . . . . 15 . . . . 20 . . . . 25 . . . . 30 . . . . 35 40
Type of event	WHAN THAT APRI L L WI TH HI S SHO U RES S OOTE

When this happens not merely do 200 *bits* of chaos transform into 200 *bits* of the most improbable order, but, with the idea of the *Canterbury Tales*, the Knight and the Parson appear and, depending on how well I know my Chaucer, the deafness of the Wife of Bath and the Shipman's 'noble monke' and up to 4.2×10^6 bits$_D$ of predictable regularity. Indeed for philologists the whole world of Mediaeval literature emerges with 10^7 *bits$_D$* and more. But if, for example, I merely analysed the letters into the frequency groups, then even for experts in Middle English, all this order would fall into more than 4 million *bits* of chaos, of meaningless ornament, just like the hieroglyphs were thought to be before the discovery of the Rosetta stone.

What I called determinacy content (D) or order must therefore be the same as, or similiar to, negentropy (N). What I called indeterminacy content (I_D), on the other hand, could correspond to chaos (S) and entropy.

4. Order as law times the number of instances

If we formulate order as being law content times the number of instances when the law applies, or more briefly, as law times instances, then we can solve three problems which, especially in the field of biological order, have hindered the application of the useful theorem: 'Information is equal to order or negentropy.' (Beyond this I do not wish to anticipate the general consequences discussed in Chapter VIII.) After this I shall start to discuss biological events.

a. *Solution of the information paradoxes*

I shall deal in turn with the paradoxes of contradiction, of reduced number of instances of a law and of increased number of instances of a law. By way of illustration I shall use examples well known in the literature.

1. *The problem of contradiction.* 'A theorem by Einstein or a random assemblage of letters both contain the same information, provided the number of letters is the same.'[19] 'If our piece of iron is now sharpened into the form of a gear wheel, the change in its physical entropy will be negligible.'[20] In a conditional way I agree with both these statements.

In the case of the letters of the alphabet we recognize the transformation of determinacy content into indeterminacy content by the shaking up of Einstein's letters so that a reader could not understand a single word of the theorem. The sum of insight plus perplexity remains constant. In the case of the toothed wheel, however, regularity is added to haphazardness, to the value of the number of $bits_D$ needed to describe the non-accidental features of its surface.

2. *The problem of reduced number of instances of a law.* 'If, for example, one given molecule of guanine in a given gene (i.e. a single decision in an enormously long genetic message) is replaced by a molecule of adenine, the information, the structural negentropy of the system is the same. For the physicist, even if the mutation is lethal, nothing has changed: the content in negentıopy has remained the same. But the mutation being lethal, the altered organism is now unable to function and reproduce normally. It has ceased to be alive.' Consequently, as we can append to Lwoff's excellent example, its negentropy vanishes.

In that case it is not merely the repetition of events (*a*) which vanishes. It is not merely the repeating of the original law which ceases (*a* = 1) so that the order in the system is reduced to the short duration of life of the mutant. On the contrary the law laid down in the message (*L*) completely ceases to apply (*a* = 0). Thus the determinacy content $D = L \cdot a = L \times 0 = 0$ and order vanishes.

This formulation even satisfies the paradox that an increase in law content can destroy the determinacy content. As Lwoff has written: 'As a consequence of the introduction of the genetic material of a virus, the negentropy of a cell-virus system is greater than the negentropy of the normal, original, noninfected cell. But the infected cell will die; that is, its information will be destroyed.'[21]

We have already established that the importance of a law does not depend on its formulation, and above all not on the length of its formulation, but on the number of instances where it applies. There are innumerable inapplicable texts of laws lying uselessly in drawers, innumerable formulae for inapplicable inventions in the archives of the patent offices and innumerable well-formulated resolutions which we have forgotten in our own lives, because they could not be realized.

3. *The problem of increased number of instances.* An excellent example, given by Linschitz, [22] will suffice: 'The inadequacy of physical entropy alone to measure the biological information content is also seen in the extensive nature of entropy, by which the entropy of two identical cells is twice that of each cell. However, the biologist and the communication engineer might both argue that little more information is present in two identical cells than is already present in one.' I completely agree with Linschitz and yet the paradox is soluble. For in the present formulation we can say: The order or negentropy of the system is doubled, but its law content scarcely changes. The solution is

obvious: $D = L \cdot a = L \times 2$. This is a fundamental concept of the identical replication of conserved law. This concept will occupy us a great deal.

The biological difference between $L_1 + L_2$ and $L \times 2$ is even greater than that between Darwin's achievement in writing the *Origin of Species* and that of the printer in putting another 40 sheets into his machine. For the living manuscripts even contain instructions for their own reprinting. The comparison is between the achievement of setting the stop switch of the printing machine to operate after 10 000 copies of the one hand and that of living 10 000 lives of Darwin on the other. The significance of this simplication for life and evolution can be imagined and we are nearly in a position to begin discussing it.

My general conception can therefore be expressed as the statement that; order is law content times the number of instances where the law applies, or, briefly, law times instances. This concept seems to solve the weightiest of the preceding contradictions. If so, we can seek to apply it to the particular complexities of living order.

b. Instance, decision, and event

I thus come to the last, and perhaps the most important, consequences of these introductory considerations. This will lead to the cause of orderly pattern, and thus to the main part of our study. It will make things easier if we remember two things — first, the simplication that we have had to make; and second, the fact that insight into law and decrease of redundancy are opposed to each other.

As to the simplication, I have proceeded as if there were a basic difference between decisions and events which justified us in analysing the redundancy problem beginning entirely with the decisions. This is a didactic simplication, so as not to confuse the flow of the presentation. It must be recognized however, that an event can never be anything else than the system of decisions (sometimes an enormous number of them) which bring it about.

As to the opposition between insight and redundancy, we can deduce the existence of conformity to law only on the basis of repetition (Section I B1e). However, the decisions repeated will come to decrease (I B2e) for reasons of economy. These two relationships are of general relevance and are particularly relevant to the living world.

1. *The identicality of decision and event.* This is not easy to appreciate in complex systems but becomes more and more cogent as we consider increasingly simple systems. We have imagined decisions (d) from the beginning as the yes-no decisions of the relay switches of a machine or source. In Nature these correspond to the decisions in the molecular realm of matter. These can be described as the entry of atoms into one or other stable state according to position and chemical bonding, with at most two dozen $bits_D$ of possible alternatives.[23]

The attempt to trace an event, produced by the decisions of a man, back to the positions of atoms is probably today not even of academic interest. It would be a different matter if, for example, we took the production of a protein as the event. Here we could already count the number of molecular decisions which need to be taken for its production. If this is true for these chemical building blocks of organisms, then it will also hold for whole organisms and even perhaps for their functions. No philosophical discussion is needed here, however. The prediction will suffice that events (E) consist by nature of systems which, likewise, are constituted by decisions (d) and these in the last analysis are of molecular type.

The difference between event and decision depends on the viewpoint. If we are interested in the end state, then we can call this an event (E). If we are interested in the intermediate states we can call these decisions (incorporation of new decisions into a system = d). I have already explained this in the introduction. As I wrote earlier:

Predecision	a	a	b	b
Final decision	a	b	a	b
Event	I	II	III	IV

In this case I contains no more than *aa*, just as a *b* implies no more than II. Suppose we equipped our source with more determinacy than previously — for example in the form of luminous figures which show I after the decisions *aa*. This again would imply a simplification since we have not described the decisions built into such a number display. For the present I do not need to say more — both decisions and events will concern us in the whole book, though the psychological causes of the distinction will not be treated until the end, in Section VIII B7*b*.

2. *The fate of decisions and of events*. Here, however, it is important to note that the fate of decisions is different to that of events. First, a decrease in the number of repetitive and thus redundant decisions (d; $bits_R$) is not necessarily connected with a decrease in the number of instances of a law (a), i.e. in the number of identically repeated events. On the contrary, if the repetitive decisions are not merely erased but, so to speak, economized by ranking and re-use, then a system arises which, by reduction of the number of incorporated decisions (d), actually favours the identical repetition of the events.

As a very simple example take the single repetition of our message with events I to IV. Before and after the elimination of $bits_R$ we obtain the following three values on repetition:

Before elimination of $bits_R$									
Predecision	a	a	b	b	a	a	b	b	} $R = 10$
Final decision	a	b	a	b	a	b	a	b	} $r = 2.66$
Event	I	II	III	IV	I	II	III	IV }	$a = 2$
After elimination of $bits_R$					x				
Predecision	a	—	b	—	—	—	—	—	} $R = 0$
Final decision	a	b	a	b	—	—	—	—	} $r = 1$
Event	I	II	III	IV	I	II	III	IV }	$a = 2$

As usual in the systemized example (below) the $bits_R$ are replaced by the memory (—) of the mechanism. We can save 8 $bits_R$ of visible decision redundancy and 2 $bits_R$ of hidden decision redundancy (i.e. of long-windedness). But the repetition of the message (a) is conserved assuming that a seventh decision can be inserted (predecision *x* with a content of about 1 bit_L).

Thus we have two different parameters for the identical repetitions that we refer to in ordinary usage as instances of the applicability of a law. (The expressions 'occurrence' and 'original plus replicas' convey the same idea.) On the other hand there is the repeated application of identical decisions which we have already met as *r*. On the other hand there is the repeated occurrence of identical redundant events (E), which we describe as *a* (the relative redundancy of events). The process of systemization (i.e. the systematic dismantling or reduction in the number of decisions) necessarily leads, when combined with the increase in size of the determinative systems (i.e. in the length of the messages), to the result that the number of identical events greatly exceeds the number of the remaining decisions ($r_{(syst)}$). Thus $a \gg r_{(syst)}$.

3. *The ratio* of these two parameters to each other depends in the first place[24] on the degree of systemization of the decisions. Before systemization, *a* is equal to the visible redundancy *r* and is still exceeded in quantity by the total visible and hidden redundancy of decisions. When the systemization is complete, however, all decision redundancy has disappeared and $r_{(syst)} = 1$. The difference between the two parameters is then at its greatest.

Thus we can describe the degree of systemization (s) by the quotient $a/r_{(syst)}$. Its values range from $s < 1$ to $s = a$. I deal with this again in Section III B2.

It is important that, when the determinacy content of a system $D_{(syst)}$ has ben fully systemized, then this also can be described as the product of the law content L and of the relative event redundancy *a* (the number of instances where the law applies).

$$D_{(syst)} = L \cdot a \tag{18}$$

This gives the composition of the systemized determinacy content. The structure of the statement of the law as well as the nature and extent of the repetitions can also be predicted from the minimal number of replaced decisions ($bits_L$). Compare, in this connection, our simple example – the twofold transmission of events I to IV.

The cause of this increase in systemization will be the basic theme of all later chapters. The following will emerge: the prospects of a system being conserved (i.e. its inner conditions being maintained under defined external conditions) increase both with an increase in the number of instances of the system that exist (a correlation which has already been proved) and also with a decrease in the number of interpolated decisions required to produce the system. For this decreases the reproduction costs, mistakes and adaptive difficulties that the eliminated decisions would have caused.

At this point, of course, this statement means both too much and too little. We need to consider the concrete aspect of biological order, so as to bring what has been said to life.

I do not doubt that with this preliminary essay I have sorely tried the reader's patience. To the biologist it may have seemed scarcely relevant and to the information scientist unnecessary. I think, however, that it will soon be evident that the problem of living order would not be soluble if I had not first tried to sort out the relevant epistemology.[25]

NOTES

1 Schrödinger's ideas were presented as lectures in 1943 and first published in 1944.
2 Well expounded by Strombach (1968).
3 Compare Driesch (1927) and Schubert-Soldern (1962).
4 Schrödinger (1969, p.73).
5 Bridgman (1941) cited in Morowitz (1968, p.3).
6 Morowitz (1970, p.169).
7 Going back to Wiener (1948), Shannon and Weaver (1949), and Fisher (1942).
8 This was very clearly expounded by Hassenstein (1966)
9 Instead of P_D, the probability of determinacy, I write here the probability of law P_L. This corresponds to the minimal number of decisions required to describe a set of lawful events. This will be justified in Section I B2. For discussion see Section I B4.
10 For background see, for example, Zemanek (1959), Hassenstein (1965), Flechtner (1970).
11 This problem is dealt with in information technology as counting through bundling and optimal group or tree selection, e.g. Zemanek (1959).
12 This was already pointed out by Weaver (p.95) in Shannon and Weaver (1949).
13 The first comprehensive presentations were published by Wiener (1948) and Shannon and Weaver (1949). Compare also Wiener (1952, 1961), Hassenstein (1966), Peters (1967), and Flechtner (1970).
14 (1949), p.12).
15 Compare e.g. Zemanek (1959) and others.
16 Linschitz (1953), Quastler (1964), Lwoff (1968); compare also Morowitz (1968, 1970), Monod (1971), Eigen (1971), Schuster (1972).
17 Recent discussion includes: Popper (1967), Woolhouse (1967), B. Campbell (1967), Wilson (1968a, 1968b).
18 Quoted from Flechtner (1970, p.74).
19 Lwoff (1968, p.84).
20 Linschitz (1953, p.261).
21 Lwoff (1968, p.93).
22 Linschitz (1953, p.261).
23 According to Dancoff and Quastler (1953) it is 24.5 *bits*.
24 Later we shall find that, in the evolution of systems, a is reduced to the benefit of L. This, however, cannot be dealt with yet (cf. Section VIII B7c).
25 The essentials of these epistemological questions can be found in Popper (1962) and D. Campbell (1966b). In particular Lorenz (1973) has endorsed the views here presented but his book, unfortunately, only appeared after the German edition of this book had been set in type.

CHAPTER II

THE DIMENSIONS AND FORMS
OF LIVING ORDER

This brings me nearer to the proper subject. For in the known Cosmos there is no other phenomenon whose order content begins to approach that of life. And there is no phenomenon of life which does not depend on an enormous structure of order. Lorenz[1] has stated that: 'Human knowledge, personal, cultural, and scientific represents but a special case of the principle by which organic life performs the miracle of developing, in seeming defiance of all laws of probability, in the direction from the less orderly and more probable towards higher harmonies of almost immeasurable improbability.' It is not only biologists, but also chemists, physicists, mathematicians, and epistemologists who agree that life is *the* dominant orderly phenomenon; it *is* order, pure and simple.

At first sight it might now seem fitting to discuss the strangest thing of all. This consists in the mechanisms which, by evading the law of entropy, allow such order, as particular universal patterns, to arise out of chaos. However, it is necessary to describe these phenomena, both quantitatively and qualitatively, as well as the associated problems, before proceeding to solve them. For, as I shall show, the wonder of the mechanism does not lie in its procedure, which is relatively simple, but in the miracle that it produces.

A. THE PARAMETERS OF BIOLOGICAL ORDER

In approaching a quantitative description of biological order, I shall first discuss the two estimates that biophysicists have already developed. Afterwards I shall derive a third estimate from comparative anatomy by the use of my theorem, i.e. order = determinacy content = $\Sigma\ bits_D$ = law × instances = law content × relative redundancy = $\Sigma\ bits_L \cdot r$.

1. Order as energy

The oldest of the three estimates describes order as energy. This was already foreshadowed in 1916 by Otto Meyerhof. In every organism continuous processes are in action which contribute to a decrease in potential energy. Meyerhof said: 'Since life requires the continuation of these potentials of energy, work must be performed continuously.'[2] And this can only happen. . . 'by the flow of energy from a source to a sink.'[3] Morowitz has recently summarized our knowledge in this field so I can limit myself to quoting his most important conclusion: '. . . A living cell represents a configuration showing a very large amount of energy as configurational or electronic bond energy relative to the amount of thermal energy when compared with the equivalent equilibrium system'.[4]

This equivalent equilibrium comes into existence at death. Simply expressed, order would correspond to the difference between the Helmholtz free energy of the living and the dead condition of the organism, of the: '. . . tension between storing energy and the decay of energy into the most random possible distribution'. Furthermore: 'The selection for stability plus the constant pumping by energy flow will lead to the largest possible degree of order'.

I cannot try to explain Helmholtz's free and thermal energy here. Also it is not necessary, because I shall not make further use of this energy concept. Instead I shall proceed by using the concept of information which is closely related to that of energy.

The connection between energy and information can again be illustrated by the paradox of Maxwell's Demon, who, as explained in Section I B3c), has been given previous knowledge of the movement of molecules. The energy which he can apparently build up, by sorting molecules, corresponds to the information which he must have about their movement... 'so that in some way information, which is a rather biological or even psychological concept, is related to purely energetic concepts'.[5] Conversely Brillouin has shown that work must be done in order to obtain information.[6] In brief: 'You don't get something for nothing — even information.'[7]

2. Order as improbability of state

The second estimate gauges order from the determinacy content and thus by the number of these determinative decisions which are necessary for its description, its stipulation or its construction. Redundancy and law content need not at first be distinguished in this connection.

It must be remembered: first that the measurements will be in $bits_D$; second, that the sum of the determinacy content of two equal, and thus equally unlikely, conditions is twice as big as the determinacy content of each one; and, third, that the probability of explaining determinative states by using accident increases with the negative power of the number of required decisions. Thus:

Type of event	1	2	3	4	1	2	3	4
$bits_D$ (cumulative)	2	4	6	8	10	12	14	16
Accidental	2^{-2}	2^{-4}	2^{-6}	2^{-8}	2^{-10}	2^{-12}	2^{-14}	2^{-16}
probability }	1/4	1/16	1/64	1/256	....			1/65,536

The determinacy content of the largest information system built by man gives a good basis for comparison, as Brillouin shows:[8]

'Let us consider, for instance, a telephone network of a size comparable to the American system. The order of magnitude of subscribers may be of a few ten millions, but let us be generous and assume one hundred million subscribers.'

The number of possible individual results in dialling is thus $E = 10^8$. Since we are sure of the determinatively working relays we can apply equation 6 and state: 'that the information content [or as we should say, determinacy content] of the whole system at each time must be of the order of:'

$$E \cdot \log_2 E = 10^8 \cdot \log_2 10^8 \approx 4 \times 10^9 \ bits_D$$

'This is a large number, but still very small in entropy units i.e. $4 \times 10^9 \times 10^{-16}$ erg °C or 4×10^{-7}. It is difficult to imagine any piece of machinery containing an amount of information much higher than in the preceding example, but if we think of living organisms we find a completely different order of magnitude.'

Nevertheless such a machine already has an accidental improbability of unimaginable dimensions. The number of trials which would be necessary to construct it by a random mixture of connections would be $2^4 \times 10^9 \approx 10^{1,204,120,000}$ which is a number with a thousand million zeros. The improbability of biological systems is very much greater than this.

The determinacy content of organisms can be estimated in the same manner. We consider the atoms or molecules in an organism which, because of their positions, are necessary for life. We then compare the number of physically possible combinations of these atoms or molecules with the very much smaller number which will actually support life. Or, alternatively we work out how many decisions are necessary to define the special, life-supporting position of these building blocks. It is not necessary to describe the arithmetical process here, nor to introduce suggested improvements to make it more accurate. The mere results will be sufficient.

For small bacteria the values are estimated at between 5×10^{10} and 10^{13} $bits_D$.[9] Even the simplest forms of life exceed the most complicated man-built machine in determinacy content by a factor of five or ten thousand and consequently exceed them in improbability by a very high power. Dancoff and Quastler[10] have made calculations for the human organism. They assumed 24.5 *bits* per atom times 7×10^{27} required atoms, of which only every tenth atom was assumed to be of definite position and obtained a determinacy content of 2×10^{28} $bits_D$. This is a quantity of stipulations which exceeds the content of all libraries on earth. Even on a molecular basis our own body has 2×10^{25} $bits_D$.

This informative study confirms another important fact. Human germ cells are reckoned to contain about 10^{11} $bits_D$, but the human gene catalogue (i.e. the pure 'germ plasm') has only 10^5 or 10^6 $bits_D$. We know that all the regularity of the human organism must be represented in the genetic code. There is therefore an increase in determinacy of between 16 and 21 orders of magnitude in passing from the genetic code and germ cell to the adult. It is an obvious question to ask where this increase comes from.

The answer which I can now give is again extremely simple: The difference consists predominantly of a — the repeated application of identical law. We need only consider how even the most specialized cells of our body occur as very great numbers of identical replicas, i.e. as identical transmissions of one and the same law content. Retinal cells number about 2×10^8, neurons 10^{12} to 10^{13} and erythrocytes in the course of a life between 2.5×10^{13} and 5×10^{15}. The parameter a bridges over the 15 to 16 orders of magnitude of difference in determinacy content between the germ cells and the adult human. Moreover, in each of these identical cells we find enormous numbers of identical organelles and ultrastructures, so that we would expect the quantity of relative redundancy (i.e. the numbers of instances of events a) to lie in fact between $a = 10^{19}$ and $a = 10^{21}$. I shall deal with this in more detail later (Chapter IV).

The important point here is that the determinacy content of the code consists predominantly of $D \approx 10^6$ $bits_L$ whereas that of the completed organism is made up of $D \approx 10^6$ $bits_L \times 10^{20} a$. This provides the key to the anatomical part of the solution.

3. Order as the extent of possible predictions

In describing the first estimate I proceeded as if we did not know how the determinacy system of the human organism was organized. Indeed we do not know the building blocks as completely as the Bell Telephone Company can know the network that it has constructed. But we already know more than we are commonly led to expect. Also the possible predictions are in the highest degree certain. An example will illustrate this.

If a tiny fragment of a human hair is found at the site of an accident the experienced criminologist can identify it from its microscopic structures. How many certain predictions could be made concerning the original owner of this microscopic structure by the cooperative efforts of anatomists, histologists, cytologists, students of ultrastructure, biochemists, and molecular biologists?

To answer this question we first need to separate predictions based on law content (L) from those concerned with relative redundancy (a). We have good information on both and also have little difficulty in separating them from each other.

So as not to repeat myself I refer the reader to what is discussed in detail in Section II B, i.e. the concepts of the anatomical plural and singular — of single individualities and the number of instances when these occur. The reader may either turn to that section to convince himself that the distinction is methodologically unobjectionable or he may simply read on.

The atlases of 'the Normal Anatomy of Man' show that there are about 10^4 predictable individual features, of the locomotory apparatus and 7×10^4 of the nervous system. Histology can add another 5×10^3 while cytology and molecular biology contribute almost as much. If these features are added together they make somewhere between 10^5 and 5×10^5 and these are $bits_L$. For they contain in the first place no redundancy. We can calculate the minimal information content of the digital decisions, assuming that for each defined feature there is only one alternative, as $D_{min} = 10^5$ to 5×10^5 $bits_L$. This should give the dimensions of the law content of our genome (specified by the improbability of the states) to within an order of magnitude.

The difference of a factor of five or ten between the maximal textbook information and the true law content of the human organism is easily bridged, as any specialist will confirm, by the difference between taught knowledge and total knowledge or, at least, by the extent of what has still not been studied.

The parameter a of identical repetitions in the human organism is even easier to estimate. It increases with decrease in complexity. The following figures for identical building blocks are known for the different levels of complexity: anatomy 2 to 10^7 (for example, symmetrical identical limbs at one end of the range, 10^7 identical hairs in a large mammal at the other end of the range); histology 10^3 to 10^{10} (for example, erythrocytes); cytology 10^{10} to 10^{16} (for example, chromosomes $\times$ cells); ultrastructure 10^{12} to 10^{20} (for example, the granula of the endoplasmic reticulum $\times$ cells); biochemistry 10^{15} to 10^{25} (for example, replicas of an amino-acid molecule); molecular biology 10^{16} to 10^{27} (for example, number of nitrogen atoms). The value of a thus certainly reaches 10^{20} to 10^{21}.

The number of possible predictions concerning an organism, therefore, despite our still limited knowledge, can reach values of $D = 10^5$ $bits_L \times 10^{20}$ a at least, to $D = 5 \times 10^5$ $bits_L \times 10^{21}$ a at most. This is a range from $D = 10^{25}$ to 5×10^{26} $bits_D$. These are extraordinary dimensions of predictability. The total quantity of order in organisms seems to have been approached by research to within one or two orders of magnitude, i.e. 25 of the 27 orders of magnitude are already documented by knowledge. This insight is encouraging for the steps that we shall take later.

Later we shall see that further dimensions of order must be added to these in considering the total phenomena of life, such as that of individuals (10^8) and of species (10^6).

B. THE FORMS OF BIOLOGICAL ORDER

I now need to make a qualitative analysis of living order. This represents a totally different type of problem. I shall have to formulate it almost anew, by a synthesis drawing on almost the whole subject matter of biology.

Up to now I have considered order merely as a quantitative phenomenon and was led to a quantitative formulation. But in discussing the qualitative aspects of order, which are epistemologically much more difficult to grasp, I consider it necessary to 'stick to my last' which is biology.

We are therefore again at a beginning, at a point where it is appropriate to look around us. So as not to beg any questions we need to return to basics and ask: What are the qualities of order? Must order possess qualities? And do such qualities have anything in common?

1. The qualitative aspects of order

At the beginning of the previous chapter I used the observation of the behaviour of a source to represent research into a still unexplained natural phenomenon. I return to this analogy again. It corresponds to a region of unspecified probability, a no-man's-land between accident and necessity. In this region order can only be recognized if there is repeatedly transmitted determinative regularity.

1. With these minimal assumptions the qualitative aspects of order is contained in the law content. Replicas of the law content contain only quantity. The law content, for example, of a continual transmission of the message '1 2 3 4 5 6 7 8' would contain $E \cdot \log_2 E = 8 \times \log_2 8 = 24 \ bits_L$ in quantity. But the qualitative aspect of the transmission would be the 'type' or 'pattern' of the regularity. This pattern could be called the 'basic law', 'exclusive nature' or the 'idea', though naturally this would not mean very much. It could however be described unequivocally as ($n = 1$ with $n, n + 1, n + 2 \ldots n + 7$). The qualitative aspects of order are very complicated and can only be clarified by repetition. Therefore they correspond to the idea of pattern as used both in cybernetics and ordinary speech.

For example, the 'basic laws' of the sine, the point, and the square differ from each other and so do the patterns that they form — being a wave pattern, a pattern of points or a checker pattern. Further, the formulae of sine wave and circle present dimensionless qualities and so the dimensions need to be added to the basic law to stipulate the pattern.

2. The existence of a qualitative essence is also presupposed by these minimal assumptions because every law content must contain an irreducible essence — irrespective of the complexity of the total law content or of the size of the essence. Thus in every determinative event there is a pattern. Otherwise we should not recognize it.

A totally different question, of course, is that of how many patterns can be expected. In the first instance only one should be postulated. For a world with only one pattern is just as conceivable as a world with infinitely many. It is one of the features of living order in our world, as I shall later show, that it contains only a small number of well-defined basic patterns.

3. The common quality of all patterns is the 'identicality of their individualities'. The reader will notice that we are already disturbingly close to the limits of scientific method and I will therefore assure him that we need go no further. For, first, this connection between identicality and individuality can be completely explained and, second, it gives a sufficient basis for study of the pattern of order.

This connection is a consequence of a. It has to be with repetition, with the number of instances, and with identical replication and indeed with adherence to law and with multiplication. It is expressed in the difference between 'the same' and 'similar'. This connection is the remarkable feature in all comparison, for when we compare we assume that behind similarity there is identicality. And that again is a matter of probability.

Thus, if after the message '1 2 3 4' we again receive '1 2 3 4' we say: 'That is the same'. We ignore the circumstances that the second message arrives at a different time, is on another part of the sheet of paper, the molecules of printing ink are totally different, and so forth. Indeed we would disregard the fact that the first message was printed in black and the repetition in red, or that one arrived by sound waves and the other by light waves in totally different parts of the brain.

We consider the action of accident and necessity. And, if too much speaks against accident and for necessity, then experience shows that we do better to compare the dissimilar and assume necessity, i.e. hold the hypothesis that, despite undoubted differences, 'the same thing' is behind the appearances.

In this way we have not only defined the qualitative aspect of biological order but have already taken the first step towards investigating it.

2. The building blocks: identicality of individualities

We must now consider the so-called homology theorem which the specialist relies upon as the backbone of the biological study of structure. This theorem is the essential part of the principles of morphology. And the principles of morphology, again, are the basis of comparative anatomy, systematics, and the study of evolution — especially of transpecific evolution which considers phenomena beyond that of interbreeding organisms. The homology theorem thus includes the epistemological basis for comparing living structures which corresponds to the problem of identicality in biology.

The importance of the homology theorem for the whole of biological research can easily be conceived. In some ways it corresponds to the causality theorem. Moreover, it is one of the oldest themes of contemporary biology, so the literature is large. It will shortly appear, however, that almost everything essential was said by Goethe, among early morphologists, and by Remane, among modern ones.[11]

By contrast with the homology theorem, the individuality problem is so small that I can deal with it at once. It is a question of specifying what shall be called a unit, a complex or a system. This refers to units which are limited both structurally and functionally and which can be recognized in large measure irrespective of their complexity. Examples are a gene, a chromosome, a muscle fibre, the biceps, the nervous system, an individual of the genus *Homo*, but also the call note of the reedling, the 'aggression system', an adenine molecule or only a hydrogen bond. The limits of units can naturally lead to much discussion but that is not our theme. It is sufficient that the units exist with the same certainty as when we recognize the message '1 2 3 4 5', under specified preconditions, as a repetition of '1 2 3 4 5'.

The purpose of this section, therefore, is to try to solve the identicality problem. In doing this I can draw on an extraordinarily wide range of biology — practically the whole of morphology, anatomy, and systematics. However, I shall have to advance in three different directions. First, homology is only a special case of biological identicality; second, a quantification of the similarity can be sketched out; and third, and above all, a quantification of homology and identicality can be suggested.

This quantification of homology and identicality is particularly important because doubts about the objectivity of the homology theorem have recently been expressed. The discussion has produced nothing useful but has undermined confidence concerning the fields of morphology and systematics. The controversy begins with so-called numerical taxonomy and revolves about 'weighting', 'reality', and 'phenetics' to which I shall return at the appropriate time. In the first instance it is necessary to take up an objective standpoint.

A quantification of homology and similarity is not a precondition for my further conclusions. It will, however, bring the benefits that go with clear definition.

a. The seven forms of similarity

I shall now turn to textbook biology for help, since we have well-defined preconceptions about the most important concepts of similarity.

The key to the similarity problem is the distinction between analogy and homology. In plain English this means the separation of outward similarity from essential similarity, as if the former had been added from outside, while the latter is thought of as lying in the essence of the objects compared. Analogy depends on immediate, direct, and functional comparison. Homology depends on logical operations where the comparison involves the whole of relevant experience.

The result of this complicated mode of comparison is known to everybody and can be expressed by saying: 'That is nothing other than' or 'That finally turns out to be'. It is important to realize that 'the first step in homologization exists in every naïve inspection.'[12] If it were otherwise, how would a child be able to call an elephant's trunk its nose, for externally it has little in common with a nose? And how otherwise could children, like primitive peoples, arrive at classifications which are largely correct?

1. *Analogies*. These are structural similarities for which we have to suppose that they arose convergently, i.e. from dissimilar origins. The mechanism which requires the production of analogies is assumed to be that of adaptation to identical functions. The criteria for recognizing analogy are the inverse of the criteria for homology (cf. Section II B2a2). Analogy is the opposite of homology. Distinguishing the two is of the same fundamental importance as distinguishing between accident and determinacy.

For our powers of insight into analogy it remains an accident which groups of tetrapods adopted the whole sea as their dwelling space and adaptation space (whether ichthyosaurs, whales, or sea cows). Or which grasshoppers were able to imitate a true beetle, which predatory fish to mimic a harmless cleaner fish, which insect a leaf or flower, and which flower developed a female copulatory trap for particular bees (cf. Figs. 3 and 4a-e).

The phenomenon of analogy will always remain a miracle, but ever since Darwin it has ceased to be a problem. Selection explains it completely. However, the extraordinarily improbable end products arise by the selection of minimal changes, for only these remain viable. The end products arise as the effect of sequences of attempts of inconceivable length. This is what is miraculous.

The concept of analogy, includes the only effects that can be predicted on the basis of evolutionary mechanisms already known in the phylogeny of organisms. For, whatever form of appendage starts to be used in swimming, it will become a paddle. Whatever shape of body has to pass rapidly through water, it will become fish-shaped (Fig. 3). However wonderful these complicated analogies may be, they are no longer problems. The true problem in living order remains that of homology.

2. *Homologies*. These are structural similarities which force us to suppose that any differences are explicable by divergence from an identical origin. Before the theory of evolution, the 'identical origin' had to be imagined as a concordant ground plan or

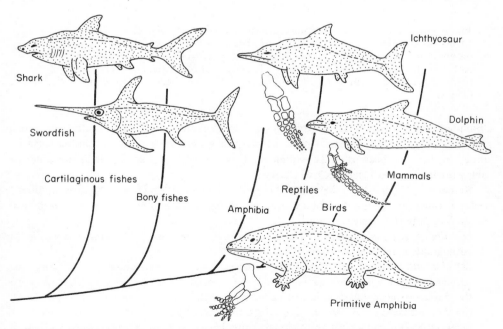

Fig. 3. Analogous occurrences of the fish shape and of the fin or flipper shape in the phylogeny of the vertebrates. Note the convergence of the ichthyosaur and dolphin flipper starting from the primitive tetrapod limb, e.g. *Eryops*. Compiled from Riedl (1970), Romer (1966) and Schindewolf (1950).

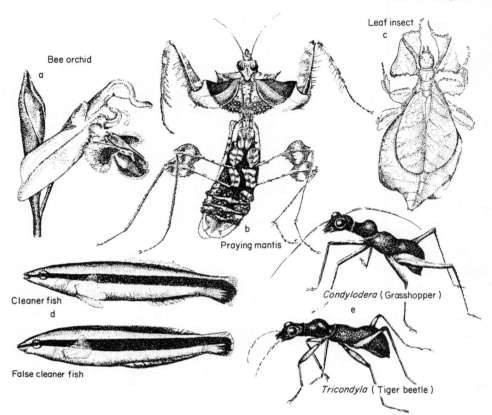

Fig. 4 a-e. Mimicry as an extreme form of analogy. (a) An orchid *(Ophrys apifera)* forms a female copulatory trap for bees. (b) The praying mantis *(Idolum diabolicum)* mimics a flower. (c) A grasshopper *(Phyllium pulchrifolium)* mimics a leaf. (d) *Aspidontus taenatus* mimics a harmless cleaner fish *(Labroides dimidiatus).* (e) A young grasshopper mimics a true valiant tiger beetle. Mainly from Wickler (1968).

'Bauplan'. Nowadays it is interpreted as 'common ancestry'. The mechanism which enforces adherence to identical patterns, despite the severest variations and changes of function, has not been causally explained. However, it is a chief factor in the order of living organisms, and I shall seek to clarify it.

Remane (1971 p.30 ff.) distinguished three principal criteria (1-3) and three auxiliary criteria (4-6) by which homology can be recognized. I shall quote him verbatim, modifying only the nomenclature.

(1) *The positional criterion.* 'Homology can be recognized by similar position in comparable sytems of features.' (Fig. 5a-j)

(2) *The structural criterion.* 'Similar structures can be homologized, without reference to similar position, when they agree in numerous special features. Certainty increases with the degree of complication and of agreement in the structures compared.' (Fig. 6a-e).

(3) *The transitional criterion.* 'Even dissimilar structures of different position can be regarded as homologous if transitional forms between them can be proved so that, in considering two neighbouring forms, the conditions under headings (1) and (2) are fulfilled. The transitional forms can be taken from the ontogeny of the structure or can be true systematically intermediate forms.' (Figs. 7a-f)

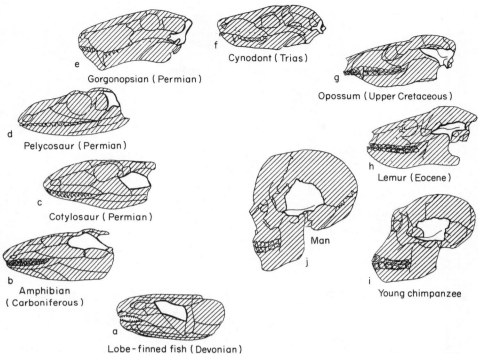

Fig. 5 a-j. Differences in form of a homologous feature as exemplified by the development of the temporal bone (not cross-hatched) from crossopterygian fish to man. Note the form and position in f. From Gregory (1951).

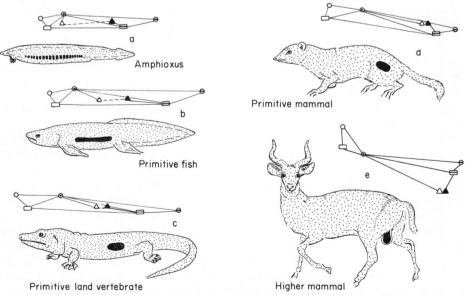

Fig. 6 a-e. Variations in position of a homologous feature exemplified by the evolution of the gonads from amphioxus *(Branchiostoma)* (a) to the ungulates (e). The gonads are shown black in the outline drawings and as triangles in the diagrams. From Portmann (1948), supplemented by diagrams of the positional relationships.

35

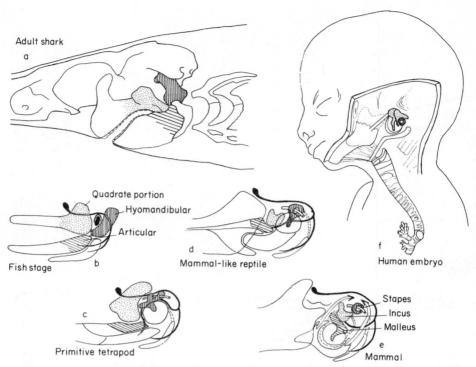

Fig. 7 a–f. The importance of transitional forms for homologization, exemplified by the evolution of the auditory ossicles (hammer, anvil, and stirrup) of the mammals (e, f) out of the primitive jaw apparatus of the fishes (a, b). Without knowledge of transitional forms (d) the identicality of the parts could scarcely be recognized. Compiled from Braus (1929) and Portmann (1948).

(4) *The general conjunctional criterion.* 'Even simple structures can be regarded as homologous when they occur in a great number of adjacent species.'

(5) *The special conjunctional criterion.* 'The probability of the homology of simple structures increases with the presence of other similarities, with the same distribution among closely similar species.'

(6) *The negative conjunctional criterion.* 'The probability of the homology of features decreases with the commonness of occurrence of this feature among species which are certainly not related.'

3. *Homoiology.* This covers similarities of structure when they include both analogous and homologous substructures. Homoiologues can also be called analogies on a homologous base. The term homoiology is, however, misleading in that mixtures of convergent and divergent courses of evolution cannot exist in individual single features. The distinction between analogy and homology remains intact. However, the term is appropriate in so far as many analogies are constructed on homologous foundations. Thus the total vertebrate features of ichthyosaurs and whales are homologous, although the fish-shaped outline imposed on these features is analogous. Likewise the basic plan of their tetrapod limbs is homologous but affected by analogous modifications to form fins (cf. Fig. 3).

4. *Homodynamy.* This refers to causes which result in homologous effects. Such causes could also be regarded as commands that are followed in identical manner.

The concept of homodynamy was formulated by Baltzer (1950, 1952) and since then has also been applied to processes in developmental physiology, e.g. the effect of the inductional commands going out from the optic vesicle to form a lens in the overlying skin. In such cases it would not be justified to use the word 'homologous', for at present the identicality can only be recognized secondarily by using the effect. (This subject is illustrated by Fig. 65–67.).

5. *Isology.* This, on the other hand, is a similarity concept drawn from chemical relationship. It is important to us as a way of specifying the limits of the homology concept. According to Florkin[13] (see also Fig. 8b): 'The biochemical compounds, molecules or macromolecules which show signs of chemical kinship, we shall call isologues. Cytochrome, peroxidase, catalase, haemoglobin, and chlorocruorin exhibit this isology, as they are heme derivates.' Isologues may be homologous or they may be analogous, although all homologues are probably based ultimately on isologous compounds. The distinction as to whether isologues are homologous or analogous is again a matter of probability.

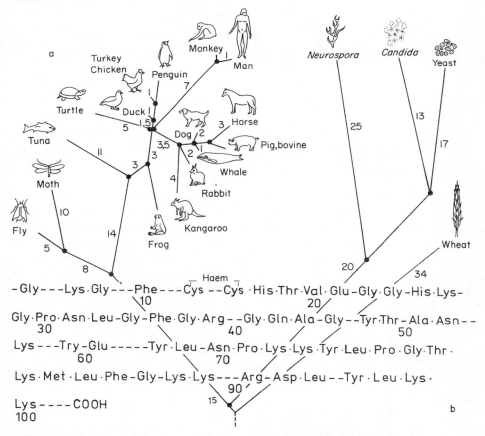

Fig. 8 a-b. The homology of isologous giant molecules exemplified by the degree of similarity of cytochrome c. (a) The phylogeny of the cytochrome c molecule on the basis of similarity. The numbers of mutations required to produce the changes are inserted between the hypothetical branching points. (b) The position of the 58 amino-acid residues which are identical between yeast and man in the cytochrome c sequence. Altered residues are replaced by dashes. From Smith and Margoliash (1964), Florkin (1966), Dayhoff (1969).

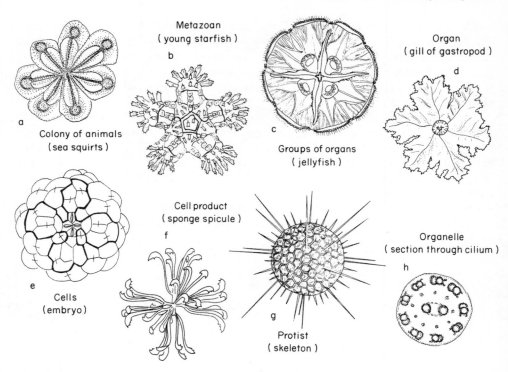

Fig. 9 a-h. Examples of radially symmetrical standard parts taken from seven planes of decreasing complexity. Even the diameters of the radial structure range in order of magnitude from decimetres (c), through centimetres (d) and millimetres (a, b) to 100 μm (e), 10 μm (f), and 0·1 μm (h).

It would be meaningless to refer to inorganic molecules of identical structure as analogous or homologous. For convergence and divergence have to do with ancestry and inheritance. Where then is the limit between the concepts of homology and isology among the organic molecules or organisms? Here Florkin,[13] the expert biochemist, reaches the same important conclusion as I shall do in anatomy when I give a quantitative solution of the homology theorem. Concerning the almost unbelievable agreement of the amino-acid sequence of the cytochrome c of mammals and yeast (cf. Fig. 8a-b) he says: 'Such a degree of isology is incompatible with chance effects.' Isologies of very high accidental improbability can be recognized as homologies.

6. *Homonomies.* These are structural similarities or identicalities between the building blocks of one and the same individual. The differences are thought of as divergences from identical basic forms, several of which occur in the same organism. Examples are the identicality of vertebrae, leaves, hairs, etc. Homonomy has also been called serial homology.

Here again I follow Remane.[14] However, he considers that homonomy is in principle different from homology on the grounds that it has nothing to do with phylogeny. I do not agree that there is a difference in principle. On the contrary, in the last analysis we are dealing with the same mechanism which is of the same fundamental importance for the formation of order in living organisms — whether such identical individualities become separated from each other to occur in different individuals or whether they replicate within the same individual.

7. *Symmetries.* These can be of radial or bilateral form and affect the axes and planes, not only of individuals, but also of their parts. For a long time they have not been seen as connected with the identicality problem. It is easy to appreciate, however, that

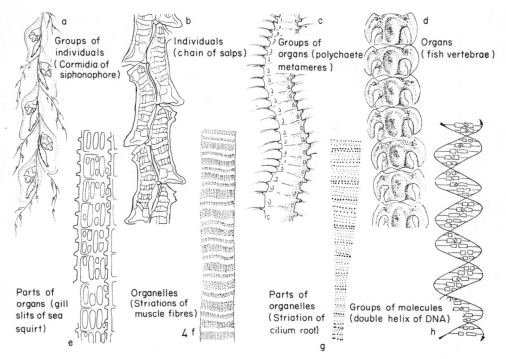

Fig. 10 a-h. Examples of serial, positional standard parts from nine levels of decreasing complexity. The diameter in these series range from centimetre dimensions (b-d) through millimetres (a), 100 μm (e), μm (f), and 0·1 μm down to an order of magnitude of 10 Å (h).

homonoms within the same living body can show not merely identical structure but also identical position. But such identical position of identical parts is symmetry, seriality, or pattern. In short it is everything which, starting from the position of a structural unit, makes everything else seem determinative and predictable with certainty.

Examples are the radial forms which range from individuals (in ascidian colonies) through organs (as in coelenterates or flowers) cells (protists) to organelles (e.g. flagella) and parts of organelles (e.g. of cilia) (see Fig. 9a-h). The same levels of complexity also exist for seriality (Fig. 10a-h) extending from groups of individuals (cormidia of siphonophores), through individuals (chains of salps), parts of individuals (segments of earthworms), organs (polychaete legs), parts of organs (vertebrae) and cells (of the notochord) to parts of cells (muscle fibres), part of organelles (cilia) down to the molecules of the genetic code.

b. Homology and identicality

After this preliminary survey of the different types of similarity we can attempt the first step in our synthesis. We assert that everything which can be recognized as similarity is connected in some way with homology.

Thus analogy is the inverse of homology. Homoiology refers to a structure which includes homology along with analogous substructures. Homodynamy is homology of effect. Isology refers to similarity whose homologous or analogous character has yet to be decided. Homonomy is the homology of building blocks within one individual, while symmetry, seriality, or pattern specify the positions of these building blocks relative to each other. All similarity is the identicality of individualities. Only analogy, which masks the principle, has a special position. For similarities are analogous on the basis of external accidental decisions. All the other forms of similarity are similarities of 'internal conformity to law', of superdeterminacy as I shall have to show. Homology is their central form.

This last assertion should for the moment be taken on trust. This will help the linear development of the train of thought, for I intend to examine the homology problem further after discussing the characteristics of the identicality of structural laws. I shall then immediately re-examine *in extenso* the principles so obtained.

I shall try later to develop an objective solution to the problem of identicality by means of this central phenomenon of homology. Before doing so I must examine two important characteristics of homologues — their mutual arrangement and their limits of occurrence.

1. *The mutual arrangement of homologues*. The term homologue refers to a homologous feature of an organism. The mutual arrangement of homologues is in every system hierarchical. This means that most homologues consist of subordinate features and conversely that several combine to form a homologue of higher rank.

An example should be sufficient to show this convincingly. There is no doubt of the homology of the vertebral column of, say, the mammals. It consists of cervical, thoracic,

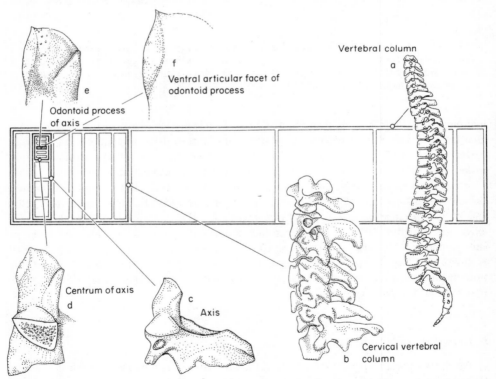

Fig. 11 a-f. The hierarchical arrangement of homologues illustrated by a hierarchical series in the human skeletal system. The series consists of five cadre homologues (vertebral column to odontoid process, shown as thickly drawn rectangles in the diagram) and a minimal homologue (ventral articular facet of the odontoid process, stippled in diagram).

lumbar, sacral, and caudal regions. Again, taking the first of these, the neck regions of all mammals are identical. Each neck region includes, as a rule, seven vertebrae and again each one of the seven is homologous. The second, for example, is always called the axis (= epistropheus) and is recognized in every mammal by means of the odontoid process which, besides other things, distinguishes it from all other vertebrae. The axis again, consists of neural arch, apophyses, and centrum, and the centrum consists of the main part and of the odontoid process. Nobody doubts the identicality (or homology) of this odontoid process in all mammals. But the odontoid process is again characterized by five parts. Each of these is a homologue as, for example, the ventral articular surface. And the entire vertebral column is only one of the homologous skeletal features of the group.

The same is true for the vascular and nervous systems, the muscular system and in short of all homologues of the mammals and all other organisms. The universal characteristic of all hierarchies can be recognized in the fact that each homology has meaning and content only if, and when, the concept of next higher rank is first named, and only when it possesses its subordinate concepts. I shall return to discuss exhaustively the phenomenon of hierarchy as one of the four fundamental patterns of order (Chapter V).

2. *Cadre homologues and minimal homologues*. Given this basically hierarchical arrangement, homologues can be distinguished according to their rank.

I shall use the term cadre homologue for all those which provide the framework or cadre for further subordinate homologues. In the above example this would be all those homologues from the concept of the vertebral column down to that of the odontoid process. They all have their own individuality because, although they differ in rank, they can all be predicted with certainty (Fig. 11a-f).

For a vertebral column of middle degree of differentiation the number of cadre homologues can be calculated as follows: 1 inclusive concept (vertebral column); 5 regions in the column; 7 vertebrae per region (on average); 4 main parts of the vertebra per vertebra (on average); 5 subordinate parts of the vertebra per main part (on average) = 1 + 5 + 35 + 140 + 700 = 881 cadre homologues.

I shall use the term minimal homologue, on the other hand, for all homologues at the bottom end of the hierarchical sequence – the minimal homologue in the example was the ventral articular face of the odontoid process. They are characterized by the fact that they cannot be further divided into subhomologues in their particular hierarchical sequence. No homologues occur beneath them but only identicalities of a different kind. This will be discussed in Section II B2b4. This lower limit is important because it limits the number of individual homologues in every system and allows them to be counted, given sufficient knowledge.

For the vertebral column in our example we count on average 5 minimal homologues for every lowest cadre homologue. Thus there are 5 X 700 minimal homologues and 3500 + 881 = 4381 homologues predictable in total.

Now I shall define the limits of countable homologues. Three such limits can be distinguished.

3. *The limit at the individual organism.* This individual limit is the easiest to describe. If in the hierarchical system of homologues the single homologues are gathered together, step by step, to form homologues of higher rank, then, wherever in the organism we begin, we soon reach the concept of the individual organism. If we were to call two mammals homologous we would be overstretching the concept of homology. But every summation of homologies leads finally to the concept of the individual organism whose identicality and individuality seem to be obvious, just because both concepts first arose by inspecting the individual organism. Our supposition that homologues are identical individualities is confirmed by summation.

4. *The homonomy limit.* This is reached on the other hand by progressive division of

homologues and again with the same regularity. If we divide a minimal homologue into its parts, as for example the ventral articular face of the odontoid process, then we reach bone cells or Haversian columns. These are likewise identical individualities, but they are represented by large or extraordinarily large numbers within the individual. And their identicality is so great that within a type of homonom they are impossible to distinguish. At this limit we pass from the individualities of the 'anatomical singular' into those of the 'anatomical plural' — from the individuality of the homologues into that of the homonoms where again we do not doubt the identicality of the representatives of the same type. Think, for example, of the similarity of the 10^{14} red blood corpuscles in a man.

Naturally the anatomical horizon or level of complexity at which the limit between maximal homonoms and minimal homologues is found depends on the differentiation (and integration) which the group of organisms has reached. We always come upon identical homonoms, however, when we subdivide the lowest level of organized homologues.

In the lowest organisms this limit lies between homologous ultrastructures and homonomous groups of molecules. In protists it lies between organelles and ultrastructures. In lower metazoa with constant number of cells it lies between cells and organelles. In other animals it lies between tissues or organs and cells. Ascending from this level of organization the concept of homonomy applies to groups of organs and metameres up to those individuals or even groups of individuals in colonies which by integration in the colony has given up their freedom as individual organisms. All this will be documented below. So also will the fact that the word homonomy can be applied in the broadest sense for all identical mass individualities in the organism.

5. *The uncertainty limit.* This is the third limit to the homology concept. Not all single structures of the organism that lie between the individual and its homonomous constituents can be homologized unobjectionably or excluded as being analogies. There is a zone of uncertainty, as might be expected from the homology criteria. Sometimes it is a question of lack of insight (into structure, positional relationships, transitions or conjunction). At other times a structure does not possess that degree of differentiation and constancy that would raise it out of the anonymity of the homonomous mass as a singular, unique individuality. Nevertheless the uncertainty limit is also well defined and this is what matters.

Two examples each may illustrate these two types of uncertainty. Thus, to exemplify lack of insight there are four groups of primitive worms which possess attachment tubes, but in comparing these, both the first and third criteria of homology fail, i.e. those of position and transition. The second criterion, that of structure, also has not helped up till now for these organs are too small to be resolved by the light microscope.[15] They await electron microscopical analysis. Again *Xenoturbella* has remained a systematic problem animal, since it has too few special structures and too few similarities with systematically adjacent animals (criteria 3-6).[16] The homologies will be resolved only by means of still undiscovered transitional forms or by the study of development.

To exemplify the difficulties caused by lack of distinctiveness, there is no doubt about the homologies of the great blood vessels of man. However, the capillary vessels, lying at the extreme ends of the system, form a huge anonymous mass of nameless homonoms. Between these two conditions there is a narrow region of small terminal vessels which are at the limits of constancy and identifiability. Again, in mammals the vertebrae are all specialized as single individualities. In fishes, however, most of them are an anonymous uniform crowd. The primitive tetrapods include the transitional forms.

6. *The constancy in number of homologues.* The numerical constancy of the homologues of a system or an organism depends on the constancy of the three limits. All three are probably in motion, moving slowly with the advance of differentiation and our experience. This movement is slow enough, however, for us to be able to calculate the total for each condition of development or of knowledge.

The variations in counting homologues within single systems, or subsystems, or in discussing interpretations, are small compared with the absolute numbers. A sceptic may leave out half the homologues that another worker has defined, but even so the factual differences from system to system lie several orders of magnitude beyond such differences of opinion.

Homologues therefore appear as countable and identifiable, hierarchically arranged, single individualities. They are bounded above by the total individualities which we call identical individuals. And below they are bounded by the mass individualities of their building blocks.

c. Individuality and law content

Before continuing I should like to sum up the position which we have now reached. In examining the orderly phenomena of life we have established enormous dimensions of accidental improbability, i.e. of the predictability of determinative decisions. Qualitatively we foresee patterns whose building blocks appear as always identical individualities, meaning structures which seem to follow the same law. The caution implicit in the last sentence is justified so long as we have not established how far this supposition may, in fact, be correct, though everything seems to point towards it.

I shall examine this question starting from homologues, which are the most critical points of structural identicality. My reason is that two identical individuals must depend on identical laws and the same is true of identical cells, whether these are separate or whether they remain together in the metazoan manner as identical building blocks. This identicality of individual organisms or of homonoms is much more obvious than that of homologues.

1. *Event, feature, and probability.* We can be fully objective in trying to judge the reign of accident or of regularity if we make the distinction by comparing probabilities. To be precise we compare the probabilities that a chain of events can be predicted by one of the two 'causes' which are possible in the world.

The probability of the reign of law (P_l), taking the number of instances into consideration (P_{la}), can be defined as determinative probability divided by determinative plus accidental probability, all to the power of the number of disappointed expectations (cf. equation 5): i.e. $P_{la} = P_D^a/(P_D^a + P_I^a)$. The accidental or indeterminate probability of an event depends on the number of decisions that the system leaves to chance.

Such an event, like 'heads' in tossing a coin or '32' in roulette, plainly corresponds to what in orderly patterns we have called an identical individuality. For we are forced to suppose, not only that the feature 'heads' represents a constant individuality, but also that it is identical, whenever and wherever it occurs. The identical single event can thus be equated with the single homologue.

Another question, as we have already seen, is how many alternatives the respective single homologues could possess. There must be at least one alternative but there could be several of them. In the first place, however, we cannot specify the number. I shall therefore be generous and proceed on the minimal assumption. I shall suppose only two alternatives as in coin tossing. The information will consequently be equal to one *bit*, irrespective of whether the decision is accidental or determinative.

2. *The decision content of the system.* This can now be calculated from the sum of the homologues that are found by applying the structural and positional criteria, i.e. from the position-structure. I shall now proceed to prove this.

Assuming that every system can only have one probability we take the second synthesizing step. The first step led to the conclusion that there was only one homology.

But now I shall establish that there is only one criterion of identicality. Position and structure (Remane's first and second principal criteria) mean one and the same thing as concerns the probability of the presence of identical determinative laws.

Our example of a hierarchical series of homologa in the vertebral column (cf. Fig. 11) is sufficient to prove this. In the structural statement 'cervical region of the vertebral column' the axis vertebra is the structural feature 'between vertebrae 1 and 3'. But in the positional statement 'axis vertebra' these same relationships to vertebrae 1 and 3 are positional features within the cervical region of the vertebral column. In the structural statement 'axis vertebra' the odontoid process is the structural feature 'cranial to the main part of the centrum of the axis and medial to the cranial articular surfaces of the axis'. But in the positional statement 'odontoid process of the axis' these same relationships to the main part of the centrum and to the cranial articular surfaces are positional features in the axis — and so forth. Position and structure refer to the same identical characters up or down the line along which homologues are hierarchically arranged.

On the other hand the third or transitional criterion of homology (cf. Section II B2a2) simply disappears in this connection. Thus if, in comparing A and C, a third object of comparison B is introduced, then all the criteria of homology are valid for the comparisons AB and BC. Thus the transitional criterion is inherent in the unified identicality criterion. This, of course, does not diminish the importance of intermediate forms — indeed it increases it.

3. *Features times number of instances.* A third step in synthesis is to establish that structural agreement is related to conjunction of occurrence as is a law to its instances, or, more precisely, as is law content to the number of instances where the law applies.

Conjunction of occurrence is understood here in the sense of the criteria of homology nos. 4-6 (Section II B2a2). It is thus understood as the occurrence of identical structures in closely related species. We could also say in the sense of a conjunction with other features that possess the same systematic distribution. In this connection we must remember the precondition for recognition of law which is the repeated independent occurrence of identical messages. And we also need to establish that a single occurrence, even of the most salient feature, would not permit any homologization.

For what can homologization be compared with, when no comparison exists? How would law be recognized without repetition? We see that homology corresponds to something more than law content. It represents an order content or determinacy content. This is law content times the number of instances or $D_{(syst)} = L \cdot a$ (cf. equation 18). Thus a homology is a homologue times its various different occurrences or the law content of the homologue's position-structure times its conjunctional occurrence.

$$D_{(syst)} = \text{Position-structure} \times \text{conjunction of occurrence (cf. equation 18)}$$

The sixth homology criterion is that of negative conjunction and is merely the converse of the fifth criterion of special conjunction. The first, second, and fourth criteria can also be formulated as reciprocals. They all define non-homology, i.e. analogy or the unexpected if the reign of internal laws or determinative laws is assumed.

Seeing that the homologues of a system can be counted, the law content can be derived from the structural and positional criteria while the number of instances is given by the general and special forms of the conjunctional criterion. Two important values can therefore be extracted. The first is the probability of the presence of determinative happenings and the second is the order content (of determinative decisions), within which law content and number of instances can be distinguished.

d. Homology and order

No general solution can be expected here. On the contrary, what follows is merely a preparation for a quantification of the homology theorem. It is worth remembering that the solution of the problem of living order does not require the quantification of homology but only the proof that homology really exists as a determinative condition. It is this proof which is important.

However, the agreement with the quantification of orderly phenomena extends beyond specifying the probability of the expectation of law. I stress this so as to develop to its limits the basic thesis that: 'Homology is determinacy'.

Apart from a few attempts, morphologists have abandoned the search for such a quantification. Remane[1][7] says: 'In principle it would be possible to make a more exact estimate. However, we are still uncertain of the basis for assigning values to the various criteria. We do not know how to establish, for example, how the value of the third criterion increases with increasing numbers of intermediate stages, or how it should be estimated for ontogenetic and morphological intermediate stages. And we are still uncertain how the various criteria mutually increase or decrease each other.' Consequently Remane considers that only a very rough estimate would be possible.

This is a most remarkable prophecy and I can fully confirm its truth. I would merely add that the third or transitional criterion disappears in a chain of comparisons, that the first and second criteria (of position and structure) prove to be identical as also are the fourth and fifth (general and special conjunctional criteria). Furthermore, criteria (1 + 2) and (4 + 5) complement each other as L and a. For each case there is only one probability of determinacy and from this point of view only one homology and there must therefore exist a single value for its estimate.

1. *The estimate* can relate to the usual $bits_D$ or required determinative decisions if we are certain of these. Alternatively it can relate to the quotient $bits_D/bits_I$ when we are uncertain. I shall first consider the former and simpler case.

If we assume, as previously, that the occurrence of a homologue has an accidental probability of ½ (only one alternative) then the improbability of explaining it in terms of accident increases with the number of identically occurring instances of the homologue as the power of 2. The determinacy content (in $bits_D$) increases linearly with this number. And indeed both numbers are dependent on whether the law content increases or the number of instances. For $D = L + R$ (cf. equation 9).

As before, coin tossing can be used to illustrate a binary accidental decision. A little system with five homologues can thus be represented by five coins while its particular structure is represented by the defined position of these coins (e.g. all five are tails-up). The accidental probability that this particular condition could occur at a single throw is 2^{-5} or 1/32. A system with ten homologues reaches an accidental probability of $2^{-10} = 1/1024$.

But the same values would be reached in another experiment in which the five coins are thrown twice ($a = 2$). The accidental probability that all will be tails-up in both tosses is $2^{-5 \times 2} = 2^{-10}$ and thus again 1/1024. This second toss, or second transmission of the identical message, corresponds to a statement that identical positional and structural situations occur a second time independently. Biologically speaking independently means, for example, in an independent genome, or in another species.

2. *The probability of law*. We have already recognized this (equation 5) as the relationship: $P_L = P_D/(P_D + P_I)$. When the event is certainly predictable assuming the reign of determinacy then $P_D = 1$ and $P_L = 1/(1 + P_I^a)$. The probability that a constant structure depends on determinative laws, i.e. that a true homologue is present, corresponds to the quotient of unity divided by unity plus the accidental probability. If accidental probability approaches zero then the quotient is approximately unity and our certainty is very large. This probability increases as the power of each identical feature and each identical repetition. Figure 12 illustrates this connection.

Position-structure (in $bits_L$) and the number of identical systems (in r or a) have the same effect on the attainable degree of certainty if these repetitions can be seen as independent realizations, i.e. each depends on its own sequence of decisions.

For the completely impartial observer (who cannot exist among man) even the most closely related member of the same species would represent an independent repetition. Also the decisions in two genomes are always separated from each other in space. But we have forgotten to be surprised that our hands are identical to our father's hands. Also we have learnt that the genome of one species is in a state of perpetual mixing. On the other hand we are still surprised that homologues occur in totally separated species such as the 'fingers' of a dolphin's fin or a bat's wing. In our calculations we can therefore be generous again and regard one species as one repetition.

The great importance of repetition has already appeared twice independently as a precondition for the recognition of law and in all the conjunctional criteria of homology. We can again confirm Remane's opinions by establishing that, in actual fact, it is small or obscure law contents which specially require consistent repetition for us to be convinced of their existence. On the other hand, suppose that our certainty of the conformity to law of a system is confirmed in a thousand species, with a hundred identically conserved homologues. Then the discovery of the thousand-and-first confirming species does not appreciably increase our certainty, for it was already virtually absolute. The probability of explaining the number of events by accident was already less than $10^{-30\,000}$ — a value with 30 thousand zeros after the decimal point. It does not signify if this number is further reduced by another case.

Only the very simplest systems, with less than five supposed homologues and less than five repetitions, have any appreciable accidental probability. But these would scarcely be called homologous by an anatomist. With all systems of identical position-structure which have greater complexity and constancy the certainty is extremely large to absolute. For them we must conclude that determinacy reigns.

3. *Law content and order content.* Once their determinacy has been established, these are easy to estimate. As shown in Fig. 12, the law content (L) of a system corresponds to the total of its structural and positional homologues while the order content or determinacy content (D) corresponds to the quantity of homology. This latter is the sum

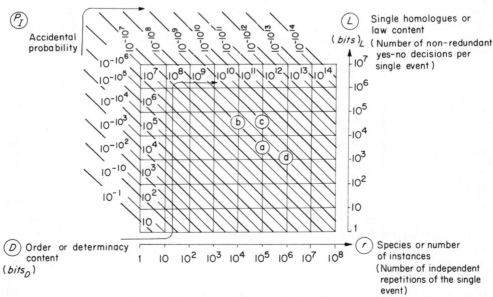

Fig 12. The probability of homologies, in the case where features are constant, according to the formulae $D = L \cdot a$ and $P_L = 1/1 + P_I$. The accidental probability (or uncertainty of homologization) decreases with the number of single homologues and the number of species that show them. (a) = vertebral column of vertebrates; (b) = nervous system of mammals; (c) = the nervous system of vertebrates; (d) = nervous system of insects. In big systems like these the accidental probability corresponds to a value with a billion zeros after the decimal point.

of the homologues times the number of species (r or a) which show these with the same position-structure ($D = L \cdot a$; cf. equation 18). Thus L, as already shown, reaches dimensions of 10^4 to more than 10^5 $bits_L$. For example, for the osteology of the vertebral column we have 4.4×10^3, for the locomotory apparatus 10^4 and for the nervous system 7×10^4. The number a, of species showing homologues, ranges from 10^4 to 10^6 (mammals 10^4, vertebrates 10^5, insects 10^6). Thus individual homologies ($L \cdot a$) reach orders of magnitude of 10^8 to 10^{10} $bits_D$. Their accidental probability (Fig. 12) would be a number following some billion zeros after the decimal point.

As regards counting the number of homologues, I emphasize the three simplifications that I included by way of caution so as to avoid overestimates in all cases. First, the cadre and minimal homologues are counted according to the structural criterion. For determining their position the homologa equal in rank to the highest one considered ought really to be taken into account within the homologue of one rank higher.

Second, I assumed that the identical minimal homologues consist of identical homonomous building blocks. It is impossible to do otherwise, but the law content of the homonoms is thereby excluded. The number of different types of homonoms involved ought likewise to have been considered, since they are all homologous and their determinative regularity is particularly certain because of their considerably increased redundancy content.

Third, up till now I have only considered identicality versus non-identicality of the supposed homologues. Taking the single homologue as the smallest unit of information, with only one alternative, justified counting it as 'one'. However, if we took degree of proportional resemblance into account, even assuming the same number of identical homologues, then clearly a further quantity of information would be brought into the comparison.

In fact, therefore, the attainable certainties and order contents are even higher.

e. *The problem of degrees of similarity*

Before finishing this discussion of the forms of similarity, a second relevant group of questions should be touched upon. Up to now I have discussed only the qualitative forms of similarity, but naturally there are quantitative aspects also — the degree of similarity within a given form of similarity.

The problem therefore is that of quantifying qualities. The main stumbling block is that a solution becomes more difficult to reach as the object becomes more complex.

This can be seen in comparing two straight lines. Here measurement is unequivocal, so long as it is compatible. But even with two triangles there are three lines, three angles, and a surface area to be compared and we first need to establish which corresponds to which and what weight the parameters should have. With irregular surfaces the possibilities for comparison become very numerous. With organic structure they increase even more.

Consequently approximate solutions have to be adopted. The more precisely the principles of comparison are defined the more reproducible the solutions become. And the principles of comparison becomes more precise the more details they consider. Obviously in considering analogies, such as the horn of a rhinoceros with that of a hornbill or of a rhinoceros beetle, there is no sense in going into much detail. With homologues, however, things are different.

We need this quantification of similarities only in dealing with homologues and I shall consider it in the framework of the hierarchy phenomenon (Section V B1*f*) where it is most needed (cf. Fig. 38a-f).

f. *Freedom and necessity*

Up to now I have only considered homologues that recur with certainty. This was sufficient for a first estimate of the extent of law and probability of law. In a world of accident and necessity, however, we must always expect the action of both. This is true for homologues also, for these identical individualities likewise show variations. This is because they now exist where once they did not, that they have come into being, that they can become widely distributed although they must once have been narrowly distributed and they are subject to transformations.

Consequently even in constant systems there will be a certain measure of freedom, of indeterminacy in a determinative framework. Patterns of different levels of freedom and determinacy are the essence of evolution, and of natural order in general, and will be dealt with in detail below. However, in the four basic patterns of living order which I discuss later the levels of freedom and determinacy are as different as the four mechanisms that cause these patterns. Consequently I shall also have to consider the dynamics of freedom and fixation in the four chapters devoted to these basic patterns. At present I shall only emphasize what all four have in common.

The ratio of freedom and fixation corresponds in expressions of probability to that of accident and determinacy. And these two values specify the three crucial insights used in the study of mixed systems, i.e. systems made up partly of accident and partly of necessity. The first of these insights is the probability with which we expect the reign of law $P_L = P_D/(P_D + P_I)$; the second is the determinacy content $D = \log_2(P_D/P_I)$; and the third is the volume of experience $I_D + D = constant$ (cf. equations 4, 8, and 17).

A simple example will show how these expressions apply. Imagine we discover three new species and compare in them a supposedly homologous system with only four positional and structural homologues, which we suspect are identical. Two of these turn out to be constant while two show gaps in the expected conjunction. We can symbolize this by three tosses of the coins 1 to 4; we obtain the following picture if we signify a positive result by H (heads) and a negative result, corresponding to an absence or an alternative, by T (tails).

H1 H2 H3 H4   H1 H2 T3 H4   H1 T2 H3 H4

For the whole system therefore:

1. $P_L = P_D/P_D + P_I$. The probability of a determinative explanation P_D is disappointed in two single events, i.e. $P_D = 1/4$ (cf. Section I B1*e*). The opposite explanation in terms of accident (P_I), on the other hand, is disappointed in four out of the six double events $0.25/0.25 + 0.0625 = 0.8$. It is therefore possible, but by no means certain, that we are dealing with identical regularities and thus with homologues.

2. $D = \log_2(P_D/P_I) = \log_2[(1/4)/(1/4096)] = \log_2(4096/4) = \log_2 1024 = 10\ bits_D$. If the system depends on identical regularities then it includes 10 determinative decisions.

3. The total experience in the current state of investigation contains $12\ bits$ which, corresponding to the not very great probability (estimated in 1), consists of $2\ bits_I + 10\ bits_D$.

This shows the evidential value of discovering further, related species, discovering a positional conjunction in additional cadre homologues, and of discovering that one of the features contains additional minimal homologues. It also shows the uncertainty which makes estimates of 2 and 3 difficult; $P_L = 0.8, D = 2\ bits$, total content = $1\ bits_I + 2\ bits_D$.

All this suggests that the potentialities of an organism in evolution consist in the ratio of freedom and determinacy in its building blocks. This ratio of accident to necessity in the individual building blocks specifies, by the interplay of all the hierarchically arranged parts, the prospects presented to the organism by both accident and law.

I shall show that a dominant portion of structural conditions and of evolutionary prospects has long been excluded from the effects of accident. This excluded portion consists of the four basic patterns of organic order.

3.   The patterns of open questions: the identicality of regularities

Where are we then? We have found that the qualitative aspect of order lies in the first place in the qualities of regularity — in the qualities of that component part of an organism which in morphology is named the homologue. However, as we saw, the identical individualities of repeated regularity exist not only one beside the other, but one within the other, in a condition of mutual dependency each with its respective individual history. Biological order consists not only of regularity in the component parts, but also in their arrangement — we can anticipate this from what has already been said concerning this arrangement.

Indeed, regularity of component parts would have no meaning without regularity of arrangement, nor vice versa. Our nature makes us think linearly and therefore it seems necessary to us that the identicality of the building block is a precondition for the identicality of their arrangement. But in fact it is only a precondition for deducing this arrangment. We are convinced of the identicality of homologous parts by accidental improbabilities so large that they are impossible within this Universe. We can therefore now start to consider the identicality of their regular arrangement.

What types of regularity do we need in order to describe the patterns that the building blocks make up? Can these patterns be distinguished and counted? Or am I constructing a problem so as to be able to solve it? I shall have to deal with all these open questions. Some of them have not been obvious previously. Most, however, have long been known and are as old as the theory of evolution or indeed as old as thought about the laws of living structures. These old questions, in fact, are so fundamental that they amount to the problem of transpecific evolution itself. This problem is the perplexity remaining when we consider the laws of phylogenetic development, whose products we are ourselves.

I shall discuss these open questions in connection with a first look at the above-mentioned patterns formed by the building blocks — the four patterns of the standard part, hierarchy, interdependence, and traditive inheritance. For the four open questions are all consequences of the four patterns of order, being instances of these four laws whose causes we shall seek and find.

It is the most surprising result of this study that the open questions of phylogeny constitute the four basic patterns of organic order.

a. The standard part

The first pattern of organic order consists in the universal occurrence of standard parts or units (*Normen* in German). These exist in a limited number of types but in an unlimited number of identical replicas (definition in Section IV A). It is characteristic of these parts that they range in dimensions and complexity through more than two dozen orders of magnitude from the biological molecule through the single individual to the colony. They are less like the symbols of an alphabet than like those of algebra which can precede brackets of unlimited content. They are arranged hierarchically within one another as explained in Section II B3*b*.

Of the four basic patterns of biological order this is the only one which, even in its subordinate aspects, has not previously been recognized as a problem in the biological literature. All the others have been swept by argument and in fact constitute *the* controversies of biology, both ancient and modern. This peculiar fact may be the reason why the total problem of the nature of biological order has not long ago been solved. The standard-part pattern is the entry point into this complex of questions. I therefore have double reason to be precise.

1. *Complexities.* The universal concept of the standard part in biology therefore has to be developed here for the first time. This is because the identical individualities are not only very diverse in extent but also vary greatly and, what is most confusing, are separated from each other in space to very different degrees. Pairs of identical molecular sequences, ribosomes, cilia, cells, organs, metameres or brothers seem at first to be so different from one another that even my meaning will not immediately be understood.

I must therefore point out, in the first place, that there are three levels — lower, middle, and upper, in this sequence of ascending complexity and with each level the identicality of similar individualities is self-evident.

The identicality of two genes of the same type is self-evident since one can be derived from the other as if by matrix subtraction. The identicality of two cells is also self-evident. For by identical subtraction from the matrix of their possibilities everything is suppressed except, say, the features of a striated muscle fibre. And the identicality of siblings is particularly convincing in the example of homozygotic twins. Our feeling of conviction is supported, in the latter case, by knowing the replication mechanisms.

However, there is no reason to doubt the identicality of the corresponding gene complexes whose decipherment produces the morphologically identical components from large molecules to organelles. Nor is there reason to doubt the identicality of the corresponding cell complexes whose genetic possibilities are directed to produce morphologically identical components ranging from tissues to metameres. We shall later recognize the matrix mechanics involved in intermediate dimensions of component parts and convince ourselves of the identicality of the determinative decisions that produce them.

2. *The individual and the problem of individuality.* In the ascending sequence of individualities three levels can be recognized as regards the way in which the individualities separate from each other. These differences in mode of separation likewise make it difficult to see the problem of biological identicality as a single whole. The three levels of separation, in ascending order of complexity, are cell divisions, reproduction, and speciation. These themselves seem to be very different.

Here, however, we are dealing with total commands which consist of determinative statements and form a gradual changing continuum. In cell division the 'punched tapes' on which the commands are recorded are replicated individually and they separate from each other, but the individuals that carry them remain adjacent and share common fates. In reproduction the punched tapes are again identically replicated and they separate, but

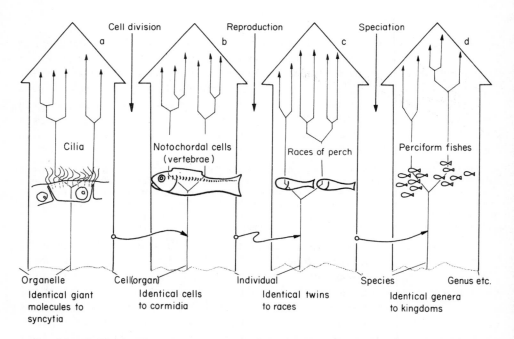

Fig. 13 a-d. The levels in the separation of identical individualities. The thin arrows show the branching of the paths of identical determinacy content. The broad arrows show the higher ranking frameworks of individuality which recur as branching paths of determinacy (thin arrows) in the frameworks of next higher rank. (a) Identical individualities in the cell; (b) between cells and individuals; (c) Identical individualities of individuals; and (d) of phyletic groups.

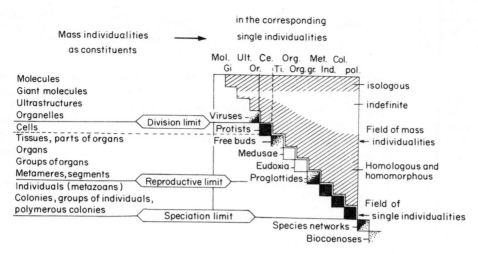

Fig. 14. The field of individualities arranged according to the hierarchical stage of complexity of the mass individualities (left) and their occurrence in the single individualities (above : Mol. to pol.). In the field each portion to which the traditional concept 'individual' applies is shown black. Their marginal regions, with examples, are stippled.

so do their bearers (Fig. 13a-d). However, there is no breakage in the exchange of commands between these punched tapes. The bearers can expect that the bases of command will remain connected. In speciation there is a separation of the punched tapes. These are replicated almost identically at first, but there is also a cessation of the exchange of identical commands. The total fates become separate.

In actual fact the system of levels of types of separation is even longer than this. Below cell division the punched tapes replicate but do not even become spatially separate. The so-called polytene or giant chromosomes are an example of this being bundles of from 10^3 to 3×10^3 'punched tapes', all of them probably absolutely identical. Above speciation, on the other hand, the identicality of commands gradually breaks down which eventually excludes the possibility of forming homoiologies – the analogous reactions of an identical inheritance to identical conditions.

These three levels in mode of separation emphasize the three limiting cases of individuality which are the cell, the individual, and the species. These limiting cases tend to mask the continuity of the phenomenon of identical individualities. Nevertheless, there are, in the strict sense, no superindividualities. One could say, perhaps, that every individuality in the long, hierarchical sequence of complexity is a superindividuality with respect to the subordinate individualities as explained under the hierarchy. The individual is a special case within this sequence of complexity (cf. also Fig. 14). For in the first place all three limits, whether of cell division, of reproduction, or of speciation are imprecise (Fig. 13a-d). And, second, the concept of the individual applies over many levels of complexity (Fig. 14).

Concerning the imprecision of the limits, the speciation limit can be seen as applying 'too late' when species unite to form hybrids. The reproductive limit applies too late when individuals fuse or fail to separate so forming colonies, or in the attached growth of male dwarfs. The cell-division limit applies too late when cells fuse, or fail to separate and form syncytia. On the other hand the species limit applies 'too early' in polymorphism or caste formation. The reproductive limit applies too early when cells, cell groups, organs or segments (proglottides) separate off as individuals in the various forms of asexual reproduction. And the cell-division limit applies too early, when parts of cells or incomplete cells lacking the 'punched tape' survive for a certain time, as with the erythrocytes of mammals (Fig. 14).

Concerning the applicability of the concept of the individual over several levels of complexity, individualities can be recognized lower than the cell concept as, for example, viruses or possibly mitochondria. Between the concepts of the cells and the individual we can recognize as individualities those tissues (e.g. buds), organs (e.g. sporangia), organ groups (eudoxia) or segments (proglottides) which are not complete individuals because of specialization in the colony or incomplete division. Above the species concept we can recognize as individualities such indivisible agglomerations of species as in lichens, for example (Fig. 14).

3. *Homomorphs and homologues.* The third seeming difficulty that might argue against a unitary individuality concept is the number of identical individualities per individual. Homology signifies identicality from species to species within affinity groups while homonomy is identicality from organ to organ within individualities. However, all homonoms would be homologous when compared from species to species. Consequently the only difference between homologues and homonoms is the number per individual. And it would be absurd to construct a difference of principle on this basis.

For example the identicality of the breasts of a woman with the udders of a mammal is not altered by the fact that they are only a single pair (as a rule).

The identicality of the genital apparatus of a giant tape-worm is not altered by the fact that it is developed *not* singly but several times, indeed several thousand times, for the animal reaches a length of 60 m.

It is only the look of the phenomenon and the phylogenetic consequences which are different, as Remane[18] already pointed out, but to that extent the concepts of homology and homonomy can justifiably be distinguished. In particular the homonomy concept has classically been limited to larger structures. However, our probabilistic definition of identicality allows us to extend the homonomy concept to the tiniest organs and cells, down to ultrastructures or indeed giant molecules. To avoid stretching the classical concept of homonomy this totality of homologues represented by several or many examples per individual can be described as homomorphs -- this word has not been overloaded by usage and signifies 'like in form'.

It was probably intended that the limits of homonoms in the classical sense would lie within the range of the optical microscope. For the number of positional or boundary features of a homonomous ossicle, for example, is less by several orders of magnitude than the features revealed by ultrastructural research for a striated muscle fibre or a cilium, or by biochemistry for a giant molecule. This is even more true for the total number of conjoined features which for cells and giant molecules reach 10^{14} to 10^{20} per individual.

4. *Anatomical singular and plural.* We thus find identicality of individualities, i.e. a dependence on identical determinative decisions, in a field that includes all living structures. This field extends to the limits of uncertainty of our probabilistic definition. It is a single field of decipherment events and extends from the smallest identical pieces of genetic code up to whole 'manuscripts'. It is indifferent to whether these occur as 'anatomical singular' or 'anatomical plural'. It is also indifferent to the fact that the anatomical singulars, or homologues, are dealt with under morphology, anatomy, and systematics, whereas the anatomical plurals or homomorphs are found in the textbooks of cytology, histology, ultrastructure, and biochemistry.

There is an extraordinarily large number of these building blocks which are known to be standardized individualities and are therefore predictable. In Section A of this chapter I discussed these qualitative orders of magnitude. In man, for example, we can predict 10^5 to 5×10^5 individual homologues which correspond to the law content, and up to 10^{20} or 10^{21} identical homomorphs, corresponding to the redundancy content. This is an order of magnitude of from 10^{25} to 5×10^{26} standard parts (i.e. a number with 26 zeros!).

Furthermore, it has been shown above that this number of standard building blocks closely approaches the determinacy content calculated from and required by molecular position. It must be remembered that, so as to be generous, I counted only one single alternative per standard part, i.e. 1 bit_D so as never to overestimate the determinacy content. This leads to an important result. For 5×10^{26} standard parts, because of this simplifying assumption, could well represent more than 5×10^{26} $bits_D$. As research into total structure advances we might reach values of 2×10^{25} to 2×10^{28} $bits_D$. In the present connection this signifies that organisms must be built up almost exclusively of standard parts.

5. *The standardization of structure.* This thus appears as a fundamental pattern of organic order and must depend on an equally fundamental mechanism or indeed be required by this mechanism despite the most various adaptational necessities. It is not at first sight obvious why Nature builds exclusively with standard parts, when standardization is not at all the aim of evolution, but merely its resultant event. A biologist will already be able to see the mechanism but it can easily be explained to the layman also.

6. *Standardization of position.* This would also be expected, especially for standard parts occurring in huge numbers.

Classical morphologists repeatedly began with the principal positions, axes and symmetries in organisms. These seemed to them to be basic and unifying and, as we shall see, in this they were entirely right.

The positional determination of molecules, from the individual molecules of the genetic code up to the giant molecules, can be understood from the laws of chemical combination. Concerning that of chemical ultrastructure and organelles, too little is yet known. The positional determination of the levels of complexity, ranging from cells to the great symmetries and axes of the metazoans and their colonies, is described by gradients which always act on the standard building blocks of the level in question. I shall later point out a unifying principle for these gradients also.

A close and necessary connection with the next pattern of order, i.e. hierarchy, can already be recognized. In practical terms they are the same thing. The relationship is like that of a letter of the alphabet to grammar, of a symbol to algebra, or of words to syntax.

b. Hierarchy

The second basic pattern of biological order is that of hierarchy. It consists in the fact that all the standard building blocks of living structures are fitted inside each other in a system of frameworks which mutually require and determine each other. There is a striking similarity to the hierarchical system of our conceptual thought and in this fact the problem lies (definition in Section V A).

A survey of the individual problems which derive from the basic pattern of hierarchy is simple. By contrast with those connected with standard parts, most of them have long been seen by biologists, have been argued about, and mostly recognized several times. To start with I can therefore limit myself to showing that five problems, although they all look very different, are all derivatives of the phenomenon of hierarchy.

1. *The reality problem.* I begin with the most basic problem. The question is how we can believe in the reality of order when it corresponds so completely to the way in which we order our thought. Do we not need to suppose that this order has merely been projected by us into Nature? It is already obvious from Fig. 14 that no single concept, out of the ten levels from biological molecule to individual, can exist without containing all the subordinate ones, nor outside the totality of the higher-ranking ones. And we shall see later that above the individual there is a further sequence of levels of systematic concepts. The identicality of the individualities cannot be doubted, nor can their hierarchical order. As soon as I have shown the necessity of hierarchy I shall have to stand the problem on its head and ask: How is it that our thought repeats the hierarchy principle?

2. *The homomorphy problem.* This is the only one which, like the phenomenon of

homorphy itself, needs to be formulated for the first time. It is implied in the question: How can we understand the limited number of types of homomorphous standard parts and their extraordinary constancy? Hidden in this question are a whole group of problems, not all equally obvious but all unsolved.

One of these problems is the discrepancy between the theoretically immeasurably large number of possible combinations of organic building blocks, on the one hand, and the very much smaller number of species realized. This is a biophysical problem. In evolutionary theory it has repeatedly been asked whether we must assume some inner principle or, from a different viewpoint, a basic plan for phylogeny, so as to explain this puzzle. The connection with hierarchy has remained totally hidden.

These problems will turn up again when considering the constancy phenomenon — a condition of superdeterminacy which is a necessary consequence of hierarchical order.

3. *The homology problem.* This has been the subject of an increasingly sharp controversy over the last twenty years. One side asserts that comparative anatomy is impossible without ascertaining homologies. The other side says that homology is a thought construct which is not taken out of nature, but thought into it, and is therefore worthless. They argue that no mechanisms are conceivable which would cause the persistence of homologues. For all genes have virtually the same random chance of being changed.

This appears to be a very conclusive assertion, able to bring the whole of classical morphology down with a crash. But the answer is equally clear and basic. Homologues are fixed simply in so far as they carry a hierarchical burden.

This burden is reckoned, as I shall show, by the number of features (or of their determinative decisions) which have become dependent on a homologue in the course of evolution. I shall not anticipate further.

4. *The problems of morphotype, ground plan, and weighting.* These constitute the converse of the above-mentioned first consequences. If the homology theorem is vulnerable then the notion of morphotype, i.e. the generally applicable characteristics of an affinity group, becomes a Platonic idea and morphology ceases to be a science. At the same time the ground plans of groups of organisms become fictitious and the weighting of features, by which an experienced worker judges affinity, becomes mere prejudice.

As against this I shall show that the position and burden of homologues determine their degree of fixation, and this latter determines morphotype and ground plan and also the assigned weighting.

5. *The problems of biological classification and of systematics.* These are indirect consequences. If morphotype, ground plan, and weighting were unscientific, then so also would systematics be, and also the concept of a natural classification as being self-contradictory. Beware the avalanche of false consequences!

In actual fact, however, all the premises on which natural classification is based are correct. Indeed they are causally required. There is therefore no need to abandon the study of phylogenetic affinity, which is one of the most remarkable syntheses in human thought, but rather we must gain insight into its causes. Indeed, the problem should be stood on its head by asking how morphology, without knowing its own causal foundations, could produce the correct synthesis which natural biological classification is.

6. *The hierarchy of features.* There is, therefore, a universal basic pattern of order. If so, it must be penetrated by an equally universal mechanism acting against the manifold adaptational demands which the environment makes on the organism. In the last analysis, not even the antagonism of analogy and homology can be understood without this mechanism which I shall explain in Chapter V. It has the same cause as the standardization of parts and depends on the same necessity.

c. *Interdependence*

The third pattern of order can be called interdependence. It consists in the mutual dependences of features and of the determinative decisions that form these features. These mutual dependences extend beyond those of sorts of standard parts or of hierarchical sequences. Mutual dependence is so universal that an interdependent feature has no meaning without its partner and would produce no meaning. This is like the contents of human concepts (definition in Section VI A).

A number of much discussed questions turn out to be sub-problems of the interdependence phenomenon. I put them into three groups.

1. *The problems of single connections.* These include the coadaptation problem, corresponding to the question: How is it that features of different phylogenetic origins develop in nice coordination with each other? The interdependent connection is obvious.

2. *The problems of directionality.* These are the problems of trend, orthogenesis, and typostasy as well as that of Cartesian transformation. These are four forms of the same problem which is based on roughly the following controversy. One side says that the directionality of evolutionary paths is so obvious that the mechanism of mutation plus selection cannot explain it; instead an internal regulator of evolution must be discovered. The opposing side says that there is no directionality but mere tendencies at most; these are nothing special and no inner principle has yet stood up to examination.

The connection with interdependence is probably not yet obvious. It will immediately become clear, however, when we remember that different features have different degrees of freedom, i.e. variations are tolerated to very different extents. This fact requires directionality in itself. But this requirement is strengthened when mutual dependences arise as necessary results of the coordination of determinative conditions.

3. *The problems of coordination.* These follow naturally. They all concern the same miracle — the functional directionality that aims at the 'complete organism'. They are the problems of regeneration (and asexual reproduction), of regulation, of homoeosis and of the organic nexus.[19] The recurring basic question is: The structures of organisms display an extraordinary degree of balance and purposiveness; how can we explain this without supposing some unknown internal regulator?

Interdependence thus likewise turns out to be a form of order penetrating the whole organism. It ranges from the control of the adaptational possibilities of individual features to that of phyletic groups of organisms and from the regulation of single dependences to the harmonious picture of the whole individual. The mechanism that results in interdependence is similarly universal.

The pattern of order produced by this mechanism is fundamental enough but cannot be made visible in the same manner as the standard part and the hierarchy. It is, as shown later, a four-dimensional 'Gestalt' along a time axis. It can, however, be seen along all possible time axes from the minutes-long axis of physiological regulation to the billions-of-years-long axis of evolution. But dynamic phenomena are no less real than static ones. Many of them are merely longer lived than their observers. Thus the orbit of the Earth is no less real than the planet Earth which follows it.

In the same way the fourth and last basic pattern of organic order is also a four-dimensional phenomenon, although its time axis is much simpler and easier to understand. Indeed it is so closely related to interdependence that it could be called successive interdependence as opposed to simultaneous interdependence. I shall name it traditive inheritance.

d. Traditive inheritance

The order pattern of traditive inheritance again consists in a universal connection. This shows itself by the fact that no organic structural condition is conceivable without its predecessors (definition in Section VII A). The pattern of traditive inheritance therefore depends on the fact that all structural conditions represent sequences of coordinations. This, once again, is so necessary that no temporary condition would have meaning without what it produces, and no end condition would be possible without all its predecessors. In the same way the sequences of letters 'padre', 'Vater', 'father', and 'père' are constant in meaning but different in form. This discrepancy can only be understood by chains of scarcely noticeable changes from the common Proto-Indo-European sequence 'pəter'. There is obviously a whole group of relevant subordinate aspects to be mentioned, especially among the open problems of physiological and morphological embryology.

I must emphasize that this basic pattern of traditive inheritance has, of course, already been recognized as a connected whole. Schrödinger's order-on-order principle[20] clearly means the same thing, although referring mainly to the physical aspect. Shrödinger's synthesis, which is already a generation old, has been held in respect. In this point, however, it has never been built upon so far as I know.

The individual problems are again very various. Indeed at first sight it will seem doubtful that all (as I shall show) can be seen as derivations of the traditive inheritance. There are about a dozen such open questions. I shall briefly survey them, grouping them according to five main types of connection with the phenomenon of traditive inheritance.

1. *The problem of old patterns.* This includes those phenomena known to the biologist as atavism, vestigialization, and neoteny. They all have in common the question: How is it that archaic character-states are so stubbornly preserved in organisms, or indeed retroactively re-established? In atavism it is bygone conditions taken from phylogeny which are re-established, e.g. a little tail in man or four nipples. In neoteny the re-established features are taken from embryonic development, being larval features occurring in adult organisms. In vestigialization there is no unanimous explanation for the obstinate persistence of features which seem to have long been functionless.

2. *The recapitulation problem.* This unites two questions: Why is it that conditions passed through in phylogeny are repeated during embryonic development (Haeckel's biogenetic law)?[21] And why do the 'building instructions' of related organisms show the same degree of relationship as the organisms themselves? By 'building instructions' I mean the pattern of biochemical compounds which, in the embryological development of organisms, are required by every structure so that differentiation will be correct in time, form, and position. The similarity of the places from which these commands proceed is the so-called induction pattern. The similarity in the effects of the commands constitutes the problem of homodynamy.[22]

I shall later discuss these problems in detail, especially as they can all be explained as necessary results of traditive inheritance which itself is necessary. At this point I have merely listed the problems connected with traditive inheritance. In one respect, however, I must anticipate: long-accepted facts come to seem self-evident. Haeckel's law, of the ontogenetic recapitulation of phylogenetic stages, has been confirmed by more than one hundred years' unmixed success in the study of relationships, so that to us it is self-evident. However, even the highest degree of unanimity about a law gives no indication of its causes. We have merely learnt to live with the unknown entity or to accept pseudo-explanations such as: 'Nature does not make jumps' or 'Everything shows its own origin'. No doubt there is much truth in folk wisdom, as this shows. But science must explain its own laws by necessities, not by proverbs.

3. *The problem of the irreversibility of phylogeny.* This belongs here also and can be expressed by the question: Why are homologues, once lost, never formed again?[23] A dolphin's fin is no longer a fish's fin, however alike they may be in external appearence. The 'calling back' of old patterns, so long as they are preserved in the archives of

inheritance, must be expected. Just as a pattern no longer in the archieves will, because of the effects of accident, never be produced again. Such a resurrection would be inconsistent with our stochastic definition of homology.

4. *The problems of the switching-on of complete patterns.* These are the centre of the complex of problems that the traditive mechanisms needs to explain. I must state here, however, that I summarize the problems knowing what the solution is. In the specialist literature only particular single problems are recognized. These look at first so diverse that even to summarize them (as is important for the argument) requires a brief glance forwards. All the problems listed here have one question in common: How can whole complexes of meaningfully interconnected determinative decisions be switched on by a single mistake in giving out the genetic commands?

The matter is particularly interesting for another reason. This mistake in the commands may depend on a mutation, i.e. on a mistake in punching the genetic punched tape. Equally it may be a so-called phenocopy, i.e. in a certain sense a copy of mutational change. This happens by experiment when a mistake is introduced into the commands transmitted during development. Indeed, such mistakes in transmission may be introduced into the system without our intervention, as sometimes happens, for example, in regeneration. I shall discuss these natural mistakes first.

Heteromorphoses. These raise the question: How can it be that in the mistaken regeneration of an organ, a meaningful structure of the wrong type arises, instead of a mere medley of features? An example is the replacement in a crab of a lost antenna by a biramous limb. Virtually the same problem arises when the mistake in development is caused by experimental disturbance. This is called a phenocopy, e.g. the doubling of the thorax with almost all its external homologues in the fruit fly *Drosophila*.[24]

The doubling mechanisms which lead to the formation of complicated systems are no less astounding. Examples are double-headed calves and doubling of the legs of beetles.

Homoeotic Mutations. These present the problem in a very similar way. It is a question of a single mistake in the genetic punched tape with complex, self-regulatory, meaningfully balanced consequences. The question is: How can single mistakes produce a self-maintained purposiveness which is wrong in position or number? Examples are the replacement of antennae by small legs[25] or of a haltere by a wing.[26] Such regulative mutations are also called systemic mutations. This name is a good indication, as I shall show later, of what really is behind them.

Spontaneous atavism. In a certain respect this is a special case of homoeotic mutation such that a stage which the mutant organism has passed through in phylogeny appears again in the correct anatomical position. The particular question is: How can a single mistake in the genetic punched tape cause a complete bygone structural pattern to appear. An example is the three-toed mutant of the domestic horse.

5. *The problem of the morphotype of a natural group in its genetic aspects.* This problem, lastly, corresponds to the totality of the single problems in physiological embryology mentioned above. The concept of the epigenotype, being the sum of mutually acting gene effects, tacitly assumes the reign of comparable and thus related traditively inheriting principles. The archetype problem, as formulated by Waddington,[27] approaches even more closely to a synthesis of the problem of traditive inheritance. It contains the hypothesis that there can be only a limited number of types of epigenetic system, each corresponding to a major phyletic group.

This points the way to a new formulation of the problem of the morphotype of a group. The morphotype is a phenomenon which we already recognized in morphology as the central focus of all the various interconnections, although its reality has recently been doubted.

Traditive inheritance, the fourth pattern of order, therefore acts as universally as the other three. It extends over all grades of complexity from the sequential dependence of single gene effects to the 'orchestration' of whole epigenetic systems, or from individual features up to the developmental types of the great taxonomic groups. A corresponding mechanism, equally universal in its action, must be found which can establish this basic pattern. This mechanism must be able to extend the pattern of traditive inheritance into every side branch of the organic world.

4. The interconnections of the patterns

At this point the reader may ask why, if we are speaking of universal patterns, there should be four of them. This is of more interest than to know whether there might be a fifth pattern or even still more. Such extra patterns would not be surprising, for in this first attempt at a synthesis in the field of general patterns of order, it is certain that some things will remain undiscovered. The important question, rather, is why the structures of life on earth are ruled by such different-seeming lawgivers as the standard part, hierarchy, interdependence, and traditive inheritance. Where is the higher lawgiver which would explain this differentiation into four?

The four basic patterns form a unity, just as the problems within each of the four basic patterns (as just listed) can be seen as subproblems of one of them. These subproblems are instances of a law, just as the basic patterns are themselves consequences of one principle. This principle can be referred to as that of mathematical or geometrical symmetries. All such possible symmetries are realized in fact.

The proof of this assertion will be given in Chapters III and VIII. In Chapter III it goes with an explanation of the mechanism that necessarily produces the four basic patterns. In Chapter VIII (Section VIII B7*f*) it is discussed on its own. In the present section it still remains to explain the structural connection of the four patterns (Fig. 15).

This complicated scheme of presentation is needed because I am near to the limits of familiar ideas. I wish to produce conceptions which do not generally exist. The thing in itself is not otherwise difficult, as will be obvious when I have described the mechanisms (Chapter III). Every universal law is simple. It is only the instances that are complicated.

As already mentioned the connection between standard part and hierarchy is like that between a letter of the alphabet and grammar, between a symbol and algebra, or between a word and syntax. The 'meaning', so to speak, of the one determines the other. Of course the grammar of various languages using the same roman letters can be as different as Cyrillic, Greek, or Latin scripts applied to one and the same language. But a word results only from the hyper-system: letter-grammar. Standard part-hierarchy could be compared, perhaps, with rank, quality, content, or structure.

The connection between interdependence and traditive inheritance, on the other hand, I have compared with simultaneous and successive interdependence. This could also be expressed as condition-history. Once again there is a hyper-system which only produces its 'meaning' when it includes both types of content. Interdependence-traditive inheritance could be compared with connection, position or function (though this would not be more than a comparison).

The total connection (of standard part-hierarchy with interdependence-tradition) is the most interesting of all, and as simple as the others. Standard part-hierarchy on its own, as Fig. 15 shows, does not define a unique connection, for within the various ranks

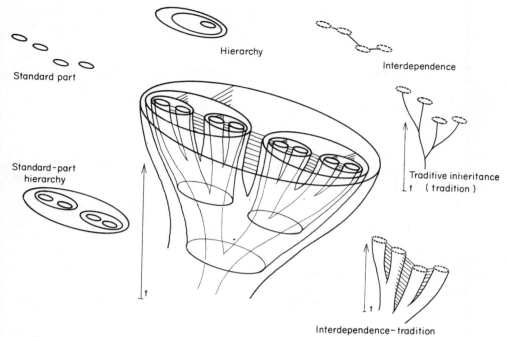

Fig. 15. A symbolic representation of the four basic patterns of standard part, hierarchy, interdependence, and traditive inheritance. Standard part-hierarchy and interdependence-traditive inheritance are represented also, as combination stages towards the total pattern. The origin along the time axis *(t)* of the three contemporary parts of the pattern is shown under traditive inheritance.

the connection would be arbitrarily variable. And connection between such undefined contents would be empty. The total connections form a whole, and the four basic patterns are its parts. They are like contents-connection or structure-function (though again this would be no more than a comparison).

However, in this summary I only need to show that the four basic patterns also form a unity in structure and function. This unity is a whole whose necessity remains to be explained.

Before taking this step, however, the problem must be approached from a second direction. Up till now I have considered what explanation could be expected of what pattern. Now I shall discuss the total problem, which equally requires solution.

C. BIOLOGICAL ORDER AS A PROBLEM

The real problem of biological order is the epistemological situation of pure morphology. Because of the truly inconceivable complexity of biological order, the causality concept of morphology is far behind that of the experimental biological sciences. This is so true that the controversy about whether pure morphology and comparative anatomy constitute a science (i.e. a causal science) has been broken off. Their study and teaching has begun to be throttled so that these huge areas of knowledge are being abandoned. This is despite the fact that one of the most profound discoveries of mankind is rooted in them — the knowledge of relationships and descent which, beyond all other areas of study, clarifies the position of man in Nature and the prospects of his survival.

Pure morphology possesses no causality concept, although such is rightly demanded of every natural science. This is the essence of the problem whose general solution I am seeking. Everything else is merely consequential.

All that, however, is exceedingly general. The reader will again suspect that I have invented a problem so as to solve it. Let me therefore be more concrete. The concrete aspect of the problem consists precisely in the three dozen questions listed in the preceding section to illustrate the patterns of order. Without doubt, therefore, the problem is both wide-reaching and already recognized.

It is evident that, behind each single problem, there stands the question of causality. Unlike most natural sciences this question relates, not to how the causal nexus is constituted, but rather to whether it should be expected at all. In surveying the problems I shall therefore arrange them according to the way in which the reign of causality has been questioned.

In this connection causality is a special aspect of order. This follows, in the first place, because order is the arithmetical product of law times number of instances (equation 18). Second, only causal laws should be recognized as laws. And, third, the controversial individual phenomena are the consequences of the four patterns of order which again are themselves the consequences (i.e. causal results) of the higher ranking systemizing principle (Section VIII A).

I shall thus consider the problem of order in its principal aspect through the controversy about its recognition.

1. The controversy of complexity

We are convinced that the reactions of molecules, like the life processes constituted by those reactions, follow causal laws completely. However, when we consider our conscious decisions, although these are made up of life processes, this conviction is greatly weakened. According to personal position we either suppose that, with increase in complexity, necessity is gradually replaced by accident or freedom, or else that the causal connections can no longer be traced. Thus the scientific problem escapes our insight in a transcausal area. Also it has been observed that in complex areas it is easier to establish mere rules while, when complexity decreases, conformities to law can be recognized. This has helped to produce the restrictive caution called reductionism.

In the present case reductionists hold that, if laws could be found at all, this would only be possible in the molecular realm. It would be impossible to pursue these laws even into the ultrastructural region where not all molecules can still be sorted out. If this opinion were right we should be in total confusion.

I shall not analyse this opinion further because it is more a question of way of life than of epistemology. The opposite tendency is represented by holism, which rightly warns that reductionism leads to an atomistic approach to Nature, to discrimination against synthesis, and to abandonment of all the biological controversies which concern us.[28]

2. The controversy of 'internal causes'

The controversy with the reductionists is a modern phenomenon, while that concerning the efficacy of the Darwinian evolutionary mechanism is a century old, like Darwinism itself. It is essentially as follows: The mechanism of Darwinism, since Neodarwinist genetics was built into it and the population and speciation studies of the synthetic theory[29] was added on to it, explains all evolutionary phenomena solely as the mutual effects of mutation and selection. Mutations are seen as random and purposeless accidental changes of the genotype, and selection, although always differential, consists of instant decisions made by the changing environment on the survival and prospects of reproduction of single individuals. No ordering directional component of external selection over various periods of time would be expected. If all this is true, where does the orderliness and directionality of evolution arise from?

1. *Does an internal principle exist?* This question has repeatedly been put by a wide range of authors. A list of the important works — some of them huge — which raise this criticism and try seriously to answer it will illustrate this fact. It includes: Baer (1876), Bergson (1907), Berg (1926), Wedekind (1927), Beurlen (1932-1937), Plate (1925), Rosa (1931), Osborn (1934), Dacqué (1935), Schindewolf (1936-1950), Meyer-Abich (1943-1950), Schmalhausen (1949), Spurway (1949), Jaennel (1950), Cuénot (1951), Bertalanffy (1952), Waddington (1957), Cannon (1958), Haldane (1958), Stammer (1959), Whyte (1960-1965), Lima-de-Faria (1962), Russell (1962), Eden (1967), Schützenberger (1967), Salisbury (1969).

These works are not all equal in scientific importance. However, it is impossible to ignore the common cause of all these endeavours. The attempt to understand this deep problem has already occupied four generations and has been seen from very different viewpoints.

Many authors have named the existence of an internal principle as the basic problem of evolutionary theory, e.g. Remane (1939-1971), Ludwig (1940), Hennig (1944), and N. Hartmann (1950). Developmental physiologists have also expected an internal principle and have considered the problem in all its complexity, e.g. Baltzer (1952-1957), Kuhn (1965), Waddington (1957).

The representatives of the 'synthetic theory' on the other hand, hold that no third causal mechanism has yet been proved and in this they are right. They fear that the search for it will open the door to such unproven entities as finality and entelechy, which need not be so. And they tend to play the problem down, which is not necessary. They even assert, which is obviously unprovable, that there is no room for a third principle. However, it is important that even the authorities of this opposed viewpoint, such as Dobzhansky (1956), Kosswig (1959), and Mayr (1967, 1970), ascribe to the epigenetic system a fundamental, although not fully understood, ordering action. And they ask whether this pattern of mutual gene effects will ever, because of its complexity, be understood.[30]

I shall consider this question later (Chapter III). In the systemic position of mutual gene effects we shall find the molecular cause of the ordering principle.

2. *Mutation or selection.* As soon as we search inside the organism for this third principle we must ask whether it has to do with mutation or selection. The answer depends on where we draw the line between the conditions facing the mutation, on the one hand, and its compatability with the ordered system of the genes and chromosomes, on the other.[31]

To my knowledge only one author[32] has proposed 'automutations', released by internal conditions, as an explanation. He did not explain the mechanism, however.

The alternative could be called 'autoselection' depending more on the systemic conditions in the organisms than on the external environment. Such selection has often been assumed. It is evident that systemic conditions in a very wide field might be effective, from the replication of code sequences to the maturation of the organism. This concept of 'internal factors' has been supported by numerous researches, e.g. Stern and Schaeffer (1943), Spurway (1949, 1960), Lima-de-Faria (1952-1962), Langridge (1958), Sondi (1961). It should also be remembered that the action of an 'internal principle' is required in developmental physiology also.[33] As regards the time when it takes effect this cannot correspond to a mutation mechanism but only to a mechanism in the epigenetic system. Indeed the authorities of synthetic Neodarwinism[34] assume a still undiscovered ordering component within the epigenetic system.

Those workers who defined limiting conditions between internal and external selection came still nearer to the point. Examples are the 'embryo selection' of Stern and Schaeffer (1943), the 'archetype selection' of Waddington (1957), the 'genotype selection' of Haldane (1958), and the 'developmental selection' of Whyte.[35]

It is astounding how little needs to be added here. The direction, the principle, and the rate have already been outlined. Only the concrete mechanism remains to be added. Nevertheless this mechanism still remains invisible and uncertain and the causal nexus is hidden. Most workers are still in doubt. No internal cause is apparent.

3. *Vitalism and entelechy.* In this situation there is still another alternative, if one is convinced of the reign of internal principles, though methodologically speaking this alternative is not a scientific interpretation. If we suppose that no causal law can be found, because the law is not causal, then we shall incline to vitalism and entelechy. This indeed is how vitalism[36] arose. Again it is a sort of world-view opposed to a mechanistic interpretation of Nature. Entelechy, a concept taken from the metaphysics of Aristotle, is assumed to be a factor that directs the individual regularities of organisms, i.e. their orderliness, harmony, plan, or goal. Entelechy would arise from the 'pre-established harmony' of living organisms. At this point, however, we depart from science and the problem can be left as insoluble.

Nevertheless vitalism confirms two of our results. It searches for the problem within the epigenetic system (as would now be said) and recognizes the plan, and sense of direction, of evolution. Indeed, in this book we even confirm the postulate of a 'stabilized harmony', though this harmony turns out to be 'post-stabilized', not 'pre-stabilized'.

3. The controversy of essential structures

Here we are not changing the subject by only the scene — although it looks basically altered. The 'internal principle' was a question of causes and functions. But the arguments about 'essential structures' concern the effects and forms of these causes and function, i.e. homology, morphotype or ground plan, and weighting. The basic subject remains the same: Are there orderly laws or not?

We should remember the mutual effects of all these controversies. If homology and ground plan are not recognized as realities, then the necessity of an internal mechanism is in doubt. If no mechanism is known which requires homology and ground plan, then it will not seem necessary to assume that these are real.

The separation of essential structures from inessential ones has been the key to the study of affinities ever since the origin of scientific morphology, anatomy, and systematics. It is also the focus of today's controversy.

Three questions will illustrate why this long-proven method, which has always lacked causal reference, seems to have difficulty in withstanding causal analysis.

1. *Weighting and features.* The accusation runs as follows: No method can be defined by which the systematic value of a feature can be specified, nor can constant value be assigned to any feature, and nor can any cause be suggested for such constancy. Weighting of features has therefore obviously been inserted by the systematists. But if relationship is worked out by features weighted *a priori* then the method is circular. According to the recipe of numerical taxonomy[37] the way out would be to foreswear weighting completely. Each feature would be of equal value.

Imagine the chaos in the study of relationships if, for example, the systematic classification of vertebrates were erected mainly on the individual skin appendages, the details of colour pattern, and measurements of every tiny feature, for without doubt these would be a majority. Against these huge numbers what would be signified by the loss of an aortic arch, the division of a heart chamber or the distinction between 'hair' and 'feather' which up till now have helped us to distinguish reptiles, birds, and mammals? Think what an enormous computer-capability of judgement is brought by man's brain to the concept of a single homologue, not to speak of the weighing of the connections between thousands of these homologues. For, in any comparison, such homologues are the only possible connection with the ground. Only by using them can total confusion be avoided.

Numerical taxonomy, which has already become a new form of literature, does right to assert that we do not know what specifies the degree of freedom or of fixation of homologues, nor how man's brain works in making a comparison. Instead of studying these two problems, however, numerical taxonomy decrees that all precausal understanding be forgotten. This is the understanding that allowed millions of affinity connections to be judged with such accuracy that Darwin could extract the law of evolution from them.

2. *Typology and morphotype.* The morphotype or ground plan, which is supposed to give the essential features of every group of organisms, has been attacked by a much larger number of workers. The accusation is as follows: The morphotype is a concept or idea belonging to morphology (which thus is 'idealistic'). It can neither be delineated in a figure nor be soundly based in method, let alone be causally explained. It therefore does not belong in a science.

Of course it is no loss to give up a word. However, it is a considerable drawback to give up the concept connected with the word 'morphotype' which is that the essential features of every natural group of organisms must have their special cause. This drawback would remain whatever sort of cause this might be. Even in Goethe's morphological work[38] the morphotype ('Typus') is defined as: 'A consequence, a rule, according to which Nature is expected to act and a metamorphosis which will always affect the parts.' In solving the problem we shall confirm this statement completely (Chapter VIII).

I shall not discuss the morphotype and its forms more fully until Chapter VIII. To many people it seems to be restricted, mistakenly, to idealistic or metaphysical concepts. This is because pre-Darwinian morphologists wisely forebore explanation, while later morphologists discovered none.

The view is increasingly held that features become fixed by becoming deeply woven into the epigenotype.[39] But the how and why remain unresolved, although the statement will turn out to be completely correct.

If, however, the hypothesis of regular conformity to morphotype is rejected, then the reality of the natural affinity groups defined by the morphotype also looks threadbare. The catastrophic consequences of this will immediately be obvious, but at this point I shall not follow the various side branches of this controversy.

3. *Natural classification and systematics.* If no certainty can be reached concerning the 'essence' of single structures, nor concerning the morphotype or ground plan which these structures should constitute, then classificatory groups are not realities laid down by natural laws; instead they would be, as the nominalists say,[40] mere aids to thought. The concept of 'natural classification' becomes a self-contradiction. Systematics, and indeed pure morphology in general, becomes meaningless, and in actual fact the subject is now in desperate retreat.

We have already established that, if the nominalist position were correct, biology would lose its basis. But it is not correct.

4. The controversy about thought patterns

We are therefore surrounded by doubts touching the existence of orderly principles or of preordered structures in Nature. If these doubts were justified where would the orderliness arise that, after all, has been described in countless books? Would it not be the orderliness of man's thought, projected into Nature? And, if this were not so, how to explain the obvious agreement between the patterns of man's logic and the supposed orderly patterns in man's environment?

As later discussed, the four basic patterns of organic order are, to an astounding degree, also the preconditions for man's thought. It is totally unlikely that this agreement is accidental. But suppose it turns out to be mere projection? What, in that case, could we learn about order in living organisms, as distinct from learning about our own logic?

Modern information theory maintains that: 'All systems that treat information, whatever field they may cover, fulfil both in themselves and between each other, the laws of information theory and thermodynamics. This holds also for the total system that includes the individual systems, i.e. for the physical universe in which entropy increases unceasingly. Followed back into the past there must have been at the beginning of all happening a condition of least entropy, and therefore of highest regularity and highest information. The first words from the *Vulgate* are: '*In principium erat verbum.*' Without blasphemous intent this could be translated as: 'In the beginning was information.'[41]

But what next? Goethe's words were: 'Geschrieben steht: "Im Anfang war das Wort!" Hier stock ich schon! Wer hilft mir weiter fort? Ich kann das Wort so hoch unmöglich schätzen, Ich muss es anders übersetzen.[42]' I also shall close the circle of this investigation by giving a different translation (Section VIII B7*f,g*). But to travel forward we need, since there is no help for it, to journey a long distance.

Up till now we have not gained much — only a few definitions and the certainty that the problems of biological order are many and concern principle. Perhaps most importantly, there is a prospect of an answer by considering probability. To this, therefore, I return.

NOTES

1 See Lorenz (1971, p.231). Compare also Whitehead (1933), Popper (1935), Needham (1936), Burgers (1965), Koestler (1968), Peters (1967, pp. 251 and 254), Popper (1962), D. Campbell (1966a), Thom (1972) etc.
2 Meyerhof (1924) quoted from Morowitz (1968, p.20).
3 Morowitz (1968, p.20).
4 Morowitz (1968, pp.7, 20, 134, 146). See also in particular Odum (1971).
5 Morowitz (1970, p.108).
6 Brillouin (1956).
7 Morowitz (1970, p.111).
8 Brillouin (1956, pp.288-9).
9 Linschitz (1953), Morowitz (1955).
10 Dancoff and Quastler (1953), Quastler (1964).
11 Goethe (1795); Remane (1971); the concept of homology goes back to Owen (1848); cf. also Hennig (1950), Troll (1941) as well as the morphological works cited in Section VIII B3*c*.
12 Remane (1971, p.28); see also the research of Diamond (1966) and Berlin and Breedloue (1966).
13 Florkin (1966, pp.6-7); cf. also Winter and coworkers (1968).
14 Remane (1971, p. 76 ff.).
15 These questions are discussed in Ax (1961), Steinböck (1963).
16 Discussed, for example, in Reisinger (1960).
17 Remane (1971, p.59). Compare the quantitative studies of Zarapkin (1943) and Olson and Müller (1958).
18 Remane (1971, p.76).
19 In N. Hartmann's (1950) sense.
20 Schrödinger (cf. 1944 and 1951).
21 This was foreshadowed by several nineteenth century embryologists but first formulated explicitly by Haeckel (1866).
22 In Baltzer's (1952) sense.
23 A biologist will recognize this as Dollo's law. I emphasize that I apply it in the revised version of Remane (1971, pp.259-74, especially p.272) incorporating a necessary restriction.
24 The 'bithorax' phenocopy.
25 'Aristopedia'.
26 The 'tetraptera' mutant of *Drosophila*.
27 Waddington (1957, p.79).

28 An exposition of the systematic thought necessary in biology is given in Weiss (1970a) and in the essays collected by Koestler and Smythies (1970) and Weiss (1971). The criticism in Whyte (1965) is also of interest.
29 Extensively treated by Mayr (1967).
30 This question was raised by Kosswig (1959) in the study already mentioned.
31 Lima-de-Faria (1962) already called attention to this.
32 Stammer (1959, p.205).
33 Baltzer (1955), Kühn (1965), N. Hartmann (1950).
34 Dobzhansky (1956), Kosswig (1959), Mayr (1967, 1970).
35 Whyte (1960a, 1960b, and 1964).
36 Driesch (1927).
37 Sokal and Sneath (1963). Contributions to the controversy are in Farris (1966, 1969), Kluge and Farris (1969), Blackwelder (1967), Goodall (1970), Simpson (1964a), Inglis (1970), Kiriakoff (1965), Mayr (1965), Margulis and Margulis (1969), Steyskal (1968), Camin and Sokal (1965), Simpson (1964b), Ghiselin (1966, 1969) as well as in Farris (1967), Cracraft (1967), Colless (1967). Bigelow (1959) is also of help. Most of these papers are in the journal *Systematic Zoology*. After this book was written in the German edition Sokal and Sneath produced a milder version of their original recipe (1973).
38 Cited from Hassenstein (1951).
39 Mayr (1967), Chapter E.
40 Gilmour (1940).
41 The final sentence of Peters (1967, p.255). The reference is to the Gospel of St. John.
42 Goethe, *Faust* Part I ('It is written: "In the beginning was the word!" I am stuck already! Who will help me forward? I rate the word so impossibly high that I must translate it otherwise.').

CHAPTER III
THE MOLECULAR CAUSE
OF PATTERNS OF ORDER

This brings us to the kernel of the matter. We need to deduce the mechanism by which the four basic patterns necessarily arise. Obviously such a basic mechanism must act at a basic position in the evolutionary process. It will have a molecular root and a morphological one and can be explained only by the two together. I shall start with the molecular root.

The question is: Why does living order always lead to the four special patterns of order which I have called standard-part hierarchy, interdependence, and tradition? This is the specifically biological question which I wish to solve in this book.

A.   ON CAUSE IN GENERAL

But another question can be recognized behind this one: Why does order in general arise, when previously there was none? This is a question in thermodynamics or statistical mechanics. The answer to it has long been foreseen by physics, theoretical chemistry, and biophysics. For present purposes it can be called:

a. The cause of the cause

The question can be put as follows: How do living systems build order up, although they are part of a universe which, following the law of entropy, passes from order to disorder? Can it be that the second law of thermodynamics does not apply to living systems? This question has been studied for about one hundred years and has led, particularly in the last few decades, to an extensive theoretical structure known as steady-state thermodynamics, non-equilbrium thermodynamics, or the thermodynamics of irreversible processes.[1]

The answer[2] is somewhat as follows: The law of entropy is not violated by living systems, but evaded. Or, more precisely: The law of entropy can only be applied to isolated closed systems. All of these, indeed, transform themselves into more probable, disorderly states. They all tend towards equilibrium. All organisms, however, are open systems. They cannot be isolated, since their very existence depends on a stream of matter and/or energy flowing through them. Like a drainage system they must all lie on an energy gradient connected both to an energy source and to an energy sink. However differentiated these may be in any individual organism, the orginal source of energy is the sun and the energy sink of the biosphere is the cold of the space of the universe. To this, after death and decomposition, everything will be lost again by nocturnal radiation. During life processes, however, there is a storage of energy which greatly exceeds the thermal energy of the equivalent equilibrium condition (of the corpse). This storage of

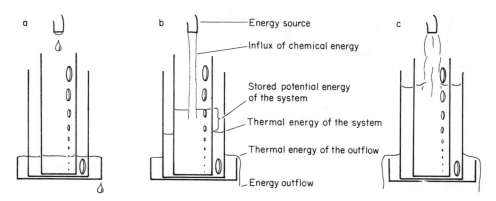

Fig. 16 a-c. Hydrodynamic model of the storage of potential energy and of increase in order. (a) Energy flow too small; the energy level of the outflow is scarcely exceeded. (b) Optimal influx; potential energy exceeds thermal energy. (c) Influx too large; potential energy is raised by the subsequent increase of thermal energy. Based on Morowitz (1968) elaborated.

energy has modes of appearance described as performance, random improbability, distance from equilibrium, or as functional or structural order. Thermodynamically speaking, order can be described as the tension between storage and random distribution of energy, between an improbable condition of balance on the one hand, and the greatest mixture of component parts on the other.

The theory further states that an optimal throughput of energy (Fig. 16b) necessarily causes the building-up of the systems that we call ordered. Or more precisely[3] : Models of such systems can be developed which result in the origin of information, or, as I would say, of determinacy. A steady throughput and the steady selection of more stable conditions necessarily cause a steady increase in order.

The simplest model of such processes is the hydrodynamic analogy of Morowitz,[4] as shown in Fig. 16a-c. Imagine two tall cylinders, one inside the other. The inner one has a series of side openings which decrease in size downwards. The outer has only one side opening, near the bottom. Both cylinders are standing in a shallow vessel over the edge of which any quantity of water can flow. Let a stream of water (chemical energy) flow into the inner cylinder. The maximal difference in water level (i.e. maximal stored potential energy) will depend on an optimal throughput of water. If the influx is too small (Fig. 16a) then the level in both cylinders will sink to that of the shallow vessel (i.e. to that of the thermal energy of the outflow). If the influx is too large (Fig. 16c) then the level in the outer cylinder (the kinetic temperature, or thermal energy of the system) will once again eliminate the difference by rising to the level in the inner cylinder. Without doubt this general subject is one of the most fascinating in science, but I must immediately leave it, so as to stay within my scope. Interested readers should consult the reference given.

The cause of what causes the forms of organic order is itself being actively investigated. The origin of order seems to be a necessary result of matter, even though the necessary conditions are themselves accidental and rare (or indeed improbable) and even though an increase in order can happen only at the expense of the order in the universe.

b. The results of this cause

These are the next object for investigation. More precisely we can ask: If order necessarily arises, why does it take on a small number of special forms?

Again I shall start with probability questions. I shall consider how far we can proceed by using the determinative decisions that are required for a definition of order. This connects information technology with molecular genetics.

This first step is justified because molecular genetics allows us to describe the mechanism in a region of relatively low complexity, before it breaks up into more complex special functions. Also the molecular solution will perhaps convince some people more than the morphological one discussed in Chapters IV to VII. But it would be wrong to believe in the primacy of molecular mechanisms; we are faced with a hen and egg situation with molecular mechanisms as the egg. No molecular determinative decision has meaning unconnected with its effects — no matter at what level of complexity these effects occur.

B.   DETERMINATIVE DECISIONS IN THE ORGANISM

First I shall examine how far the punched-tape model applies to the arrangement of the decisions that are laid down in the inherited material. A glance at the basic ideas of genetics[5] and of the transmission of information will be helpful.

As is well known, genetic information is laid down in the chromosomes. It is codified in a one-dimensional chain of four molecules, in a script somewhat like a strip of morse code.

The DNA (desoxyribonucleic acid) is a very long aggregate of numerous nucleotides which differ from each other by having a particular one of the four bases — G, A, C, and T (guanine, adenine, cytosine, and thymine). The bases therefore correspond to the holes on a punched tape while the rest of the DNA corresponds to the paper. The code is read off in groups of three nucleotides (triplets) beginning from a fixed starting point. This gives definite codons which are triplets of nucleotide with $4 \times 4 \times 4 = 64$ possible combinations. The decoding mechanisms recognizes most of these combinations as meaningful triplets and translates them into the 20 different amino-acids. This is rather like translating the three Morse symbols (.-/) into letters. To complete the analogy, an amino-acid can act as starter and a meaningless codon necessarily functions as a terminator. This is like the spaces and punctuation of a script.

The collinearity of nucleotide sequences can be compared with the sequence of determinative decisions in a Morse transmitter. And the amino-acid sequences of the coded polypeptides can be compared with the rows of letters that constitute the translated words. Furthermore, the universality of the genetic code indicates its origin from a single ancestral language. It is only the interweaving of these decisions — what I shall call systemization — which transcends far beyond the non-linear component of script, as poetry transcends information. I shall discuss this later.

1. The importance of the single decisions

This depends on two different possible effects that the decisions may have. From the static point of view there is the identical replication of part of the total system of the individual to the extent shown to be necessary in the ancestral line. Dynamically speaking there is the adaptability of the same part, if a relevant change should bring an appreciable benefit to the total system. Accuracy of copying versus mutability of single determinative decisions, together with their antagonist 'selection', form the mechanism of adaptability. This is the mechanism of evolution as known up till now.

a. *Adaptability — the designer plays dice*

Despite much search, no mechanism has yet been found by which the decisions in the genome could be informed about an adaptational demand, however pressing. This preposterous and indeed unbelievable and catastrophic circumstance (catastrophic for the billions of organisms removed by selection) fuelled the controversy between Neolamarckism and Neodarwinism for many decades. However, it now seems that such retrospective action has never been possible. Genome decisions cannot be meaningfully influenced by the environment. Their effect is unidirectional. This dogma must be accepted, but we know that such a unidirectional manner of looking at causes cannot take in all the connections.

We too shall find no retroaction of Lamarckian type. But this is no reason to accept that creation happens in a random manner. Einstein[6] found it repugnant that God should play dice with molecular phenomena. The supposition that man arose by pure accident, that God plays dice with evolutionary phenomena, is equally repugnant. However, as we shall see, order has less to do with 'meaning' than with 'self-meaning'.

If mutations are accidental, we need to know how much a genome is altered by a mutation, how often a mutation happens, how great are its prospects of success and what are its effects.

1. The genetic extent of a mutation is usually small. Moreover, the prospects of success increase with smallness. To use our analogy, it is generally a few letters that change, in the text of a monumental tome.

Genetic deletions — the loss of pieces of code of various lengths — are almost always lethal, causing the death of the bearer. Point mutations are most important. These alter a gene, or more precisely a cistron, which is a portion of a DNA chain which determines a continuous polypeptide sequence (or, in our analogy, a word). The sequence of 'letters' or of triplets will only be altered as from the codon ('letter') which the mutation strikes. If an extra nucleotide is added to a codon, or lost from it, then the reading-off of triplets goes out of phase, the reading raster shifts, and all subsequent readings will be wrong. But if a nucleotide is altered, rather than lost or added, then the rest will remain in phase. Only the affected triplet or 'letter' will be wrong.

2. The frequency of change in a gene (i.e. the mutability) is not high under natural conditions (this is the so-called spontaneous mutation rate). One mutation per gene in 10^4 cases counts as a high rate. Mayr[7] says: 'One may estimate that in the higher vertebrates the average mutation rate per individual per generation is somewhere between 1 in 50 000 and 1 in 200 000 per locus.' Obviously some mutations must be even rarer than this: 'Besides many have been observed only once so that no statement about their probability is possible.' On the other hand there is good information about the maximum rate, which is all that matters to us in the first instance. Adaptation of a feature, even the most pressingly required adaptation, must wait on the next accident, and this remains improbable for at least 10^4 tries.

3. The prospects of success of a mutant are also not high: 'Improvement by mutation is as unlikely as the enhancement of a good poem by a printer's error.'[8] This illustrates also, what we have already established, that small changes have a greater prospect of success than large ones. Only a few per cent of mutants have a prospect of being passed by selection for: 'The sorting of hereditary factors has been so thoroughly worked over that accidental changes are seldom tolerable.' We shall discuss this more fully later. The process of adapting a feature must wait not only on a rare accident, but on a still rarer happy accident.

4. The result of a mutation almost never affects a functional whole. The belief that one gene specified one feature was long ago abandoned. Gene effects are interwoven with each other in two different ways.

First, most mutated genes result in alterations to a whole series of features — this is called polypheny or pleiotropy. Second, no functional system in an organism depends on a single gene only. At least a few genes are always involved, and usually many — this phenomenon is called polygeny. Formerly this seemed strange. However, existing genes must have been fitted together gradually, in the same way as the features that they control, and by the same circuitous route. This makes the interweaving of gene effects not only comprehensible but to be expected as a necessity.

The process of adapting a functional unit thus waits, not merely on a happy accident, but on an accumulation of happy accidents. Adaptability, on which the prospects of advancement mainly depend, quickly becomes more difficult as the number of required changes in determinative decisions increases.

b. The holes in the punched tape – accidental programming

We have still not considered a fundamental point in the comparison. If the nucleotide bases of the DNA chain are compared with the holes in a strip of Morse code then we need to note the remarkable fact that producing and placing the holes is done entirely accidentally.

Let us suppose that the decipherment of the sequence of holes produces a volume of poetry (individual) which has up till now sold well in the market (in the environment). The proceeds of sale (selective advantage) were sufficient to keep up the printing (reproduction) of the edition (species). But now the tastes of the market (environmental conditions) change and requests, profit, and printing all diminish. A licensed edition (population) will prosper if a particular alteration of the text (a particular feature) adapts it to the new fashion (new environmental conditions). But the only change allowed is a mistake (mutation) in punching the tape that controls the printing.

In this situation there can be no generosity in setting out the holes (or determinative decisions). It is difficult enough to put a single additional hole through pure accident in the right position while not making holes that are not needed. This explains a peculiar but essential difference between technical and biological punched tapes.

In the technical punched tape it has proved useful only to enter the 'Yes' in the binary code of 'Yes-No' decisions, though the positions of the missing holes, i.e. the 'No' decisions are indicated by a second uninterrupted row of holes each of which marks the position of a decision. Marks adapted for finding mistakes have also proved useful e.g. a third row of holes with a test hole at each tenth decision. Nothing corresponding to these mechanisms is known in biological punched tape.

2. The advantage of dismantling redundant decisions

As already mentioned, in determinative systems we always have to reckon with the occurrence of redundant decisions. This has been investigated in Section I B2 where it was shown (I B2e) that redundant decisions can only be avoided completely in a regime based on finalistic or on teleological principles. Redundant decisions do not matter so long as we do not assume an economical principle in the system.

The term 'redundant decision', in my sense must not be confused with the term 'genome redundancy' which has recently attracted attention. The latter simply means repetitive DNA sequences.

a. The principle of economy

Such a principle, however, is in all organisms an absolute precondition for existence. When all decisions have been materially established as molecules or molecular positions, there arise with every decision costs, sources of error, and difficulties of adaptation. The dismantling of every redundant decision must bring profits – evolutionary or adaptational advantages (A). This will be obvious even in very simple systems.

Let us use our well-tried example of a system of determinative decisions with a range of 1024 numbers which establishes the events I to VIII ten thousand times. Here we have a law content of only 23 $bits_L$ as against a redundancy content R_{max} of 799 977 $bits_R$ (cf. Section I B2d).

1. The costs of conserving redundant decisions are incurred first by conserving the structure and position of the nucleotides carrying the decisions. These energy costs due to 'maintenance' will probably increase linearly with the number of decisions. In addition there will be the cost of storage which will also increase appreciably, though not linearly.

Thus, in our example, instead of storing 10 metres of punched tape carrying the pure law content, a 400 km strip would have to be stored if the redundancy content were undiminished. Besides the costs of maintenance and storage there will be the costs of replication. These can be worked out exactly.[9] Every replication would have to copy and reprint about 1.2 km of DNA, instead of 3 cm. This cost also would increase proportionate to the relative redundancy.

2. The susceptibility to error will also increase with relative redundancy. For we must expect that, if the determinative decisions laid down as molecules are doubled, they will be affected by double the number of replication errors. To illustrate the difference, let us suppose that each decision will be wrongly reproduced every ten thousandth time. If a new copy is complete at about 20 $bits_L$ of law content, then a single error will be contained only in every five hundredth copy. But if a new copy has the full redundancy content, then every new copy with 800 000 decisions will carry 80 mistakes.

This full level of redundancy in no way increases the adaptability of a system, but only its liability to error. For undiminished redundancy increases the number of non-acceptable alternatives only. Assuming that, in a balanced system, only one alternative is adaptive, then the rejection rate will increase as the power of the relative redundancy. It is as if a cell in the middle of the eye were specified not only as being a lens cell or a cell of the vitreous humour, but also had the choice of being a bone cell, gut cell, blood cell, or enamel cell.

3. The adaptability of a system certainly decreases as the power of the decisions required to produce the system and thus with the relative redundancy. A numerical estimate of all the losses connected with redundancy still needs to be worked out in detail, but an estimate of the restriction of adaptability can be given. This single quantity, among those yet measurable, is enough to show that redundancy necessarily has drawbacks.

b. Adaptability and redundancy

An important point arises as a result of estimating the value of systemizing the genome by dismantling redundancy. How does redundancy arise and what does its dismantling signify?

1. *Why redundancy arises.* I discussed this problem in Section I B2e but in general terms (considering inorganic determinacy also) I left the answer open. As concerns organic determinacy a conception can be formed.

A message is determined by decisions. The number of decisions which can be left out will depend on what we expect from the message. The estimates of redundancy content in Chapter I depended on the simplifying assumption that the messages had to remain the same. Such a precondition, in fact, gives the greatest possible redundancy content (R_{max}).

From living messages, or organisms, however, evolution demands adaptive change also. The simplest assumption is to expect that all single decisions could change independent of each other, because an alteration of any single decision might come to be required. This stipulation would no longer apply as soon as two decisions become to some extent functionally dependent on each other. They could then not be altered separately but only together (discussed later in Chapters IV to VII).

For example:

Decision number	1	2	3	4
Preliminary decision	a	a	b	b
Final decision	a	b	a	b
Event number	I	II	III	IV

Preliminary decision 2 (in italics) can only become redundant if events I and II become dependent. It could be left out if the deciphering system 'remembered' the preliminary decision 1 until changed by preliminary decision 3. This, however, has been discussed already.

We know that maximal redundancy content reaches huge values even in very simple systems (cf. Section I B2d). This indicates that optimal redundancy content under conditions of adaptation must also be important in biological systems for these are much more complicated.

2. *The significance of the dismantling of redundancy.* This can now easily be shown. We must remember two genetic parameters — the mutation rate and polygeny. The mutation rate is a low probability (P_m) of 10^{-4} or less. Polygeny, on the other hand, implies that the shaping of each individual functional system of an organism is determined by more than one gene. The prospects of adapting a functional system by means of accidental changes must also be a probability. This will decrease with the number of mutations required.[10] If the probability of two required mutations is for each $P_m = 10^{-4}$ then the chance of their accidental conjunction is: $P_{m_1} \cdot P_{m_2} = 10^{-4} \times 10^{-4} = 10^{-8}$.

An important question now arises. How large is the selective advantage A (i.e. the increase in the probability of a change being accomplished A_a) if one of the two required mutations can be avoided by dismantling, as being a mutation of a redundant determinative decision? In the present case the prospect of accomplishment increases from 10^{-8} to 10^{-4}. This means that 99 990 000 attempts can be dispensed with, i.e. the number of attempts is diminished to 1/10 000. This means that

$$A_a = 1/P_m \text{ or } A_a = P_m^{-1} \tag{19}$$

The selective advantage in adapting a system, gained by being able to omit one of the required mutations, corresponds to the reciprocal of the probability of that mutation.

This advantage is very large. We should expect that evolution would have sought a way to use it. The difficulty, or accidental improbability, of finding a way might perhaps exceed that of a single mutation by a factor of 10 000. Even so it is likely that a way would have been discovered long ago by evolution. And, in fact, it *was* discovered. The procedure can be called systemization and will now be examined in detail.

c. *The necessity for systemization*

This can be deduced from the selective advantages, since these can reach inconceivably large values. This is immediately obvious when we remember that the improbabilities increase as the power of the number of mutations. In a single system the selective advantage A_a has the redundancy content in $bits_R$ as the power. That is:

$$A_{a\ max} = P_m^{-R} \tag{20}$$

The maximal accomplishment advantage (or increase in the probability that a change can be accomplished $- A_{a\ max}$) that can be reached by systemization thus corresponds to the reciprocal of the mutation rate to the power of the number of redundant decision dismantled in the system.

Even the simplest system, such as a single transmission of events I to VIII out of a range of only eight numbers, is determined by 14 $bits_L$ and 10 $bits_R$ (Section 1 B2d). The maximal selective advantage with complete systemization i.e. complete dismantling of the 10 redundant decisions. is:

$$A_{a\ max} = P_m^{-1 \cdot R} = 10^{4 \times 10} = 10^{40}$$

And with a somewhat less simple system, such as the transmission 10 000 times of I to VIII out of 1024 possibilities (Section I B2d) then:

$$A_{a\ max} = 10^{4 \times 799\ 977} \approx 10^{3\ 200\ 000}$$

This would be a number with three million zeros. Both of these values, of course, are maximal. We shall soon see, however, that real selective advantages are likewise extremely large.

It follows that the advantages of saving even a small number of required mutations rise very steeply. This steep increase makes it totally unlikely that the mechanism needed for systemizing the genetic determinative decisions should not yet have evolved. Indeed it is, as we shall see, a necessary requirement for organisms.

C.   THE SYSTEMIZATION OF DECISIONS

By systemization I therefore mean the process by which the action of determinative decisions is differentiated out of a condition of uniformity. The simplest differentiation would be the ranking of one decision above another. This simplest model causes high-ranking decisions to affect lower ranking decisions in a one-sided manner. But, at the same time, in the evolutionary process it also produces a retroaction affecting the high-ranking decisions, as I shall show later. This feedback is the basic precondition for systemic effects and for bidirectional causality.

These concepts are already current, and indeed self-evident. In the language of genetics they are basic features of the gene interactions which have long been called the epigenetic 'system'. In the design of apparatus they correspond to the obvious fact of the wiring, which arranges the individual switches in ranks. The agreement of the genome with our systemic model has been implied ever since the hypothesis of 'one gene, one feature' was abandoned.

The individual types of genetic switching action necessary for systemization have mainly been studied in very lowly organisms. Their presence in all other organisms is assumed,[11] however, and is gradually being proved.

1. The model and its molecular realization, Part I

Pursuing the comparison of the systemic model with molecular genetics, I shall first examine the conditions that make one decision outrank another. In a piece of apparatus a switch is inserted in the circuit between the source and the effector. A general or mains switch is connected in series, as in every household. In the genome it is the production and spread of groups of molecules which must have an analogy in the wiring model. Superposition of ranks is obtained when one unit of command, cistron, or synthesized compound can switch others on or off. Biologists will know that this does happen in the genome. But a more exact and basic question is: What are the simplest and most basic elements that can be switched? These elements are, in fact, on-off switches and change-over switches. Both are fundamental for systemization and well known in the genetic system. Moreover, they are the cause of two primary patterns of order — the standard part and hierarchy. The systemic model implied by these switches will now be compared with the genetic system.

a. The repeat switch — 'repeat on demand'

The repeat or on-off switch exists in every apparatus that is capable of repeating. Its 'yes-no' alternatives are 'go' and 'stop'. Its effect depends on the chain of subordinate commands which it repeatedly causes, or allows to be repeated e.g. a minute-light, an electronic flash, a radio alarm, or in changing a disc in a record player or replaying a tape in a tape recorder. It allows a whole sequence of stored determinative decisions to flow once more.

In the last analysis all depends on the wiring of the subordinate decisions. These, as we know, have been stored and can be dealt with in connected sequence, one after the other, and when they have finished are switched back to the starting position. This prevents the avoidable repeat decision which would otherwise appear as visible redundancy.

The selective advantage of the on-off switch is so enormous that we cannot even imagine a machine without this mechanism. Indeed it is not easy even to think without this obvious feature. The mere storage and sequence of decisions, indeed do not in themselves make a machine. Only the power of repeatedly starting them off will do that. Without being able to use them repeatedly their regularity could not be recognized, nor even be reconstructed for examination. As shown in Section I B this repeatability is the basis of empirical knowledge.

Naturally, storage and sequencing have two preconditions — both the nature and the sequence of the preserved decisions must be recorded. Repeatability, however, has the advantage that nature and sequencing do not need to be reinvented accidentally. This advantage has already been estimated as R'_{max} (see Section I B 2d). In terms of single events (E) even a single repetition of the series saves $E \cdot \log_2 E = bits_R$ (cf. equation 11). With 16 single events, therefore it is: $16 \times \log_2 16 = 16 \times 4 = 64\ bits_R$. But how large is the selective advantage A'_a when even a single bit of visible redundancy is saved? It is at least as large as the reciprocal of the mutation probability, i.e. $P_m^{-1}\ bits$.

This can be shown as follows: Even the alteration of a single base is more than 1 *bit*, because there are four alternatives. Furthermore, for each cistron there is a whole chain of such decisions, perhaps hundreds, though we do not know precisely how many. What we do know is how often a mistake happens to a cistron, in one or other of its decisions. This happens P_m times and thus with a frequency of 10^{-4} at most. The P_m of the individual decisions is therefore perhaps 100 times less, or 10^{-6}. It is therefore generous to say that a cistron equals at least one *bit*.

However, this generosity is intentional since we need to compensate for the degree of uncertainty. Indeed, we may have achieved two or three orders of magnitude of generosity, for P_m itself ranges from 10^{-4} to 10^{-7}, or even further. More important still, in this respect, is the fact that A'_a ranges over dozens or even hundreds of orders of magnitude.

Correct adaptational decisions, as already mentioned, can only be discovered by accident. Each such 'discovery' that can be avoided as being unnecessary raises to a higher power the selective advantage of dismantling redundancy i.e. the increase in the prospects of accomplishment A'_a of the adaptation.

$$A'_{a\ max} = P_m^{-[E \cdot \log_2 E(a-1)-x]} \tag{21}$$

Nevertheless we do not know how difficult it is to discover the decision which overranks the others — the 'go'. Let us suppose, in the first instance, that this degree of difficulty (x) is of the same order of magnitude as the others, i.e. P_m. We can then make a numerical estimate as follows. In the replication once only of four events, each out of four possibilities, systemization would lead us to expect an accomplishment advantage $A'_a = 10^{4 \times 7} = 10^{28}$. Even this is a huge number.

In the tenfold replication of the system with merely 16 individual results out of a range of numbers of 16, systemization gives an accomplishment advantage as follows:

$A'_a = 10^4$, to the power of $E \cdot \log_2 E \cdot (a-1) - 1$.

The power can therefore be reckoned as: $16 \times 4(10) - 1 = 639$. Consequently $A'_{a\ max} = 10^{4 \times 639} = 10^{2556}$. This is a number with 2556 zeros and is too big to be conceivable.

However the discovery of the 'go', even if it were a million times more difficult than this, would possess even in the first replication of the simplest possible system a selective advantage of $A'_{a\ max} = 10^{28-6} = 10^{22}$. This is a number with more than 20 zeros. Obviously, therefore, the 'go' will have been discovered by evolution. It would certainly have developed in the very earliest phase of the evolution of living matter.

Thus the elimination of redundant decisions — the elimination by which visible redundancy comes to be excluded from the play of accident — is a basic feature of living matter. It is as basic as its structuring to form an open system, the way it is driven by energy, or its storage of determinative decisions, i.e. the build-up of information.

It is easy to see that this institutionalizing of 'go' necessarily leads to the order pattern of the standard part. The molecular proof follows immediately. The morphological proof is given in Chapter IV.

b. The nucleic-acid systems

It is easy to show that the 'go' has in fact been discovered in the molecular biological process and to show how it is realized there. It is an obvious feature rooted in the DNA system and, in several different ways, in the RNA system. The 'go' mechanism is as basic for genetics as in designing apparatus or in recognizing regularity in general. It consists in the processes of semi-conservative replication and transcription together with the phenomena of gene reinforcement, division, and reproduction.

1. The thread-like molecule of DNA carries the original sequence of determinative decisions. It is a double structure. Each purine base is in apposition to a complementary pyrimidine base. Along its whole length, therefore a thread of DNA carries its own template. When the thread divides, the template DNA forms an original DNA thread while the original forms a template (Fig. 17a). More than 10^{15} copies, for example, are therefore produced for the cells of the human body.

2. Besides this, in cells that have a high protein requirement such as egg cells or gland cells, there are giant chromosomes. These are made up of hundreds of DNA threads packed together like a cable. These threads are reproduced, but remain together to help the mass-production processes of the cell.

3. The chains of decisions are sent out from the nucleus into the cytoplasm by the thread-like molecules of messenger RNA (mRNA). These again are copies (transcriptions) which reproduce in large numbers the information contained in individual pieces of DNA.

(The copying process moves at about 30 nucleotides per second which is chemically very slow. Measured in number of decisions transmitted per second it corresponds roughly to the rate of human speech or of typing by a good typist.[12])

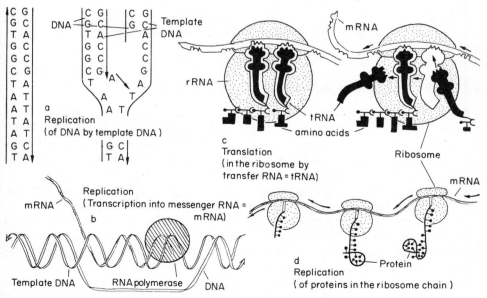

Fig. 17 a-d. The three replication mechanisms of the nucleic-acid systems. These are shown schematically and in correct sequence in (a), (b), and (d). Figure 17c illustrates the process in the ribosome. (a), (c), and (d) after Bresch and Hausmann (1970); (b) after Watson (1970).

4. The instructions for building the decipherment devices are then copied from the original, producing the ribosomal or rRNA. 'It is interesting to work out how many identical rRNA molecules are synthesized simultaneously on the genetic assembly line of the ovum.'[13] It is 5×10^7 rRNA molecules.

5. These decipherment devices (the ribosomes) are themselves a mass product, therefore. The mRNA threads are pulled gradually through them (Fig. 17c) and, with the help of a third type of RNA (transfer or tRNA), the 'proteins are crocheted together, amino acid by amino acid.'[14] 'This all irresistibly recalls an assembly line in a machine factory.'[15] The ribosomes are like molecular printing works taking in messages which are copies of the original strip of Morse code. By transfer they gather the individual letters from the type case of the 20 amino-acids and produce the translated text as protein by the process called genetic translation. The ribosomes are not merely produced in huge numbers. Very often many of them read and 'publish' simultaneously from the same mRNA message (Fig. 17d). When doing this they appear as rows of ribosomes – the so-called polyribosomes or polysomes.

Besides this, genes have been identified which control the switching-on of the replication process. It is supposed that an initiator, produced by a regulator gene, reacts with a **replicator**.[16]

Many details are still unknown. For example, we still do not know how the highest-ranking control switch that sets the copying process in motion is recognized by the polymerase that switches it on. There is no doubt, however, about the systemization. We can forsee that extensive avoidance of redundancy would be possible by means of the structural genes in the original DNA.[17] Notice also the production of huge numbers of molecular standard parts on 3 to 5 levels, each level having a multiplying effect on the parts.

The standard parts produced in this manner range in complexity from molecules (bases, nucleotides) up to complete individuals and are also found as large and giant molecules (amino-acids, polypeptides, proteins, complete DNA 'manuscripts'). Morphologically, standard parts can also be shown to be necessary (Chapter IV) at all other levels of complexity. The mechanism for producing them is always the same in principle.

c. The selector switch – 'applies until further notice'

Change-over or selector switches form part of every electrical apparatus. Their alternatives are not 'go' and 'stop' but 'a' or 'b'. Their effects again depend on the extent of subordinate interconnected commands – compare local and long-distance telephone calls, for example. It is obvious that the switches can be arranged in ranks, one subordinate to another. Thus in the kitchen we have: washing machine (not the cooker): drum (not the spinner): cold rinse (do not heat) etc. These selector switches pick out the 'index letters' under which whole sequences of subordinate decisions have been stored.

Again, in the last analysis, everything depends on the wiring. For this ensures that a preliminary decision, standing for a whole row of subsequent decisions, does not need to be discovered accidentally whenever it is to be used. This prevents long-windedness in the sequence of commands and allows the dismantling of hidden redundancy.

Indeed the explanation is once again so self-evident that it would occur to nobody to arrange the sequence of commands differently. It is not a question of building a new switch into the system but only of ensuring that the position of the switch is noted by a number of sub-switches, until it is changed. The selective advantage in terms of ease of accomplishment (A_a) associated with saving a single *bit* of hidden redundancy is large.

The caution implied by rating a cistron as only one *bit* is therefore justified. I shall use the same rating here.

By way of explanation I refer to the previous result. The selective advantage of the on-off switch (A'_a) is that building the switch into the system increases the applicability of stored decisions which are excluded from the play of accident. The same is true also for the present case. However, the on-off switch has only a quantitative effect since it specifies the repetition of a sequence of commands. With the change-over or selector switch the effect is qualitative — building the switch into the system shields from the play of accident the making of alternative changes in the stored decisions.

1. *The single case.* Let us take the simplest case first and consider the advantage of interpolating single change-over switches into the system. The interpolation of such switches implies the dismantling of what we have called hidden single redundancy (cf. Section I B2d). This advantage can be written A''_a. For example take the transmission of the events (or realization of the features) I II III IV. At first $E \cdot \log_2 E = 4 \times 2 = 8\ bits_D$ (a or b) will be required. We can write $aa = 1$, $ab = II$, $ba = III$, $bb = IV$, where the two italicized preliminary decisions can be recognized as redundant if the message is to be repeated unchanged. If a unidirectional change occurs as for example the change of I II III IV into I II I II then both the fifth and the seventh decisions must mutate from b to a. If however the preliminary decisions are systemized, i.e. made to apply until cancelled, then this change can be achieved by a single mutation in the fifth decision. The advantage of saving this mutation can be recognized as $A''_a = P_m^{-1}$ and the maximal systemization is known from equation 12. These $bits_R$ depend on the number of events E and affect the achievable advantage as a negative power:

$$A''_a = P_m {}^{-(E \cdot \log_2 E - \sum_{i=1}^{\log_2 E} 2^i)} \tag{22}$$

Taking our old example in Section I B2d, then even in the transmission of 8 events out of 8 possibilities, 10 $bits_R$ can be saved. Consequently $A''_a = P_m{}^{-R(bits_R)} = 10^{-4 \times -10} = 10^{40}$. The systemization of only one of the first preliminary decisions (e.g. the first 'a' decision, cf. Section I B2d2) saves 3 *bits*. It therefore gives a selective advantage of $A''_a = 10^{4 \times 3} = 10^{12}$. This advantage corresponds to a number with 12 zeros.

Thus, even in very small groups of features, the systemization of preliminary decisions produces huge advantages. We should therefore expect that the mechanism of the molecular code would long ago have evolved the required technique.

2. *The serial case.* The advantages of a series of change-over switches are considerably greater. These advantages always exist when it is possible to exclude a number of alternative choices (e) of the possible programme (whether statements or features) so that only a selection of events (E) is chosen. Applying equation 13 (in addition to the advantage A''_a) we can work out the serial advantage:

$$A'''_a = P_m^{-(E-1) \cdot [\log_2 (E+e) - \log_2 E]} \tag{23}$$

For the systemization of every preliminary decision saves $E - 1\ bits_R$.

Even in the case of quite simple systems with only 8 E out of 1024 possibilities $(e = 1016)$, as in the example in Section I B2d an additional 49 $bits_R$ can be saved. For this example, therefore,

$$A''_a + A'''_a = P_m^{-4x - (10+49)} = 10^{-4x - 59} = 10^{236}$$

Let us take, by way of illustration, the simplest possible case, which would be the switching-off of a programme '1—8', or its switching-over to the programme '9—16' or '513—520', or whatever. The systemization of almost all the preliminary decisions will then avoid 7 $bits_R$. The reign of accident through lack of systemization will result in hopeless chaos as can be seen by comparison with an electrical device. If the channel selector switch of a television set were wired up to seven subordinate decisions, with only two alternatives in each position, then if the device is fully systemized, the chance of getting the right programme by turning the knob at random is $P = 2^{-1}$, i.e. ½. But, with an unsystemized device, in which we had to play eight switches at random, then $P = 2^{-8}$, i.e. every 256th attempt would be successful. It would be impossible to use such a device blind, i.e. without knowing the switches.

We can estimate how impossible such a transmutation would be in the genome. For the prospects of success would be still further reduced so that only a single one of the switches would change after 10 000 attempts to switch on. Moreover the switching-off or switching-over of a message would require consideration not of eight single events, but probably of a multitude. We know the method of dismantling hidden redundancy of the genome, i.e. of reducing long-windedness in the formulation of laws. But if we did not we should have to presuppose its existence. A biologist will appreciate that it has long been operating.

Perhaps it is not yet apparent how this mechanism of the change-over switch necessarily leads to the establishment of the hierarchical pattern of order. I shall prove this connection in the next section (III D). I shall first discuss the molecular-genetic mechanism involved while the morphological mechanism will be dealt with in Chapter V.

d. The operon system

A mechanism for reducing long-windedness in the statement of laws requires only the ranking of determinative decisions, one above the other. More precisely we should expect to find early-acting switches (i.e. of higher rank) that set the signals for a greater number of later-acting switches (i.e. of lower rank). The preliminary decision must be remembered by all the subsequent decisions until it is superseded by the appropriate alternative. Precisely such a system has been discovered by molecular geneticists in the form of the operon system.

Research on this matter is, of course, still in full progress, and only the simplest aspects are near to being explained. These are the operon structures. Two results, however, are beyond question. The first concerns the old conception of the genome as a string of beads made up of equal-valued determinants such as eye colour, number of bristles or wing shape, scattered haphazard. This conception has now been completely superseded. The second result is that the operons probably take in the whole genome and together represent a complex system of grouped controls and transdeterminant controls at all levels of complexity. This means that, as already suggested, the nucleotide chains of the genetic message are comparable, not with the Morse code of a verbal script, but with algebraic script.

1. *The operon.* In the simplest case this is a little sequence of genes in which an operator gene (or a promotor and an operator) switches on a sequence of adjacent structural genes. The operator genes correspond completely to the preliminary decisions of our model of determination flow, while the structural genes correspond to the final decisions. The agreement with expectation is complete (Fig. 18).

The operator gene, like all the others, is a fairly usual chain of codons of some 20 nucleotides in length (ranging from 10 to 100) and the promotor is also of about this size. The adjacent row of structural genes is only short. Either the operator is the place of attachment of the suppressor that switches it off (see below) and of the RNA-polymerase which cooperates in forming the mRNA templates of the consecutive structural genes. Or else a special promotor, peculiar to the operon, forms the place of attachment for the RNA polymerase. Genetics has naturally concerned itself particularly with the regulator problem, which I shall not deal with until Section III C3. At present I only need to prove the existence of ranked switching.

Mutations of individual structural genes do not influence the preliminary decisions. But mutations which do affect such decisions, producing the so-called constitutional mutants, make it impossible to control any of the consecutive structural genes of the operon. Furthermore it transpires that: 'every operator affects only those genes with

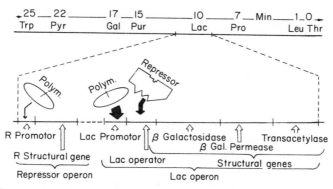

Fig. 18. The operon system as illustrated by the lac operon and its regulator gene. Above is a larger portion of the chromosome map of *Bacterium coli*. From Bresch and Hausmann (1970).

which it is structurally connected.'[18] These crucial conclusions are already twenty years old[19] and are in the textbooks. Other conclusions, however, are still tentative. In connection with the requirement of the present theory they can be listed as follows:

2. *The principle of the group key.* In molecular genetics[20] this corresponds to our further requirement that preliminary decisions, as soon as they are developed, must in themselves form a ranked system. Bresch and Hausmann refer to this as: 'Repeated pattern formation by using a corresponding switching scheme.' They then elaborate: 'For it is no doubt possible to set up many different switching schemes with the same elements, and thereby produce totally different regular effects.' And moreover: 'The possibility of group switching should be borne in mind, by which single specific repressors act on *a group of several operons* or single effectors alter a number of different repressors.'[21] (My italics.)

3. *The principle of transdetermination.* This corresponds to the last of our requirements which is that even large complexes of features can be altered meaningfully by a single mutation of an early acting and therefore high-ranking decision — meaningfully in the sense of how the system of the relevant portion of genetic code is organized. This alteration would be much like the change in our last example (Section IIIC 1c) by which the message 1-8 could transform into the message 513-520, or 513-1024 could be switched over to 1-512.

In the replication of cells, not only the structures are copied but also the controlling conditions. If a change occurs in one of these conditions then the clone of cells from an antenna anlage can by error produce a leg, or the clone from a haltere anlage can produce a wing.[22]

The error may breed true, or even be cancelled out by a back transdetermination. I cannot go into the details here. The problem will be dealt with more fully later under the headings heteromorphosis, spontaneous atavism, and homoeotic mutation. For the present we are only interested in the genetic consequences.

The complex details have, of course, not yet been resolved, but Bresch and Hausmann summarize the situation as follows: 'The results suggest, nevertheless, that the whole state of determination depends only on a few molecular switching events, or possibly on one only. If it were otherwise no transdetermination could occur or else the determination of a whole clone of cells would become unsettled simultaneously and remain unstable. This point gives hope that we shall soon understand the situation at a molecular level.'[23]

We have therefore shown that ranked switching must necessarily exist, and proved that it is realized on a molecular level. It will be a small step to show that the hierarchical pattern of order is a necessary result.

2. The first consequences of systemization

Even folk wisdom asserts that nothing is got without being paid for. This could count as a general rule of living, but here I shall prove it numerically by a careful reckoning of evolutionary prospects. The systemization of redundant determinative decisions leads to a momentary increase in freedom, but this must later be paid back to evolution by a loss of freedom. This is the first seemingly paradoxical consequence.

In considering these prospects the question arises: What is the point of this to and fro of advantages when there is finally no gain accomplished? However, this question is teleological, for no evolutionary path can predict in any way how its future phylogenetic bank account will stand. What group of animals could, at its origin, know its own prospects of dying out? The question whether, with subsequent loss of advantages, any selectional profits would remain at the finish to drive the mechanism likewise has no biological meaning. Selectional advantages are valid for the instant. The paradox is purely fictitious. Evolution lives from hand to mouth.

a. Burden and canalization

This brings us to the concept of burden (not 'genetic load'). By burden I mean the responsibility carried by a feature or decision. I intend to show that, with systemization, the functional burden carried by decisions increases and with this a new lack of freedom called canalization also increases. This burden of decisions belongs to the realm of genes and molecules. But its counterpart, the corresponding functional burden of features, is predominantly a morphological phenomenon and will not be discussed until Chapter IV to VII.

The change in the prospects of success with increasing systemization provides the key to the problem of burden and canalization.

The term 'functional burden' of the genome, in my sense, has nothing to do with 'genetic load'. The latter was introduced by Muller (1950) as a measure of the reduction in fitness of a population caused by an accumulation of subvital genes (see also Mayr, 1964, Chapter 9). Confusion is only too likely, for genetic load has been translated into German as 'Bürde' (= burden).

1. *The advantage of systemization.* Whatever sort of redundant decision is dismantled, we have found in Sections III B and III C1 above that there is an increase in the prospect that a change can be accomplished. This increase, or accomplishment advantage, is a probability similar to the reciprocal of the mutation rate (P_m^{-1}). The biological reader will have noticed, however, that we have left one parameter out of the account. This is the prospect of success that a mutant possesses, whatever its frequency (P_m) may be.

2. *The chance of success of a mutant.* This again is a probability — the probability of a beneficial effect — which will be written P_e (e = Erfolg = success). In individual cases it can be measured empirically and ranges from zero to fairly large values. The zero values represent the near certainty of death, corresponding to the so-called lethal factors. The fairly large values approach almost within an order of magnitude of probable success (10^{-1}), especially if total viability should count in the first place as success.

There is now much information about the success of mutations.[24] Two facts are particularly interesting here. First, the prospect of success decreases with increase in the extent of mutational change, i.e. with increase in the features or individual events altered by the mutation. Second, the maximal visible prospects of success are no more than a few per cent. We can with fair certainty assume that the mean prospect of success of a mutation of a single decision affecting only a single event, is about 5–10 per cent, or in any case between 1 per cent and 50 per cent. The increase, as a result of dismantling redundancy, in the prospect of successful change (A_{ae}) is therefore less than P_m and less than P_e, namely $A_{ae} = P_m \cdot P_e$.

What then is the prospect that two events (features) altered independently by mutation will be accepted by selection? We can assume a probability of P_{e1} for the first feature and P_{e2} for the second. The mutant that bears these two alterations will have an even smaller prospect of being accepted by selection. It will correspond to the product of the two individual probabilities, say $P_{e1} \cdot P_{e2} = 1/10 \times 1/10 = 1/100$. The prospects of success (P'_e) of a mutant will correspond to the products of the individual prospects of success of the number (n) of altered events (E) or features. Thus $P_e = P_{e1} \cdot P_{e2} \ldots \ldots P_{en}$

Thus $P_{en} = P_e^{En}$

$$\text{(24)}$$

Let us suppose, however, that the adaptive pattern demanded by the external environment agrees with the systemized pattern that the genome has achieved. For example, if two events (features) are changed in the same sense by the mutation of a single preliminary decision, then their prospects of success will likewise be tested in the same sense and will not decrease. They will amount to P_e^1 rather than P_e^2. Here again, therefore, systemization will increase the prospect that a subsequent change will be successful (A_e). This increase could equal the reciprocal of the prospect of success for a single change $(1/P_e$ or $P_e^{-1})$ to the power of the number of further dependent single events (E').

$$A_e = P_e^{-E'} \qquad (25)$$

The overall success of systemization, or the positive alteration in the prospects that a mutant will be realized and will be successful, can be called the accomplishment and success advantage of systemization (A_{ae}). At a maximum it will correspond to the product of the altered prospects of accomplishment $A_{a(max)} = P_m^{-R}$ (cf. equation 20, Section III B2c) and of the altered prospect of success $A_e = P_e^{-E'}$ (equation 25). Therefore:

$$A_{ae} = P_m^{-R} \cdot P_e^{-E'} \qquad (26)$$

3. *The failure of systemization.* The advantages of systemization will disappear at a characteristic point, when the pattern of adaptive changes demanded by the environment no longer agrees with the systemization pattern of the genome. Here I must anticipate somewhat. I shall show later that systemization patterns copy the functional pattern of features, and thus the environmental requirements (Section III D2), and I shall consider the changes in these environmental requirements in Chapters IV to VII. At present I shall only sketch out the feedback pattern.

The advantages of systemization hold when the pattern of requirements continuously corresponds to the pattern of systemization.

Assume a systemized genome with the following characteristics ('–' signifies 1 bit_R):

Number of Decision	1	2	3	4	5	6	7	8
1st preliminary decision	a	–	–	–	b	–	–	–
2nd preliminary decision	a	–	b	–	a	–	b	–
Final decision	a	b	a	b	a	b	a	b
Event (feature)	I	II	III	IV	V	VI	VII	VIII

Suppose that the genome enters an adaptive niche demanding the features I II III IV *I II III IV*. Four events (E) must therefore change. The systemized genome would only need to mutate the first preliminary decision (no. 5) while the unsystemized genome would need to mutate the first preliminary decisions, nos. 5, 6, 7, and 8. Consequently $R = 3$ and $E' = 3$. The exponent in $A_{ae} = P_m{}^{-R} P_e{}^{-E}$ will therefore be positive. If $P_m = 10^{-4}$ and $P_e = 10^{-1}$ the change in the prospect of successful accomplishment will be positive compared with the unsystemized genome. Thus $A_{ae} = 10^{-4 \times (-3)} \times 10^{-1 \times (-3)} = 10^{12} \times 10^{15}$. This is an enormous adaptive advantage, corresponding to a number with 16 places in front of the decimal point.

The disadvantages of systemization occur when the pattern of requirements (the external pattern or the environmental conditions) no longer corresponds to the internal or epigenetic system.

Let us take the same example but suppose that it enters an adaptive niche which demands the features I II III IV *I* VI VII VIII. In this contrary instance only the first preliminary deicision (no. 5) needs to mutate in the identical but unsystemized genome. But in the systemized genome, decisions 6, 7, and 8 must also mutate as independent single decisions. The change in prospects of successful accomplishment, again compared with the unsystemized genome, will be negative. Three decisions which up till now have been dismantled as redundant must be interpolated again ($R = -3$) with the help of accident ($P_m = 10^{-4}$) and their dismantling cancelled out. Moreover it may be that the prospects of success ($P_e = 10^{-1}$) of three up-till-now correlated features will be tested separately ($E = -3$) by selection.

In this instance, therefore, the change in the prospects of successful accomplishment will reverse in sign $A_{ae} = P_m{}^R \cdot P_e{}^{E'} = 10^{-4 \times 3} \times 10^{-1 \times 3} = 10^{-12} \times 10^{-3} = 10^{-15}$. This corresponds to an enormous adaptive disadvantage for the prospect of successful accomplishment will have 14 zeros after the decimal point.

4. *The burden of a determinative decision.* In the systemized genome this depends on the number of other decisions that the decision in question implies, and also on the number of events (features) that it results in. If the required adaptive pattern agrees with the systemization pattern this 'latent' burden will never be felt. On the contrary, systemization will increase the prospects of adaptive success. However, as soon as the two patterns depart from each other, the burden will lead to a drastic decrease in the prospects of success, to a disadvantage in terms of realization or success $A_{ae(neg)}$. This is the reciprocal of the previous advantages.

$$A_{ae(neg)} = P_m{}^R \cdot P_e{}^{E'} \tag{27}$$

We can confidently neglect P_e as less important and difficult to estimate. The mere difference between $A_a = P_m{}^{-R}$ and $A_{a(neg)} = P_m{}^R$ is enough to show the extraordinary advantages offered by systemization of the genome and the equally enormous disadvantages which may result from the burden of determinative decisions.

5. *Canalization of evolutionary prospects.* The result is a narrowing of evolutionary possibilities. This narrowing does not stifle the process of adaptive change uniformly. Instead the change will conform completely and characteristically to the pattern of the burden, this latter being a 'metamorphosis' in Goethe's sense. These burden patterns correspond to the systemization patterns of the determinative decisions in the genome.

b. *Freedom, determinacy, and superdeterminacy*

It goes without saying that freedom, even in evolution, is a relative thing. How should we judge freedom within the limits of the determinative process?

In everyday life, freedom within the framework of law consists in those transgressions which are either tolerated or not noticed, or merely in a certain measure of confusion in the legislature and the executive, or in those who are expected to observe such laws. As concerns the transmission of law in genetic determinative decisions we describe this degree of freedom in the first place as the mutation rate. This asserts that to a certain degree, in at most every ten thousandth instance, each point of the law text is granted the 'freedom' to alter. It also asserts that the kind, place, and time of the alteration is decided by accident and that this accident will necessarily be based on the molecular conditions of this form of law. Such freedom is the failure rate of this particular determinative process. But in addition there is a monitoring mechanism, being the limits of the prospects of success. This specifies that, even among the tiniest textual alterations, at most between a tenth and a hundredth will be tolerated.

In principle, therefore, every point in the text, and thus the text overall, would have the prospect of being completely changed only at every millionth replication (10^{-6}). Without doubt this precision is absolutely astounding or even unbelievable, compared with machines or the other laws of life. However, if we compare the result with the number of reproductive occurrences and chances of alteration over at least a billion years (10^9) it becomes an unbelievable imprecision. Assuming an average of one reproductive process per year, which is very cautious, we should expect that every feature in every line of ancestors of every recent organism would already have transformed completely one thousand times ($10^{-6} \times 10^9 = 10^3$). Comparison between organisms would therefore be impossible, which is in no way the case.

Certainly there must have innumerable features which have changed one thousand times. (Imagine for example the changes in the markings of the skin in man's ancestors, starting with the predecessors of the fishes.) At the same time a large number are preserved virtually without charge, e.g. from the central canal of the spinal cord to the tails of sperms, to the ribosomes and to nucleotide bases.

This determinacy exceeds the precision of the basic mechanisms at least one thousandfold, and probably one hundredthousandfold or a millionfold. It is a superdeterminacy or superprecision which could not be reached either by the mechanism of single determinative decisions, nor by their monitoring process. Instead it depends on the systemic conditions whose elements I have described as systemization, burden, and canalization. It is not merely the patterns of organic order which demand this explanation but also the stability shown by this order.

c. The building-up and dismantling of decisions

One more question remains. The determinative decisions of the genome differ in rank. In which rank therefore would the build-up of new decisions occur? For it is both required and proven that the law content of the genetic manuscript expands with the evolution of its bearers.[25] Even in this connection the consequences of rank and burden permit a conclusion.

The nature, place, and time of a new decision is a matter of accident, just as with a change in a decision, so that its prospects of success can be estimated as for a change. As we saw (equation 24) this prospect (P_e) diminished exponentially with the number of single events (E) affected by the decision. New decisions will therefore be expected to arise in the lowest ranks and, within these, most commonly as new groups of nucleotides in the structural genes. They will only gradually be taken into the higher ranks.

The same must hold for the dismantling of decisions that leads to systemization. The loss of a high-ranking preliminary decision has only a vanishingly small prospect of being tolerated. The build-up and dismantling of decisions must therefore tend to affect the lower ranks. The movement of decisions into higher ranks is a result of the general increase of systemization which, according to the theory, would be expected to occur in every genome.

3. The model and its molecular realization, Part II

We are therefore faced with a molecular code of determinative decisions, and the prospect of survival depends on appropriate adaptation of the whole by deciphering mistakes in the code. In such a case we should expect two further types of dependence to arise. I have already called these simultaneous and successive dependence, i.e. interdependence and traditive inheritance (Section II B4).

I shall deal with the systemization of these two kinds of dependence here, having already dealt in Section III C2 with the 'first consequences' connected with the standard part and hierarchy. I choose this approach because simultaneous and successive dependence are themselves affected by the 'first consequences'. In principle, however, they are merely two further forms of switching action — 'synchronous' and 'sequential' switching. I shall now develop a model of these and explain how they are realized at a molecular level.

a. Synchronous switching — 'If N, then M'

Synchronous switching is an obvious feature of electrical devices wherever two originally separate events are supposed to function only together. Thus, in a slide projector, the lamp cannot be lit unless the blower is running — the blower can probably be switched on separately but the lamp only when the blower has been switched on already. The aim of this is to prevent the lamp overheating by a mistake in switching.

We should expect by analogy that the functional interdependence of features (e.g. organs) in organisms would result in synchronization of the decisions on which the features depend. In a movable joint, for example, the switching-on of the determinative decisions to produce one articular surface might also switch on the production of the other surface, for this would give a great selective advantage. It would be at least as large as P_m^{-1}. In fact, however, it will probably always be considerably greater than that, because the modest hypothesis of 'one feature, one alternative', which I have used up till now, will certainly not apply when more complex unities are being altered. The number of alternatives will increase with the individual events involved, as also with the required precision.

Thus assume only 10 single features for either articular surface of a joint (E' and E''). With independent switching we should expect $E' \cdot E'' = 100$ different alterations. If only one of these can be accepted by selection the prospect of success will be as small as $1/(E' \cdot E'') = 1/100$. But if synchronization decides not merely the time but also the nature of alteration then the prospect of success rises from $1/100$ to $1/10$, i.e. by a factor of E. The selective advantage of synchronous switching would then be $A_a = E \cdot P_m^{-1}$. If quantitative coordination is also included the advantage would be even greater, by a factor x.

We can speak of the increased probabilty of realization or realization advantage under conditions of interdependence (A_{ax}). This will be

$$A_{ax} = X \cdot E \cdot P_m^{-R'} \tag{28}$$

X would correspond to the precision of the required dependence and R' to the redundancy which thus arises.

Redundancy will be dismantled even with this synchronous switching. A certain number of the determinative decisions ($bits_D$) required for separate working will become redundant if the events they determine become functionally interdependent. And, as soon as this pattern of functional interdependence of the events is copied accidentally by coordination of decisions, then the redundant decisions will be dismantled which will bring a selective advantage.

The advantage of such switching, assuming 10 degrees of precision (X) and ten single events involved (E), will be about 1 million $(A_{ax} = 10 \times 10 \times 10^4 = 10^6)$.

With a non-linear code of this sort it is important that individual groups of decisions can be synchronously switched over any number of single decisions. Biologists know that this requirement is fulfilled in the most remarkable fashion by the 'chemical messages' of the genetic system.

b. The regulator-repressor system

In considering the requirements of the 'selector' or change-over switch we found that, starting from an operator gene, the conserved message sequence of adjacent structural genes can be called up (cf. the operator system, Section III C1*d*). But a synchronous switch would be expected to control the activity of distant operons also.

These synchronous switches have been recognized in the form of *regulator genes*, while their messages take the form of *repressors*. To start with, we must distinguish between, first, the switching action itself and, second, how it leads to synchroneity.

1. *The switching action.* The action of the operator gene was structurally dependent, being limited to the *cis*tron (or operon) in the *cis*-configuration. Genes with a distance effector, however, can be *cis* or *trans*. This fundamental difference depends on the fact that the messages from these regulator genes are sent off in quantities into the plasma, which is like an unsorted in-tray. These 'telegrams' all contain only the command 'yes' or 'no' along with the precise address of an operator gene. As soon as one of the huge number of sent-out messages reaches the appropriate addresses the command will be transferred. Let us suppose, for example, that it is 'no'. The regulator molecule looks for the operator molecule as if key to lock and if it finds a fit it fixes over it and shuts off the operon function like a lid (Fig. 19a-b).

We need not go further. Molecular genetics is particularly concerned with how the formation of the complicated protein molecules known as enzymes is controlled. It is also concerned with allostery which is the double specificity of the repressor, i.e. the alteration of the message from 'yes' to 'no' or conversely. Thus an effector can alter 'no' into 'yes', in the induction of catabolic operons (Fig. 19a) and 'yes' into 'no', in the repression of anabolic operons (Figs. 19b). [26]

In addition to these 'negative controls', systems of positive controls are beginning to be discovered. In these an activator, allosterically controlled by an effector, activates the previously dormant process of reading the gene. [27]

2. *The synchroneity of the switching action.* The mechanism just discussed is potentially capable of synchronous switching. There is only one additional requirement.

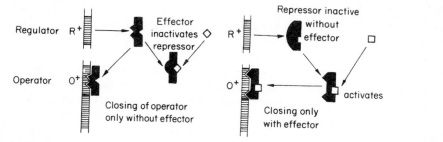

Fig. 19 a-b. The regulator-repressor system. An example of negative control of the activity of operons involving co-operation with an effector.. (a) The induction of catabolic operons where the repressor is inactivated by the effector. (b) The repression of anabolic operons where the repressor is activated by the effector. Compare Fig. 18. From Bresch and Hausmann (1970).

This is that the repressor should not be specific for one address only, among the countless number that exist, but for more than one, as would not be surprising. This requirement is likely to be satisfied, not only because of the molecular conditions, but because the selective advantage so attainable make it probable to the point of certainty.

Indeed the process is already a solid constituent of molecular-genetic theory. I have already discussed the principle and how it shows itself when I discussed the *group key* and *transdetermination* (Section III C1d).

All this refers to the molecular-biological aspect. In morphological and developmental physiology an even greater quantity of material can be understood by assuming synchronous switching. I shall deal with this in Chapter VI.

c. Sequential switching — 'N only after A'

This last wiring instruction likewise seems self-evident, for all the three types of switching action so far described appear to be impossible without its help. Thus it is obvious that in a record player the playing arm cannot swing in and come down on the record until after it has been lifted up. In a washing machine the heating switches on and the drum turns only after filling with water. There are hundreds of other sequences which are every day so obvious that they need no explanation.

In the genetic system, however, there is a complicating circumstance which is just as necessary, but less obvious because it is not immediately visible in everyday life. This is that the building instructions are copied from one piece of apparatus to the next along with the operating instructions. Every piece of apparatus has its own history so that we should expect the building instructions (even for a washing machine or a record player) to contain an account of their own course of development. To prove the absolute necessity that even historic, archaic decisions must be kept available, I shall discuss a simple example quantitatively.

1. *The impossibility of not conserving old decisions, a quantitative example.* For illustration consider Fig. 20a–f. A letter of the alphabet, regarded as analogous to an extremely simple technical or biological system, is adaptively built up and modified step by step. The modern phenotype O has evolved through the sequence I L C E F A N O (Fig. 20c). Adaptive modification to form Q will happen much more easily by repeating the whole series, than by way of the much shorter and teleologically simpler series I L C O Q which omits the detour E F A N.

For purposes of calculation the playing rules are as follows: There is a framework of squares with positions for ten bars (features). At each reproductive process each feature can with equal probability ($P_m = 10^{-4}$) disappear or appear in an unoccupied position, i.e. could be wrongly placed in 9 out of 10 possible positions ($P_e = 10^{-1}$). The prospect of success of each step is therefore $P_m \cdot P_e = 10^{-5}$ and this applies also in passing from O to Q. The prospect of building up E features in the correct position and sequence is $(P_m \cdot P_e)^E$.

There are theoretically two ways of changing the phenotype. The organism could either follow the teleologically shortest path, omitting detours. Or the organism could repeat the detours, which are large but specified by determinative decisions, and then introduce a single, new, accidental decision. The selective advantage of the second way over the first is the difference in the probability of successful mutation assuming a principle of traditive inheritance (A_{aet}). This is equal to $1/(P_m \cdot P_e)^{(E-1)}$, i.e.

$$A_{aet} = (P_m \cdot P_e)^{(1-E)} \tag{29}$$

This is the selective advantage (taking into account the prospects both of realization and of success) under conditions of traditive inheritance. In our example we can

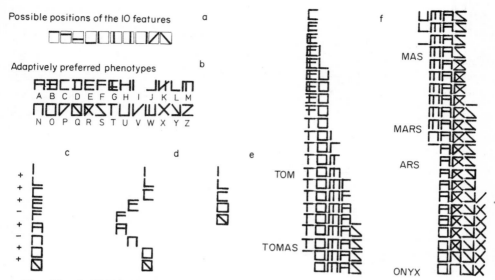

Possible positions of the IO features   a

Adaptively preferred phenotypes   b

Fig. 20 a-f. The order-on-order system illustrated by adaptive change in letters of the alphabet. The ten features (a) that constitute the letters (b) change from above downwards by addition or subtraction of a single bar (+ or −; c). They pass through detours which teleologically could be avoided. Compare d and e. (f) The 'ontogenetic' stages *TOM* to *ARS* of the definitive stage *ONYX* to illustrate a fairly simple case.

theoretically either transform the determinative sequence I L C E F A N O into I L C E F A N O Q by accident, or into I L C O Q by accident. The selective advantage of the first case with $E = 5$ is $A_{aet} = (1/P_m \cdot 1/P_e)^{(E-1)} = (10^4 \times 10)^4 = 10^{5 \times 4} = 10^{20}$. This is a number followed by 20 zeros.

Another possibility with the same example would be to shorten the detour by a mutation in the embryonic stage C (Fig. 20d) so that the determinative decision C → E disappeared and C → O arose. But then the organism would have to await the coincidence of this mutant with a second mutant O → Q. This chance of shortening exists but is 100 000 times smaller than that of the longer detour ($A_{aet} = 10^{-5}$). With increasing complexity it too will disappear completely (Fig. 20f).

Such simplified models give little conception of the requirement that new development can only be undertaken after passing through all previous stages of development. For example, the wheel with pneumatic tyres is only conceivable by way of the solid-tyred wheel, the spoked wheel, the disc wheel, the cylinder wheel, and the rolling cylinder. I shall discuss this exhaustively in the morphological section (Chapter VII). At present I shall continue to consider the characteristics of systemization by sequential switching.

2. *Time and redundancy.* Sequential switching brings the temporal dimension into the discussion for the first time. Indeed this mode of switching is the temporal component of the other three wiring patterns. As a result certain ideas such as redundancy and long-windedness, already familiar within a single time-section, show new characteristics.

We have already recognized that the essential feature of systemization is the avoidance of repetitions and prolixity in printing determinative decisions. *This is because order, developed on the basis of accidental decisions, is the easier to modify adaptively the more it can be excluded from undesirable accident.* This is the kernel of the matter. It is, so to speak, the paradox of the evolutionary mechanism. Repetition and long-windedness, and indeed cancellation and circuitousness, are necessary in building order up. But they have no further role to play as soon as the decisions behind them move away from the unreliability of adaptational accident, i.e. as soon as they become required history.

Thus, to refer to the example (Fig. 20c), the decisions which cause the events I L C E F A N O Q to be repeated are redundant in the sense that those which lead to O no longer need to be shielded from accident. This is true no matter how 'long-winded' the historical path may seem.

Long-windedness in this sense is the invasion of accident into determinative laws. The to-and-fro of environmental conditions forces the self-coding system into crooked adaptive paths. This to-and-fro, so far as our powers of insight are concerned, is a game of chance. The inorganic aspect of history has no meaning. It merely exists.

d. The order-on-order system

The way in which sequential switching is realized in the molecular realm is likewise obvious. Indeed it is so transparant that little needs to be said. For how could any determinative decision be effective except in a system that is already almost completely excluded from accident? It needed the creativity of a man like Erwin Schrödinger[28] to see beyond the obviousness of the facts.

His 'order-on-order' principle showed that order necessarily depends on order, as has since been fully confirmed. And it also showed, as we can add, that accident which complements order, and from which order must arise, can only be allowed a vanishingly small part in the system. However, in morphology the consequences of sequential switching are an extremely complex and much more opaque area (Chapter VII). In studying it, many arguments and results can be added to Schrödinger's.

To look back, it is virtually certain that the systemization of genetic determinative decisions will be arranged in four patterns. This assertion is based on the enormous selective advantages which accrue from these four basic patterns of wiring, for all four greatly reduce the extent to which adaptive instructions depend on accident. The four switching patterns mutually condition each other in many ways. Indeed they presuppose each other and in this respect form a unity and a functional whole. They represent, as shown later, all the symmetries of dependence possible. Having proved how these switching patterns are realized at a molecular level, we can now move forward. We now need to ask whether, and how, the molecular switching patterns will produce macroscopic patterns of order.

D.   PATTERNS OF SYSTEMIZATION AND PATTERNS OF FEATURES

1. We already possess the key to this connection. It consists in the concepts of burden, superdeterminacy, and canalization (cf. Section III C2). It amounts to the principle of 'lack of freedom tomorrow for freedom today'. The mechanism consists in the connection between a preliminary decision and the number of the events or features E which depend on it — a number which raises the prospects of successful adaptation to the power^E (equations 24 and 25). The mechanism ought already to be obvious in morphology.

2. But should we not expect that today's disadvantages, which pay for yesterday's advantages, would also wipe out the originally successful pattern? On the contrary, *the advantages will probably be lost again as a sort of payment, but the burden pattern of the molecular realm will be re-emphasized in the morphological realm.* Molecular causes have morphological causes as true partners, connected with them by a reciprocal two-way

feedback connection. These morphological causes will be presented in Chapters IV to VII. At this point I shall only indicate the principle involved.

1. Patterns of features are systemization patterns

I shall start with the antithesis between phenotype and genotype as being the easier of the two approaches. What consequences would the four systemization patterns possess in the morphological realm?

a. *Repeat switching, nucleic acids and the standard part*

1. Repeated switching of identical DNA sequences and their morphological utilization must be the cause of every standard-part pattern of order in the morphological realm (Section II B3*a*). Furthermore, standard parts occur at all levels of complexity, from that of proteins to that of complete individuals, while the replicated sequences are not in principle limited to DNA. This indicates a complete agreement between the wiring pattern of the replication switches in the realm of the gene, and of the standard-part pattern in the realm of morphology.

2. The canalization of standard parts in the phenotype depends on their conservation — on the considerable difficulties which selection raises to every departure from the standard. I shall deal with this in Chapter IV. I shall only anticipate by saying that the degree of fixation of standard parts is connected with the number of different positions that the particular standard part takes up in the organism and with the burden that the parts have to carry.

By way of analogy, consider the standard screw fitting of electric lights. Changes in this would always be rejected by selection in the market until a whole industrial group had, by accident, altered a sufficient number of types of sockets to correspond (cf. Fig. 30c).

b. *Selector switching, operon and hierarchy*

The pattern of hierarchy is less obvious than that of the standard part. At least I personally made many mistakes before I saw the connection. Correspondingly it is more difficult to explain and I want to be as precise as possible.

1. I shall begin with the operon. As explained already, this consists of one or two preliminary decisions (promotor and operator genes) which give the signal to a set of final decisions (the structural genes). If a preliminary decision is stopped then all the subsequent decisions are also stopped. This establishes the basic structure of hierarchical order in decisions, as already explained in considering the dismantlement of hidden redundancy (cf. Section I B2*d*). Functionally it is the same connection as between 'the last preliminary decision' and the 'final decision' as discussed above. Its establishment is compelled by the same extraordinary selective advantages that are offered by the dismantling of hidden redundancy.

The molecular biological mechanism involved in this ordering of determinative decisions into groups strongly suggests that periodic patterns will be formed. That is to say that decisions (promotor-operon systems) will be established which over-rank several operons (the group key in Section III C1*d*). Indeed we should expect that these decisions would form systems of enormous complexity, with preliminary decisions of several ranks, as in the transdetermination phenomenon. There would thus be a series of preliminary decisions ranked one above another so as to reduce hidden redundancy more and more effectively, as selection more and more insistently demands.

Thus all conditions for the hierarchical pattern are fulfilled, i.e. the triangular units of preliminary and final decisions and the superstructure of further preliminary decisions. The genetic and somatic patterns of hierarchy undeniably correspond to each other.

The opacity of the problem, however, consists in the difficulty of recognizing the morphological derivatives of the hierarchy phenomenon (as described in Section II B3*b*) for what they are. The whole of Chapter V will be devoted to this question.

The ubiquity of the hierarchical pattern of decisions is caused by its enormous selective advantages. Its stability is maintained by the burden, which increases exponentially with the rank of the preliminary decision. The burden, in turn, causes the prospect of successful alteration to decrease exponentially.

2. This leads to the same crucial question. Canalization will have disadvantages, for it will make features less adaptable. Why, therefore, does selection at the somatic level not wipe the canalized pattern away? But, in fact, the opposite happens. *Adaptive modifiability disappears as the burden increases. The hierarchical pattern, however, is only strengthened by the burden, since the functional burden of features completely corresponds to that of their determinative decisions.* Gene and feature form, for selective purposes, a whole. The hierarchy of the gene is only semantically distinct from that of the feature. The factual agreement between the two hierarchies exists because their mechanism is one and the same.

As I shall show at length in Chapter V, the hierarchy of decisions and that of features are mutually dependent in the way that they are built up. The genes decide the possible ways of simplifying the switching system, while the features decide the contents and limits of the hierarchical structure. This structure thus consists in the fact, that both decisions and events have content only through their subordinate features, and have meaning only through their features of higher rank.

For illustration I recall that the final decisions are the content of the last preliminary decision, but its meaning is decided by earlier preliminary decisions. The content of the vertebral column is the vertebrae, while its meaning is contained in the vertebrate ground plan. The contents of the concept 'car' is the kinds of car; but it has meaning only within the concept 'vehicle'.

c. *Synchronous switching, the regulator and interdependence*

It is easy to see that the structural pattern of synchronous switching in the molecular realm agrees with that of interdependence in the somatic realm. In symbolic form we can write: 'If N, then M also'. It is then obvious that the structural difference is no greater than that between decision and event.

1. The real question about the agreement is not the similarity in principle of the two patterns. It is whether they are identical. Even here, however, certainty is easy to reach. For what is synchronously switched at the genetic level will show connected, dependent alteration at the morphological level. And such connected changes of several features through alteration of a single decision are well known as the phenomenon of pleiotropy. A synchronization must have occurred wherever features are affected which arose separately in phylogeny and which therefore could only have become synchronized later.

Obvious examples are the spindle-fingered mutant of man, where spindle-shaped fingers go with lens-shaped eyes, or the gl-mutant of the mouse, where pelt colour and bone structure are correlated. In both cases the features affected arose independently in phylogeny. I shall discuss pleiotropy (or polypheny) in detail later.

We should expect that selection would tend to synchronize all those decisions whose resultant features (phenes) were functionally dependent on each other. And that it would remove all synchronizations between phenes which needed separate adaptive modification. In this sense we should expect the interdependence pattern of genetic switching and of phenetic function to be one and the same.

2. But even interdependence, though promoted by selection so long as it leads to a meaningful switching of functionally dependent features, will lead to canalization. And the canalization will still hold when the primitive functional dependence of two phenes ought to be altered, given up, or even avoided, because one of the partners has changed in function. If the genetic synchronization cannot be backmutated, because of the burden that it has in time acquired, then yesterday's advantage will again become today's canalization and today's disadvantage.

The adaptive advantages of interdependence will lead to canalization if one of the interdependent features change in function. But this canalization may remain in force because of burden and selection (cf. Chapter VI).

The pleiotropy phenomenon seems, at first sight, to contain a majority of such deleterious examples (see above). They are not truly a majority, however, but merely the most striking among all interdependent single mutations. Beneficial alterations of originally separate features are no less astounding, but are less unexpected. Examples are the mutually adaptive alterations of the pelvic bones and sacral vertebrae, of the ear ossicles and the ear drum, and of the last molars in the lower and upper jaw. Synchronization only becomes surprising where it concerns the coadaptation of originally independent features (Section II B3*c*). Here it is a well-known problem.

d. *Sequential switching, order-on-order, traditive inheritance*

It is a similar problem when the patterns of sequential switching and traditive inheritance are considered. In the realm of decisions the statement 'N only after A' must give in principle the same pattern as do the events which follow the decisions.

1. The first question, again, is that of identicality. If alterations are successful, even to the slightest degree, and if selection will only accept those series of gradually added decisions whose increasing consequences of events are functionally coordinated with each other, then the identicality of the two patterns will be virtually certain. Established phene sequences will be switched by sequences of gene effects. And what is switched in this manner, must have its equivalent in identical sequences of phenes.

Traditive inheritance only becomes a problem when the detours in a developmental series of phenes become so large that they no longer seem necessary (illustrated symbolically in Fig. 20f). In other words, when the functional necessity for a process or a structure no longer seems obvious. This sometimes happens in the study of behaviour[29] and the ethology of cultures.[30]

The problem becomes an enigma when yesterday's advantages of repetition have become today's disadvantages of canalization — when a whole species is driven to death by sticking to outdated ways.

2. Why then is the traditively inherited pattern not everywhere dismantled when it leads to a drastic limitation of the adaptive possibilities? We have already seen why (Section III C3*c* and Fig. 20a-f). *Canalization by tradition is unavoidable as soon as the prospects of successful change disappear. This happens through a high burden of decisions in the sequential pattern.*

Naturally, traditively inherited patterns do not remain untouched by selection, which in some cases has attempted to modify them for a thousand million years or more. I shall consider this in Chapter VII. This often leads to simplification, generalization, or to a symbolic language of structures. The more complex of the traditively inherited structures, however, are never completely disssolved.

2. Systemization patterns are patterns of features

To summarize, I have asserted that the four basic patterns of structural order coincide with the four basic patterns of systemized switching in the genome, and very probably are causally identical with them.

1. This assertion will be judged on the consequences that we draw from it. We could say that: 'The patterns of change in events correspond to the patterns of change in the determinative decisions.' But this would be merely self-evident. It is probably more interesting to say that: 'The orderly patterns of the phenotype are a consequence of the systemization patterns of the genotype.' For this necessarily means that the causes of the orderly phenomena are grounded in the self-designed systemic conditions of organisms. They are never imposed by external conditions. There is no special kind of selection that promotes orderly patterns. There is only one kind of selection. And it is the possibilities or impossibilities of the storage and decoding mechanism of determinative decisions which specify the formation of such special systemization patterns under the pressure of this single kind of selection.

This recalls the 'inner principle' which has been demanded often and energetically to explain the results of evolution (cf. Sections I A1a and II C2). For the language is the same, in whatever theory it is presented. But I hesitate to speak of an inner principle, when 'inner' and 'outer' would signify no more than 'organism' and 'life', or 'structure' and 'function', or 'object' and 'constancy'. It is only the systemic conditions that matter and which we are interpreting. Mutation and selection do nothing except the tasks that we have already seen for them.

2. I have stated that the patterns of order of the phenotype are a consequence of the systemization patterns of the genotype. But this is not all. For the orderly patterns of the genotype must vice versa be a consequence of the systemization patterns of the phenotype. We have already seen this reciprocal effect several times.

It would be naive to assert that so complex a whole as the evolution of organisms could have only a single cause e.g. a molecular one, and that mammals, man, and Michelangelo's Moses were its unidirectional effects. It would be equally naive to assert that only function could be the cause of structure, that only the egg could be the cause of the hen.

Again we could say that: 'The possibilities of change in decisions correspond to the patterns of change in their events.' But this is also merely self-evident. It is more interesting to say that: 'The orderly patterns of the genotype are a consequence of the systemization patterns of the phenotype.' For with this we assert that the structure of the genotype and the epigenetic system must be related to the functional patterns of the phenotype and must contain their history.

The epigenetic system must include the same primitive features of the early morphotype as the organism does and, like the organism, it must contain a shortened history of its own origin.

a. *Copying of functional patterns by the epigenetic system*

This can be expected because the same selection conditions will always hold when a synchronous feedback is established (Sections III C3a and III D1c). To be more precise, the decision that establishes the new connection will as yet possess scarcely any burden. It could still be backmutated, and lost without harm. Consequently selection can only act successfully on the decisions through the functional connection of the synchronized group of features (events of phenes). If the relationship between the pattern of features and that of decisions is a true functional connection, demanding a simultaneous adaptive change in both, then the mutant will have a definite selective advantage. But if the relationship between the parts demands the retention of the greatest possible adaptive independence then the synchronization will, to the same extent, result in selective disadvantages.

Consider the feet of ungulates, for example. In the evolution of the cloven-hoofed ungulates a synchronization of the third and fourth toes would be advantageous, while in the odd-toed ungulates it would be disadvantageous. The harmonious ch..nges of proportions known as *allometries* include numerous quantitatively defined examples of this sort.

However, among the accidentally produced synchronizations, selection will systematically promote the functionally appropriate ones. Consequently, the pattern of synchronous switching will more and more copy the then valid functional patterns. Correspondingly, dependences will arise with repeat switching. *The epigenetic system copies the functional interdependencies of the phene system.*

It is a different matter when such interdependence acquires burden, becomes almost inalterable, and by the canalization effect remains anchored in the deeper layers of the epigenetic system.

b. Conservation of the original pattern

This can be expected because patterns of decisions that have a certain degree of burden have no real prospect of being fully dismantled (as already shown in Section C3c and III D1d). They can only be built upon. Standard-part, hierarchical, and interdependent patterns are all affected in the same way. The consequence is *that the epigenetic system, in its developmental-physiological course, will contain a recapitulation of its own history, though in increasingly symbolic form.*

In the morphological part (Chapter VII) I shall prove this and answer the relevant open questions which we have already recognized as subproblems of the phenomenon of traditive inheritance.

Thus we should expect imitative and recapitulatory potentialities and processes. If so, a further consequence would follow. This is especially important because it makes it possible to test the new theory methodically. We should expect that the ontogenetic functional states of the epigenetic system would represent a simplified recapitulation of the functional states that are run through in phylogeny. We should equally expect that the physiological states would correspond to a sequence of meaningful events and thus to the pattern of previous positions in which commands are given out. Furthermore, the direction in which the commands flow would correspond to the positions and directions in which they were established millions of years ago.

I shall try to show that all these expectations hold.

———————   ———————

A short halt is desirable before starting to consider morphological complexity — which, being better known, is more extensive. The epistemological and molecular-biological results so far achieved (Chapters I, II and III) are in themselves not totally convincing.

The origin of four patterns of systemization indeed now seems highly probable on grounds of selection. I have also demonstrated a parallel with four molecular genetic mechanisms which probably correspond to these patterns of systemization. Moreover, I have shown a parallel between these four patterns of decisions (interactions of genes) and the four patterns of events (or features).

Indeed one could assert that this parallelism is so likely to represent a causal connection that, if the orderly patterns of event were not already known, they would be required to exist. I dare not ask the reader to agree on that basis, but a great mass of evidence and proof will follow.

Finally all holists know, and some reductionists suspect, that in an evolutionary mechanism neither accidental decisions nor necessary events can be the exclusive cause of the other. The evolution of living organisms can be understood only as a system. Since Bertalanffy's courageous work this has become a biological theorem.[31] Egg and hen can only be understood as mutual causes and mutual effects of each other. This corresponds to the view that causes must necessarily be multi-directional, forming feedback loops that interact with their effects — a view that has been self-evident in physics since Galileo and Newton.[32] In biology, however, it still has to be convincingly argued, however difficult that may at first seem given the complexity of living organisms. In short, I must now turn from the molecular aspect to the morphological aspect of the same object.

NOTES

1 Monographic treatments in Prigogine (1955), De Groot and co-workers (1962), Katchalsky and Curran (1965), Glansdorff and Prigogine (1971) etc.
2 In a simplified way I follow Morowitz (1968).
3 Eigen (1971) and Schuster (1972).
4 Example from Morowitz (1968, p.141).
5 In this chapter I follow the surveys of molecular genetics given by Bresch and Hausmann (1970) and Watson (1970).
6 From Einstein and Born (1969) quoted from Wickert (1972).
7 Both quotations from Mayr (1967, p.143); see Bresch and Hausmann (1970, p.63).
8 Hadorn (1961, p.47).
9 The calculation assumes energy transfer by ATP (adenosine triphosphate) and gives about −7000 cal, for every mole of the replicated nucleotide in the strip of genetic code. A clear exposition is given by Lehninger (1965) or more recently by Klotz (1967).
10 The classical example was given by Simpson (1955, p.96). He showed that the accidental conjunction of five mutations is absolutely improbable. I shall discuss this later at length.
11 Compare Britten and Davidson (1969).
12 Bresch and Hausmann (1970, p.221).
13 Bresch and Hausmann (1970, p.307).
14 Bresch and Hausmann (1970, p.224).
15 Monod (1971, p.137).
16 Jacob and Brenner (1963).
17 It has only recently been shown that structural genes tend to occur singly (cf. Sullivan *et al.*, 1973 and the four relevant papers, all 1972, cited therein). It is significant that other genes (let us say 'switch genes') are known as groups of more than 200 identical copies. In connection with ribosomal, histone, and transfer RNA see respectively Birnstiel *et al.* (1970), Kedes and Birnstiel (1971), and Morell *et al* (1967).
18 For details see Bresch and Hausmann (1970); the quotation is cited by them on p.272.
19 Monod and Cohn (1952).
20 Monod and Jacob (1961).
21 Bresch and Hausmann (1970, p.295).
22 Hadorn has surveyed these phenomena (1966a, 1968).
23 Bresch and Hausmann (1970, p.300).
24 Compare Hadorn (1961), Mayr (1970), and Dobzhansky (1951).
25 Compare, for example, Britten and Davidson (1969) and the facts compiled by them.
26 The details of these discoveries, which have been made over twenty years beginning with Jacob and Monod, are surveyed by Bresch and Hausmann (1970).
27 Sheppard and Engelsberg (1967); see also Gross (1969).
28 For the first time in *What is life?* (1944).
29 Surveyed by Eibl-Eibesfeldt (1967).
30 See Koenig (1970) and Lorenz (1972).
31 See especially von Bertalanffy (1948) and his later works (1952, 1968, 1970). See also Koestler (1968), Weiss (1969, 1970a), Lorenz (1971) and the collective works edited by Koestler and Smythies (1970) and Weiss (1971). Thorpe's conclusion to the volume edited by Koestler and Smythies is particularly short and clear.
32 For physics this was already expounded by Eder (1963).

CHAPTER IV

THE STANDARD-PART PATTERN OF ORDER

A. INTRODUCTION AND DEFINITION

The standard-part pattern of order is the first to be considered, because without it even recognition is impossible. In presenting it I shall have to consider histology and cytology as well as anatomy. It will be easy to document and establish it. *The standard-part, or normative, pattern is evident from the observation, or occurrence, of events (such as structures) agreeing so well with each other in constitution and mode of occurrence that no doubt remains of the presence of identical determinative laws.* We are dealing with what is called 'the same thing', with classes and standards, or, in the sciences, with building blocks, units or identicalities. We are dealing with repetition whose content is known in information theory as redundancy. Standard-part order is extraordinarily universal. It reigns at all levels of thought and in the lawful and predictable external world.[1]

Consider how the concept of the standard or norm, is applied in algebra and printing, sport and justice, in Communist labour laws, in petrography, social sciences and medicine, but particularly in economics, science, and technology.

a. A fantasy world without standard-part order

A world without standard-part order is unthinkable. It is unthinkable even to fantasize without standard units. This is so amazing that I must ask the reader, if he will, to experiment on himself. In doing so he should remember that every concept that can be drawn from this book, every word printed in it, every letter put in printer's ink on the paper, each of the downstrokes of an 'm', is recognizable only because of its repetition or repeatability. Redundancy of observed phenomena is a precondition for all knowledge. I showed this already in Chapter I.

In Fig. 21a-d I have tried to trace the dissolution of standard units but of course unsuccessfully. The simplest standardized order conceivable (Fig. 21a) is somewhat like a crystal. When standardization of position disappears, description becomes considerably more long-winded. When standardization of structure disappears (Fig. 21c) each individual symbol would need to be described. When identicality of size and thickness disappears (Fig. 21d) the figure becomes even more confused, but symbols are still present as standard units. If this class also disappears, then we lose even the concept of this collective, but the situation is still thinkable. One or two further steps, no longer picturable, and even imaginability will fail. For imagination fails at the limit of the utmost conceivable standard part.

But if recognition and imagination are unthinkable without standard-part order, how can we be confident that such order objectively exists? Is it possible, as already touched upon, that the supposed standard properties in Nature are in fact thought-standards that we project into Nature, so as merely to be able to think about Nature? The simple solution to this confounding question will be recalled from Section 1 B2: What is not

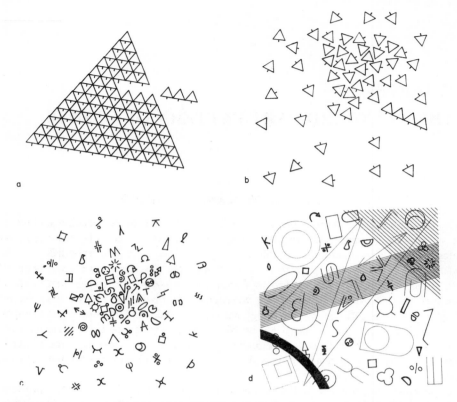

Fig. 21 a -d. A graphical attempt to dissolve standard-part order. (a) Standardization of structure is complete and standardization of position is almost complete. (b) Standardization of structure is complete, but standardization of position has almost disappeared. (c) Standardization of structure is disappearing, remaining only as the standard characters of 'symbol', 'size', and 'line-thickness'. (d) Even standardization of size and line-thickness has disappeared.

identically repeated, we do not understand. Wherever we can make predictions or recognize rules or law or meaning, then determinative happenings must redundantly occur, and standard-part order must reign.

b. Masses and classes

I shall make a few more remarks in general before concentrating on the standard-part order in living organisms. For organic standards are only a special form of universal normative order.

1. *Distribution.* The distribution of standard units extends from elementary particles through atoms, molecules, crystals to celestial bodies. And from universal concepts through words back to elementary symbols. Standards are stable or more probable states. Standard parts are identical units of regularity which, under defined conditions, exist long enough to be observed.[2] They extend from cells and organs through people and through family norms and social norms into every part of everyday life. They extend from cars, through TV channels and down to pins.

2. *General features.* These are obvious if we widen our stochastic theorem of homology, taking this theorem as an instance of a more general law. We recognize

identicalities if, under the same conditions (positional criterion) the same thing (structural criterion) is always and exclusively (conjunctional criterion) to be observed.

The positional criterion can be shown by the simple statement:

⊢ ⊏is sou–euoe !ᴧ ⊏ɐɹp ɟo ⊢ɐǝd

There is a position and a place (conjunction) for everything. If, for example, the dumbbell shape occurred in a molecule, a spore, in a gymnasium, or in a galaxy, then, however similar the shape, we should rightly conclude that here was mere analogy, not identicality. Think, for example, of the similar orbits of elementary particles and of planets.

Structure is convincing in proportion to the extent of the features. We do not confidently identify two moving points of light in the night sky; but we do identify two complex structures as, for example, two Boeing 747's. 'Metamorphoses' or basic morphological transformations occur both in the inorganic and organic worlds as modifications of identicalities such as ℬ.B.℔.ℬ.ℓ.b. The transitional criterion (cf. Section II B2*a*2) applies in the inorganic world also; thus the identicality of the Phoenician ⊿ with our **R** becomes obvious by way of ᑭ ᑭ R of archaic alphabets.[3]

3. *The fates of standard entities.* These also show general features, which can be covered by the concepts of production, collectivization (stereotyping or deindividualization), systemization and individualization or reindividualization. If a unit of regularity exists in a particular framework of conditions then, among all possible units, the identical ones will have the greatest prospect of likewise existing, or surviving.

Think, for example, of the restricted existence of the innumerable breakers on a shore line. There will arise identical individualities by the action of like parameters on the same place, so long as the conditions of sea and coast are the same. Consider also the deindividualization of unlike waves to identical standard waves appropriate to the roughness of the sea. The same standardizing conditions of formation must extend from an elementary particle to a living individual. Both, as we say, are 'conditioned', 'selected' or 'tuned' towards the condition of the great mass of individualities, because of increased prospects of stability.

But the mass of individualities in turn produces new conditions. Not only are there mutual dependences and interactions between the standard parts to form systems. There are also reindividualizations and diversifications of standard parts under these interactions. I shall give no further general examples here but treat them, because of their significance, in the morphological part (Section IV C3).

The standard or normative order of biological structures therefore seems to be only a special case. However, I shall now leave this generality, with its philosophical and indeed political implications, so as to go more deeply into the regularities of this special case.

B. THE MORPHOLOGY OF STANDARD PARTS

As in the later chapters (V to VII) I wish to separate facts carefully from theories, although the theoretical solution of the remarkable facts is temptingly obvious from what has already been said. I shall be methodical and explain the manifestations of the standard part first, and its causes later.

1. Complexity, quantities and transformations

First I shall describe the single individualities of the standard parts.

a. The limits of identicality

I have already considered the limits of identicality of biological standard parts in Sections II B2 and II B3*a*. Essentially the homology theorem implies the probable existence of identical standard parts. In other words, their existence is shown by the improbability that systems true to the positional-structural criterion and the conjunctional criterion might depend accidentally on different determinative regularities and thus be different in origin.

1. *Degrees of complexity.* Cell types in metazoa and individuals in species represent the degrees of complexity where identicality is least in doubt. For we know that individual organisms depend on identical commands, while in cells the total laws of such commands have been suppressed except for a special identical section of them.

This choice of identical commands in many cells of the same organism (which all must have received the same total laws) is the process of embryological induction. The latter implies, for example, that in forming the lens of the human eye, a substance goes out from neighbouring tissues and permits in the region of the future lens only a single highly specialized cell structure and cell position. The genome of these cells is identical and the inductive substance is also identical, so it must be identical commands that act on the individual cells.

We must also postulate the identicality of the tiniest organelles and ultrastructures, such as the ribosomes. How otherwise could different ribosomes translate a strip of DNA into identical proteins (cf. Fig. 17a-d)?

The standard parts in the smallest units of organelles, organs, and colonies answer the most stringent epistemological demands that can be made on them. The identicality of the determinative decisions that they depend on is therefore certain. There is no reason to doubt the standardized identicality which constitutes them in such a visibly congruent manner. Consider, for example, the cilia of an epithelial cell, the hairs of the human head, or the cormidia of a siphonophore (cf. Fig. 10a-h).[4]

2. *Limits of identifiability.* There is a limit to the identification of standard parts only in the lowest submicroscopical region. It exists where the particles are so small that sufficient structural details cannot be produced by the electron microscope to establish a high enough accidental improbability. If the complexity is still further reduced, so that actual molecular structure can be worked out, then identicality reappears (as indicated in Section II B2*a*). This is the case when the degree of isology in macromolecules is so large (i.e. their chemical similarity) that no accident could explain it. Thus, in the example of the cytochrome c of mammals and yeast (Fig. 8a-b), we must assume the reign of identical regularity — the existence of identical standard parts on the basis of identical commands by identical genes.

There is thus only one zone of uncertainty in which, for the moment, morphological structure is too small to be resolved by our methods and the molecular structure too complex. Normative or standardized events, therefore, extend from colonies of animals down to chains of polypeptides. Biological standardized decisions, as opposed to standard events, extend from the 20 amino-acids down to the four bases of DNA.

b. Complexity and quantity

The numbers in which the standard parts of organisms occur range from two identically formed examples (e.g. the lungs, kidneys, or eyes of vertebrates) through 10^{14} identical cells and 10^{18} or more identical giant molecules. There is a broad connection between the number and the complexity of the building blocks in an organism. The number of standard parts usually increases with decreasing complexity, because each standard part of a given complexity is built up of numerous standard parts of the next lower level of complexity. I have already discussed these levels in Fig. 13a-d and in Section II A3.

With regard to this correlation it must be remembered that the degree of complexity of organisms themselves can differ, from bacterium to man, by at least 12 orders of

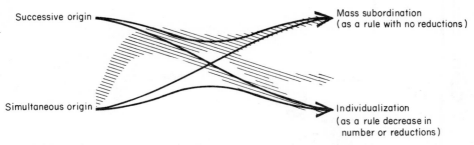

Fig. 22. The origin and fate of standard parts. Shading indicates evolutionary paths which have been followed particularly often by standard parts from their origin to their appropriate phylogenetic final condition.

magnitude (10^{13} to 10^{25} as shown in Section II A2). In the larger order of magnitude the maximum number of identical standard parts is differentiated. With increasing complexity of the standard parts the variation in complexity of the groups of organisms in question neverthess decreases (e.g. Metazoa, Bilateria, Chordata) and the range in number of similar standard parts also decreases by 10^3 or 10^2. The range in number of organs and metameres decreases similarly.

c. Origin and fate

The origin and fate of standard parts follow two partly crossing paths. There are two ways of arising and two ways of finishing (Fig. 22).

1. *Mode of origin.* Two ways of origin are conceivable — successive and simultaneous formation. So far as our knowledge of phylogenetic relationships allows a reconstruction, the following path seems to be preferred.

The first anlagen arise in many cases simultaneously. This is certain for the origin of vertebrae, teeth, scales, and the individuals of colonies and can be assumed for cormidia and gills (of chordates), metameres, parapodia and coeloms (of the Articulata). Perhaps the first phase of morphological formation has always been simultaneous.

Actually we do not know this, for example, as regards the standardized cell types of metazoans or for the organelles such as cilia. It must be remembered, however, that the same principle of identicality exemplified by homonoms within an individual is also shown by 'correspondence' in the individuals of a species, and by homologies in the individuals of a phyletic group. This shows that standards or norms in the wider sense will probably always arise simultaneously. The tiny molecular first appearance of a mutant will, if successful, become distributed until its further elaboration becomes visible.

After its origin as an anlage, every standard type seems to be able to swing into a phase of successive increase in number. Examples are the lengthening of the rows of cormidia in the growing stem of a siphonophore, of the chain of proglottides in a tape-worm or the metameres, parapodia, and gills of the polychaete worms (cf. Fig. 10c). Other examples are the replication of vertebrae (which number 435 in the giant snake *Python molurus*), of fin rays, of brain cells, cilia or ribosomes. All this is obvious, but I stress it since it is important for the ways in which selection acts on the standard part (Section IV C).

2. *The fate of standard parts.* This lies between two extremes. Most standard parts reach enormous numbers and, as mass building blocks, become subordinate to overranking systems. There is extremely little differentiation into subordinate standards in this case, seeing the numbers of building blocks, the number of species that have them, and their age. Consider, for example, the cilia, the retinal cells, the striated muscle fibres,

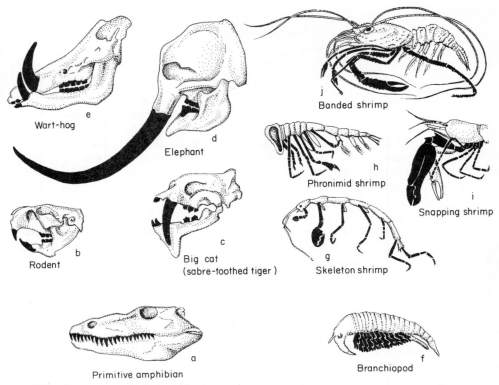

Fig. 23 a-j. Individualization of standard parts. Two examples are shown: homodonty (a) turning to heterodonty (b-e) in tetrapods; and homopedy (f) turning to heteropedy (g-j) in crustaceans. (a) *Eryops,* Permian-Carboniferous, (b) *Castor;* (c) *Eusmilus* (d) *Elephas;* (e) *Phacochoerus;* (f) *Branchipus* (g) *Phtisica* (h) *Phronima;* (i) *Alpheus;* (j) *Stenopus.* (a-e) after Gregory (1951); (f-j) after Riedl (1970).

the alveoli of the lungs, the glomeruli of man's kidneys, the ciliated chambers of a sponge etc.

On the other hand, a certain number of standard parts show reductions, decrease in number and, what is specially interesting, a process of differentiation which can be called individualization (cf. Fig. 22). Classical examples are the teeth of mammals and the limbs of most crustaceans (Fig. 23a-j). In mammals the teeth emerge from anonymity and identicality (as in primitive tetrapods, Fig. 23a), reduce in number and begin to develop the special differentiations of the individual teeth of the orders of mammals as in carnivores, ungulates, elephants etc. (Fig. 23b-e). In crustaceans the almost identical limbs of Anostraca (Fig. 23f) become so individualized in function that, in the higher groups, every appendage can be distinguished (Fig. 23g-j). The metameres of articulates, the vertebrae of mammals, the plumes of birds, and many other standard parts follow the same path from anonymity to individuality.

Contariwise this individualization can disappear again if a new change of function demands it. Thus heterodonty has disappeared in some whales.

This individualization or 'weakening' of standards does not lead to their disappearance. It cannot, however, be regarded as an exception to the rule, for it happens too often. This regular occurrence is important. For it emerges that differentiation of individuals happens in those standard types which occur in relatively small numbers and which, above all, are situated at the distal ends of functional series. I shall return to this later.

2. The placement of standard parts into systems

The next question concerns the positions in which types of standard parts are found to be inserted into organisms. More precisely, what position do they take up anatomically and in functional chains and what correlation is there between their position and their fate?

a. *Positional standard parts and symmetrical standard parts*

The alteration of symmetry relationships with increasing differentiation has long been of interest to morphologists. Translated into our terms it is a question of differentiating those positional standard parts which are the largest constituent building blocks of the organism. It is found that, with the progress of evolution, the axes of differentiation-polarity increase in number and the possible planes of symmetry between identical standard complexes correspondingly decrease. This can be seen as an individualization of what had been positional standard parts.

Thus spherical symmetry, without definable axes, is mainly found in pelagic protists and in sponges; radial symmetry, with one axis, is mainly found in coelenterates; and bilateral symmetry, with two axes and one plane, is found in all higher animals. The correlation between degree of symmetry and lowliness of organization should not be exaggerated, however.

Subordinate positional standards, such as implied by the bilateral symmetry of the primitive tetrapod hand, are dismantled equally often. By contrast, new symmetries may arise, as in echinoderms and in the formation of colonies. And some symmetries of low-rank component parts, such as cilia (cf. Fig. 9h) are conserved throughout the whole realm of living organisms.

b. *The substrate for single homologues*

It is easy to understand the position of structural standard parts in organisms by remembering the position of the homonomy limit (Section II B2b). In progressively breaking up a homologue we found this limit to lie just beneath the minimal homologues. At this limit the single individualizable identicalities of an organism always pass into mass identicalities; the features of the anatomical singular pass over into those of the anatomical plural. In fact no single homologue is conceivable that does not consist of standard parts, and nearly always of several levels of complexity of standard parts.

It is a universal characteristic of the plant kingdom that this limit lies very high in the plan of construction. Even in the most evolved forms, as in the angiosperms for example, the level of standard parts will be reached (with branches, twigs, flowers, and leaves) after only one or two steps of analysis. This is a universal feature of plants. It applies almost equally to the primitive sessile marine animals which were once called zoophytes.

Even in the most highly differentiated organisms, however, as with man, the limit is reached after at most five or six analytical steps. Taking the example used in Section II B2b (Fig. 11) it lies at the ventral articular facet of the odontoid process of the axis vertebra. But beneath this relatively deep-lying limit there follows a considerably more extensive, hierarchically layered substructure of standard parts. Thus one of the chains of standard parts making up this minimal homologue would be: Haversian pillars; layers of osteoblasts; osteoblasts; mitochondria; mitochondrial cristae; membranes of the cristae; enzymes; giant molecules (proteins); peptides; amino-acids.

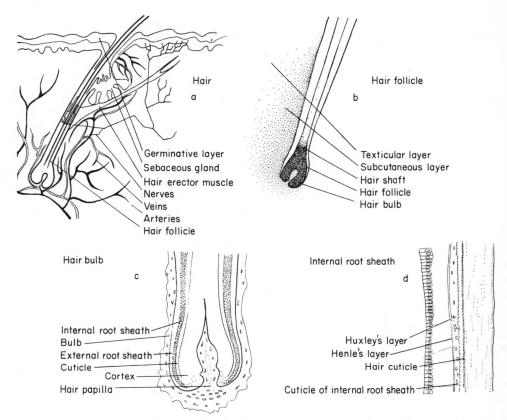

Fig. 24 a-d. The complexity of a standard part at the level of a small organ, i.e. a human hair. To illustrate the four histological levels of complexity: (b) shows details of a feature in (a); (c) shows details of a feature in (b) and so forth. After Patzelt (1945) simplified.

There are therefore five or six hierarchical levels above the minimal single homologues (Section II B2b), while beneath the single homologues there are at least nine or ten hierarchical levels of standard parts. This is already indicated by the fact that the individualized levels range over two orders of magnitude, or three at most (*Homo* > 1 m, ventral facet of axis < 1 cm). The standardized levels, on the other hand, range over more than six orders of magnitude. (Haversian pillars > 1 mm, amino acids < 10 Å) which is a 10 000 fold greater difference. In plants these hierarchically arranged standardized levels extend in general over the total plan of construction, from the dimension of 10 m down to 10 Å. This is a span of more than 10 orders of magnitude, i.e. a factor of ten thousand million.

Even in the smallest organ this great standardized substructure is never lacking, even when the organ, as with a human hair (Fig. 24a-d), has almost no superstructure. One of the chains of standard parts that constitutes the standard part 'hair' would be: hair follicle; hair bulb; inner root sheath; Huxley's layer; cells of Huxley's layer; mitochondria; and so forth as above. Thus there are 11 to 13 hierarchical levels of standard parts.

Consider one of the smallest homologues which can be observed, such as the flagellum of a uniflagellate flagellate. One such chain would be: basal portion; tubules; subtubules; arms; connecting fibres (Fig. 25a-d); giant molecules; peptides; amino acids. Even here we have eight levels of standard parts.

This is all remarkably like the arrangement of individual and standard parts in a building. But the analogy, even if we extend it to cover the planning of an entire town, would scarcely reach half the orders of magnitude or of complexity of a mammal. For example, the individualized constituent parts (single homologues) of a residential district are the individual buildings such as the Town Hall, a

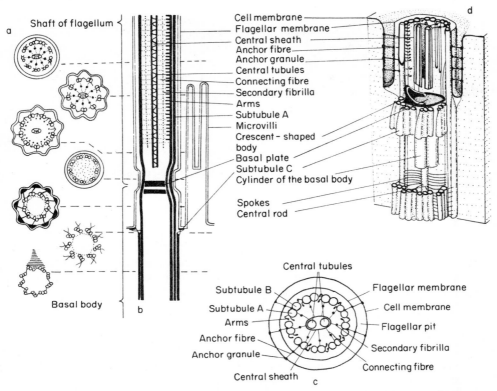

Fig. 25 a-d. The complexity of a standard part at the level of a large organelle, illustrated by the flagella of a bivalve mollusc (a, b) and of a flagellate (c, d). (a) Sections through (b). (c) Section through the upper part of the reconstruction in (d). (a) and (b) from Sleigh (1962); (c) and (d) after several authors and more highly magnified (about x 10,000).

department store, a cinema, a gymnasium, etc. But even the blocks of flats are standard parts and so are their staircases, doors, and chimneys (being below the homonomy limit). Roof tiles, pipes, and lamp sockets are certainly standardized not to speak of the nails for the roof tiles, the screw-fittings of pipes and lamps and their materials.

Even in a one-windowed garden shed (a flagellate) the lowest single homologue (window) has four levels of standard parts as in the sequence: casement; individual window frames; hinges; hinge screws.

Single individualities are always made up of several hierarchical levels of normative or standardized order. These are the substrate of all living order. There is level upon level of standard parts, each part repeated to million or thousand million. An animal tissue or the leaf canopy of a huge tree show this at a glance.

c. Diversification of placement and function

A third general characteristic of standard parts should be considered, additional to their position and their arrangement into levels. This is that a single type may occur in totally different and functionally fully differentiated systems. This again is anatomically self-evident and one example will probably be enough to show it. Nonetheless this characteristic is remarkably important for the selective conditions that act on standardized order.

For example, the cilium is a standard part from a middle hierarchical level — the level of the organelles. The cilium must have evolved as a locomotory organ on the surface of the most primitive marine animals of Precambrian seas. But since then its position has basically changed and diversified. In man it transports the sperms and in epithelia it moves the fluids in the tubes, the secretions in the alveoli of glands, cleans the auditory canal, moves the cerebrospinal fluid, is concerned with sensitivity to sound, covers the olfactory membrane, and transmits the sense of balance. Likewise, the striated musculature shares in singing, running, breathing, and hearing (stapedius muscle and the tensor tympani).

Again there is an unmistakeable analogy with the standard parts which technology has poured into civilization. Thus at one extreme screws function in holding railway bridges together while at the other extreme they are used in installing the balance wheel of a lady's watch. It will later emerge that this technological comparison is more than mere analogy. For the same principles of standardized order occur in thinking as in civilization.

3. Burden, change and constancy

I shall now establish an important correlation. This is the connection that exists between the functional position of a standard part and its constancy. This correlation is important because it will soon lead us (Section IV C) to the selective mechanism which requires standardized order even in the morphological realm.

a. *The forms of burden*

The concept of burden has been introduced already. I have pointed out that it applies both to decisions and to events. The degree of burden is genetically specified by the number of subsequent decisions that depend on a preliminary decision or by the number of single events (or features) functionally dependent on a preliminary decision or on a fundamental event (or feature).

The functional or hierarchical position of a feature therefore plays a large role. With mass standards the number of identical individualities and the diversification of their functional reference is also important.

1. *The hierarchical position.* It is easy to recognize this as an indicator of burden. To use our old example of the levels of standard parts which make up the ventral articular surface of the axis, it is obvious that the burden of the standard parts will increase along the series: Haversian columns, osteoblast layers, osteoblasts, mitochondria, mitochondrial cristae, crista membranes, enzymes, proteins, peptides, amino acids. It is self-evident that, if there is a gross defect at any of the levels, then all preceding or overlying levels will be defective, while all succeeding or lower levels will be undisturbed.

In the same way, if a roof is too small it has no effect on the roof tiles or roof-tile nails that have been delivered. But if the wrong size of nails has been delivered then not even the roof tiles can be fixed up.

2. *The number of standard parts in a standardized collective.* This increases the burden additionally. It does not increase the burden of the individual standard part but only the collective burden of the principle on which the standard part is constructed, and this burden depends on the higher-ranking single systems in which the standard part functions. If striated muscles fibres were represented only in the inner ear then a defective mutant of these fibres would merely be deaf (i.e. subvital). Since, however, they also provide the thoracic musculature, the mutation is lethal, since the mutant could not breathe.

3. *The number of functions.* If a standard part is involved in a number of different functions, its burden will be increased considerably. For a mutational change in the

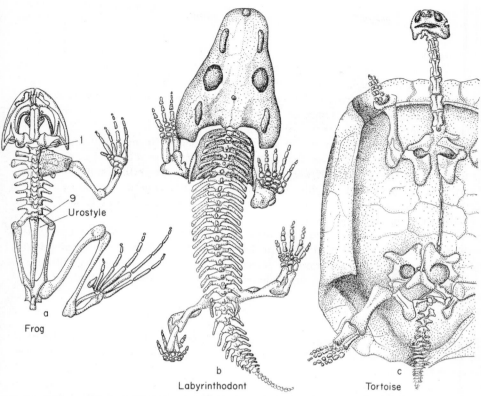

Fig. 26 a-c. Transformations of homonomous organs within a phyletic group, as illustrated by the vertebrae. (a) *Rana temporaria;* reduction of the vertebrae to ten with the pelvic and caudal vertebrae fused into a urostyle. (b) *Trematops,* a primitive tetrapod from the Permian. (c) *Testudo pardalis*, opened on the ventral side; note the thoracic vertebrae, which are slender ventrally but expanded dorsally to form part of the armour. (a) and (b) after Gregory (1951).

standard part may by accident be advantageous in one functional connection but disadvantageous in others. This will happen the more certainly, the more numerous and valid the functional connections are. (Think of the different functions of cilia in man, for example.) I shall discuss these selective conditions more fully later.

It is obvious that the burden carried by standardized order can increase very steeply — as steeply as the selective advantages of standardization did previously. In evolution, wins are paid for with losses (Section III C2*a*) as already said.

b. The forms of freedom and of change

We can now predict the extraordinary constancy of burden-carrying standard parts. Monstrous dimensions of superdeterminacy are involved, as discussed below. Changes *are* observed but these are so characteristically canalized that it is worth giving some examples.

The cilium is a standard part of low hierarchical position belonging to the organelle level. It shows no changes in principle in any of the four kingdoms of cellular organisms whether in protists, plants, fungi, or animals. Only the flagellum of bacteria is differently constructed. The cilium can be extensively suppressed, but there seems to be no class of metazoan animal in which it has totally disappeared. Thus thread worms (nematodes) are entirely surrounded by a cuticular sheath and their sperms have no tail. But even here deeply sunken cilia have recently been found in an internal sense organ.[5]

The hair is higher in the hierarchy of standard parts. It was 'invented' at the root of the mammals and is probably lacking in no mammalian species. Even whales still possess a few bristles in the upper lip, which may be a remnant of the whiskers of carnivores and perhaps serve to detect currents. But the various 'metamorphoses' or transformations of hair are unmistakable – the spines on a hedgehog, or the great 'horn' of a rhinoceros which grows up from a huge number of hairs all stuck together.

The vertebra is still higher in the hierarchy. Consequently its transformations are already quite considerable.[6] Even the number of free vertebrae varies as 1 : 44. (It is ten in a frog (Fig. 26a) as against 435 in the giant snake *Python*.) Nevertheless the identicality of vertebrae is always unmistakable. The primitive vertebral column was almost solely of mechanical importance. Even in tortoises the vertebrae are preserved but have become incorporated in the shell (Fig. 26c).

The freedom of standard parts is small. It decreases with lowness of hierarchical position, with number, and with diversification of function. Changes are extraordinarily slight; they almost always affect only superficial features of the standardized collective in question and hardly ever alter the principle on which the collective is standardized.

c. The degrees of constancy and fixation

Standard parts regularly attain inconceivably high degrees of constancy and fixation. Thousands of standard parts, especially those of low hierarchical levels, have been preserved totally unchanged since they first appeared in Precambrian seas.

This must again be emphasized. I have in mind almost all those features of living organisms described in thousands of papers in textbooks and handbooks of general biology, cytology and genetics, for these features hold for all organisms. Bacteria in some ways are peculiar. But the latest common ancestor of the other four kingdoms, to which these identical structures must go back, lived more than a thousand million (10^9) years ago.

1. *The degree of constancy.* This can be specified as a quantitative connection between change and time. We merely count the years during which an identicality has not altered beyond the framework of the defining features. (This is discussed further in Chapter V.)

Thus nearly all standard parts up to the level of giant molecules are 10^9 years old, from DNA bases to proteins. This is also true of many organelles and ultrastructures, up to mitochondria and cilia. But a large number of the standard parts which have arisen in individual classes have been completely preserved e.g. nematoblasts, cross-striated muscle fibres etc. The septa of corals, the metameres and spinal ganglia of vertebrates and many other features are 4 to 5×10^8 years old. Even standard parts, like hair, with a peripheral position in the body, are as old as the mammals, at 1.8×10^8 years.

2. *The degree of fixation.* This, on the other hand, is the measure which should be used when superdeterminacy is to be specified. It measures how far the constancy of a feature exceeds the determinacy which, knowing the mutation rate, would be expected on average. This average degree of precision of the organic determination mechanism, we have already found to be the product of mutation rate and the prospect of success $(P_m \cdot P_e)$ and to be about 10^{-6} (i.e. $10^{-4} \times 10^{-2}$). This means that, on average, every ten thousandth reproductive process will be affected by a change and, of these changes, every hundredth may be successful.

The mammalian hair, for example, is possessed by every recent species of mammals, inherited from its ancestors for 1.8×10^8 years. Let us assume only one reproductive act per individual after four years and only 10^6 individuals per species. Each species will have in its mammalian ancestry $(1.8 \times 10^8/4) = 4.5 \times 10^7$ generations with 10^6 individuals and therefore $4.5 \times 10^7 \times 10^6 = 4.5 \times 10^{13}$ reproductive acts. Without superdeterminacy, the feature 'hair' would have been changed successfully every millionth time and thus $4.5 \times 10^{13} \times 10^{-6} = 4.5 \times 10^7$ which is forty-five million times. But, so far as the basic principle of hair is concerned, this has not happened. Indeed, taking all recent mammals together $(3.7 \times 10^3$ species) it has not happened. Thus the superdeterminacy is $4.5 \times 10^7 \times 3.7 \times 10^3$ which is more than 100 thousand millions (10^{11}).

The cilium, one of the most archaic features, has certainly existed for 10^9 years. For each recent species this would be at least 5×10^9 generations with at least 10^8 individuals. This gives a degree of superprecision of $(5 \times 10^9 \times 10^8 \times 10^{-6}) = 5 \times 10^{11}$. For all recent species (certainly more than 2×10^6) the cilium has a superdeterminacy of $(5 \times 10^{11} \times 2 \times 10^6) = 10^{18}$. This is a superregularity of astronomical dimensions.

This is undoubtedly one of the most astounding phenomena of life. The order content of life represents in itself an inconceivably improbable balancing act by matter. But some of life's systems reach a precision which may exceed even the probability that matter will conform to law. Consider, by comparison, the half-lives of radioactive atoms (uranium = 4.5×10^9 years, radium = 1580 years, mesothorium = 6.7 years), or the duration of elementary particles.

We should expect that very basic laws would be involved in compelling the formation of standards and fixating them to such a degree. In the molecular realm these laws were the necessity of systemization and the burden carried by the decisions which determine living matter. I shall now discuss the macroscopic equivalent.

C. STANDARD-PART SELECTION

Many of the mechanisms of evolution may still be hidden. Nevertheless we know where to look for the cause of the establishment and fixation of standardized order. It must be some form of selection. For mutation cannot be responsible nor can a third mechanism be found.

Here, however, I must point out that the solution will lie in a two-sided analysis of what is commonly called selection. Special conditions of selection can be interpreted only by the measuring rod of special objects subject to selection. As a process, selection can be understood from the confrontation of external conditions in the environment with the internal conditions of the organism. It is therefore not surprising to find the standardized systemic conditions of organisms as a product of selection.

1. The advantages of standardization

These have long been familiar in everyday life. When buying nails it is scarcely necessary to state more than the length and number required. When bolts run out, however, we need to state the diameter, pitch and whorl section also, or even better hand over one of the remaining nuts (the definitive selection conditions). Indeed the selection conditions are specified in percentage departure from nominal, or tolerance. Woe to the firm that does not keep to the expected standard! It will lose all chance in the market.

a. The prospects of success for blind accident

These, as in throwing coins, are at most ½. This would be good enough for a player who was in the position, like evolution, to lose on an extraordinary number of occasions and to take an inconceivably long time. However, with increase in complexity of the features or the rules of play, the prospects of success decrease exponentially. As shown in Chapter I they become infinitesimal. Even the largest populations (all the matter in the biosphere transformed into tiny players) ready to lose almost indefinitely (over periods of cosmic length) would not be able to compensate.

The whole human population of the Earth could not throw a few hundred molecules into a particular position (cf. Section I B1c) by chance. But even this situation is some orders of magnitude less complex than that of the simplest organelle.

What part therefore could accident play in evolution? We know that the fate of every species depends on it, but it only has a meaning where the prospects of success for the individual accidental events are large, i.e. where accident is given the smallest possible room for action. This is in the narrow alley between firmly established regularities. To attain evolutionary success, as many holes of the roulette wheel as possible must be closed.

b. The prospects of success for established facts

These, on the other hand, lie in the prospect of re-establishing established facts again and again, which is the conservative or reactionary property of evolution. An organism is an improbable condition. If, improbably, it can endure for a certain time in the special condition of its environment then, out of all other conceivable living structures, the best prospect will belong to the one that is most like the original.

1. *Adaptation.* It is therefore rightly considered that evolution has the best prospect of success when it takes the smallest steps. This, however, is only one aspect of the matter. Another aspect, which is here crucial, is by a giant stride to produce the same improbable condition once again. The prospect of success, therefore, does not consist only in giving little space to accident, but at the same time in giving much space to law. Or, in other words, in applying law again — we have already seen (Section I B4) that order is law times the number of instances where it applies.

We now have everything to hand. We have already seen that the individual organism is only one of many levels of identical individuality; it is only one of the identical, hierarchically arranged standard units from which the world of organisms is made. The solution to the problem is identical replication of standardized order at all levels. The selective advantage (equation 21, Section III C1a), in which the number of features (E) and replications $(a - 1)$ appear as a product in the exponent, must hold for standard units of all sizes.

Thus the ability of a protist to form a new locomotory flagellum, after having lost the old one, would be as important for the persistence of its genome as its ability to divide and identically replicate itself. Indeed, when the disaster happens, the ability to form a new flagellum holds more promise of success and is the more obvious solution. Replication is always the best compensation for loss. Its selective advantage is specified only by the importance of the loss.

2. *Fitting in.* The crucial advantage of the identical replication of standard parts must be to create structures with the greatest prospect of fitting harmoniously into that improbable state of matter which an organism is.

Remember what an extraordinary range of regulations and preconditions an organism contains. A crab's claw is a complete and perfected device, but if it had been invented as an appendix on a sponge it would be a functionless absurdity that could not be fitted in.

If identical replication of component parts, at all levels, were unknown in organisms, then it would be required to exist. We found this already in considering the molecular

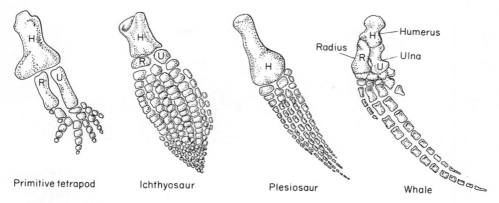

Fig. 27. Change in number of hómonomous bones in the tetrapod hand. Ichthyosaurs, plesiosaurs, and whales all show an increase in the number of finger bones (polyphalangy). Ichthyosaurs show an increase in the number of fingers (polydactyly) while whales show a decrease. After Romer (1966).

realm of genetic decisions, and it recurs in the realm of events. Replicative switching would be expected at every level of complexity, so it is not surprising to find identical structures at each level.

Examples to illustrate this are probably not really needed. For organisms as a whole consist of such identical structures. This is self-evident everywhere, from protein molecules, through muscle fibres, even to organs and metameres.

Consider, however, the case of terminal addition where structures are added to the distal ends of series of like structures. Examples are found in vertebrae (snakes and fishes), in fingers on the hand of ichthyosaurs (polydactyly), in the finger bones of whales (polyphalangy; Fig. 27), in the replication of proglottides in tape-worms, of metameres in annelids, or the somewhat absurd-looking replications of the pharyngeal and genital apparatus in turbellarians (Fig. 28a-c).

Two requirements are crucial to the prospects of success for such additional parts, irrespective of the level of complexity — maximal adaptation to the external conditions and to the internal conditions. No mechanism could fulfil these requirements so well as a universal principle of identical, standardized replication.

c. Economy and the increase of order

This principle can be illustrated from a different aspect. Order, on whose increase survival will often depend, is the product of law times the number of instances. Assume a replication mechanism that works more reliably than the mechanism that increases the law content. In such a case selection will demand an increase, whether of individuals (in a population expansion), of metameres (as in annelids and snakes), of organs or of cells (gigantism). Without doubt it will often be easier to increase quantity than to enlarge the law content harmoniously.

Think of a specialized brickworks. To begin with, the workers and owners need only two pieces of information — material and dimensions times the number of bricks. But suppose each customer comes to want only one brick and each states unpredictable, complicated requirements for the bizarre shape of his particular purchase. If the price can rise then the brickworks will become an art-work foundry. If not, then the brickworks will go out of business. However, the market more often needs cheap mass goods than valuable craftwork. For this reason lower forms are as successful in evolution as in industrial civilization. Quality is always the more mature condition.

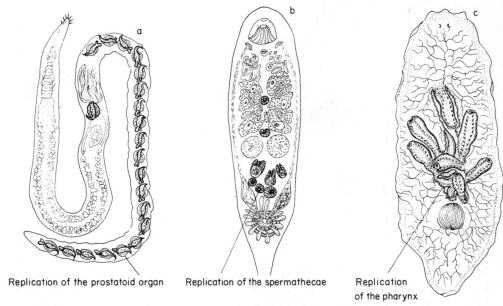

Replication of the prostatoid organ   Replication of the spermathecae   Replication
of the pharynx

Fig. 28 a-c. Identical replication of organs as shown by the turbellarians. (a) *Polystyliphora filum* (about × 15); (b) *Oligochoerus limnophilus* (x 40); (c) *Crenobia alpina montenigrina* (× 10). (b) after Ax and Dörjes (1966); (a) and (c) after Beauchamp (1961).

As previously stated (Section II A1) the determinacy content of the human organism is between 10^{25} and 10^{28} $bits_D$. But the determinacy content of our germ cells is only 10^{11} $bits_D$ and that of our gene catalogue only 10^5 or 10^6 (Section II A2). This difference of several thousand million can be explained by the great repetition of the number of instances (redundancy of events). There would not be room for a complete formulation of our determinacy content in a sperm cell, just as there would be no room for the data for each single article produced in the archives of a mass-production industry.

d. Quick breakthroughs to new forms of organization

These are another selective advantage of parts that can be replicated to a standard. This can be concluded from the small number of transitional forms observed in going from a few standard parts to their employment *en masse*. The change often happens very quickly and with great chance of success. The analogy of mass-production industry again applies.

Think, for example, of the cells of the Metazoa. Except for the peculiar Mesozoa these cells immediately appear in huge numbers. Or of the ciliated epithelial cells of bilateral metazoans which, except for the strange Gnathostomulida, immediately show a large number of cilia. Or of the number of cilia in the Ciliata where, as often happens, a reduced number indicates a secondarily derived condition. Or the occurrence of special cell types, from nematoblasts (sting cells) to glia cells, or the occurrence of metameres and parapods in the primitive Articulata.

There can probably now be no doubt that the production and extensive application of standard parts is greatly promoted by selection. But this presupposes that the standards of phene and event also exist genetically in the form of units of decisions behind them, and that these units can be switched on by a single command. In the molecular-biological part I have shown that this is possible or highly probable (Section III C1*a* and *b*). I shall prove it by considering mistakes in the regeneration process, in the transmission of commands, and by considering mutations (cf. Chapter VI).

2. Canalization and fixation

As already implied, the extraordinary advantages of employing standard parts have to be paid for. The evolutionary freedom gained yesterday in this manner, is largely lost today by canalization. I have said this already. I shall now demonstrate the selective mechanism that brings it about.

a. *The prospects of successful alteration*

For standard parts the prospect that an alteration will be successful depends on the functional burden that they carry. In the molecular-biological chapter I specified the burden for single decisions and described it in terms of the number of single events (E') dependent on the decision (equation 25, Section III C2*a*). I showed that the prospects of success (P_e) will sink exponentially with the number of dependent events ($P_e E'$) (equation 27, Section III C2*a*).

For the single event (the phene or morphological building block) burden signifies the number of mutually independent events which are nevertheless dependent as a whole on a pattern of decisions. This dependence of standard parts is familiar to us by considering standardization in civilized life. It shows three quantitative dependences, which is basically true of biological standard parts also.

1. Suppose that a mutation in the power station of a town changed the frequency from 50 to 500 Hz. The electrical clocks, fluorescent tubes, some radio sets and some motors would go out of service. The mutation would be subvital, for the town, although weakened, would survive. But if the mutation changed the current from alternating to continuous, then in addition all alternating current motors and devices, most electric heaters and the transformers would be put out of action. The mutation would be sublethal; the town would be scarcely able to survive. If the tension changed by an order of magnitude then the town would be lethally affected, for everything would break down immediately. This means that the more fundamental or hierarchically basic the feature, the greater its burden will be, i.e. the more catastrophic will an alteration be and the smaller the prospect of any alteration being successful. I call this the *position effect* (hierarchical position).

Naturally the analogy does not apply in every detail, but the mutation from alternating to continuous current is one level of hierarchy more basic than the mutation from 50 to 500 Hz. The standardized single events (or features) would correspond to the millions of standardized terminal outlets in the municipal grid. This grid would correspond to the transmission of commands from the site of the mutation to all the identical standard parts of a class.

2. Suppose that the production of the power station was not connected up with all the stationary energy users of the town. For example, suppose it excluded industry and went only to residential buildings or households, or only supplied the night current of households. The effect of the disturbance, given the same hierarchical position, would decrease according to the extent of these exclusions. This means that the collective burden of a standardized category increases with the number of dependent but functionally different systems of the organism, or of the town. I call this *the grid effect*.

3. Suppose that the power station supplied not just one town but a whole country. The prospect of adaptation and survival, for a given hierarchical position and grid, would

diminish. Suppose, on the other hand, that it supplied only a single village, a single farm, or a single do-it-yourself workshop, then the prospects of successful adaptation for the dependent functions would become more and more reasonable. This means that the collective burden of a standard-part category is dependent on the number of identical standard parts (identical terminal outlets) in the system. I call this the *size-of-collective effect*.

The difficulties, or the sizes of the catastrophes, thus depend on the parameters: hierarchical position, grid and size of collective. The comparison between a town and an organism depends on the inverse relationship between the prospect of success and the degree of disturbance in both cases. As the degree of disturbance rises, so does the ruthlessness of selection. As the improbability increases of improving the running of a town, so does the improbability of gaining a selective advantage for the organism.

b. Burden and selection

The selective conditions which are now to be described would be expected to explain the extraordinary degree of superdeterminacy that is characteristic for states of standardized order. We should expect a sort of superselection which exceeded conventional selective conditions to the extent that the superconstancy of standard parts exceeds the average constancy of features. Looking in organisms for the selective equivalents of the three conditions just mentioned, then we find the following (reversing the previous order):

1. *The size-of-collective effect.* This is least obvious. Mutations are heritable and of evolutionary importance if they affect the germ cells. If in these cells the determinacy content of a replicable unit is changed, then all identical copies will contain the same change. If the number of copies is small, then the prospect that the change in the collective will be accepted by selection is relatively large. Indeed it is almost as large as the prospect of success for a single feature. This is because the number of required changes in fit is not large. If, however, there is a considerable increase in the number of altered standard parts, then by the same token the prospects will decrease that all will harmonize with the fit-requirements, which likewise will have increased in number.

If, for example, accident, or mutation, changes an edge piece of a jigsaw puzzle then there is a slight prospect (say $P_e = 10^{-2}$) that the whole picture will be improved. But if the mutation colours all the edge pieces (E) then the prospects sink mightily ($P_e' = 10^{-2 E}$)

If a handyman accidentally bought a mutated bolt then, with luck, he might find a suitable nut in his junk box (say $P_e = 10^{-2}$). But if an industry received a million of these mutants then the prospect that they could be used is nil. (If, however, the nuts mutated with the bolts then both would conform to the same standard. The fit-requirements would be calibre etc.)

With living standard parts, consider the high speed of change in small collectives (fingers of tetrapods, joints in arthropod limbs); the slower speed of change in middle-sized collectives (segments in Articulata, limbs of arthropods); and the constancy of large collectives (feathers of birds, the ambulacral feet of echinoderms).

At the hierarchical position of cells there are standard parts with only a small number of copies (size of collective) among the rarer types of tactile corpuscles. An example is the disc cells in the Grandry nerve corpuscles (Fig. 29a). Their low constancy goes with their restriction to the tongue and lores of waterbirds and is shown by the fact that they are differently developed even on the edges of the mouth. Again the inner capsule cells are completely lacking in the related lamellate corpuscles of Pacini in mammals (e.g. in man, Fig. 29c). Compare with this the correlation between mass-occurrence and constancy in striated muscle fibres. For these must be older than the whole fossil history of animals, being represented in vertebrates, molluscs, and arthropods and always in principle identical.

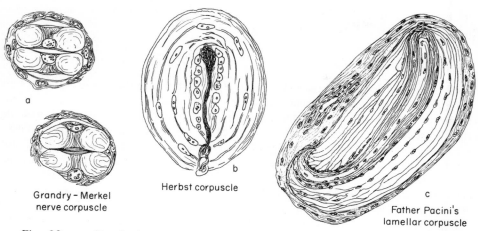

a

Grandry – Merkel
nerve corpuscle

Herbst corpuscle

b

c

Father Pacini's
lamellar corpuscle

Fig. 29 a-c. Standard parts of limited size of collective as illustrated by special tactile bodies in the skin of birds and mammals. (a) About x 500; (b) x 250; and (c) x 150. All figures from Patzelt (1945).

2. *The grid effect*. This is obvious enough. Suppose the individualities in a standardized category have somewhat different functions (fitting requirements) in the various organs to which they are connected. The prospects of success for an alteration in the collective will decrease exponentially with the extent of functional diversification.

Imagine, for example, that a type of bolt was used as a collective mass product by the car industry and that by mutation the size of the head was doubled. In a rare case this might improve the total product – if fashion in the market accidentally demanded big ornamental screws and if the outcome only depended on such ornamental screws. However, the outcome depends on everything. It seems certain that one of the other functions for the bolts would fail – the tank or the cylinder head would leak or the carburettor could not be fitted. The mutation would be lethal for the firm.

The cilium is an organic example. A change in the collective of cilia might improve hearing but prevent sperms from functioning. Or improve olfaction but destroy the sense of balance (cf. occurrence of cilia as discussed in Section IV B2c and the structure of cilia Section IV B2b). Ultrastructural research shows that any changes in cilia must not affect the principle. A replacement for cilia is also inconceivable, since all their different functions would have to be substituted successfully at one and the same time.

3. *The hierarchical-position effect*. This is self-evident. Whatever feature is adapted to the preordained function of a standard category will cease to function if the standard changes.

For example, a car industry might survive a mutation from steel to plastic if this only affected the blinker switch. It would be subvital if it affected all the levers, lethal if it affected the coachwork. And it would be an absurdity if all steel were turned into plastic, for it could not be further adapted before the market had given its verdict.

The number and demands of dependences are likewise judged by selection. Peripheral standard parts are characteristically always changing. The cormidia of siphonophores, the metameres of articulates and the parapods of polychaete worms vary from species to species. Change in vertebrae is considerably more difficult. It is damped down by the fact that ribs, dorsal nerve cord, vessels, the whole dorsalis musculature, the exit canals for all spinal nerves, and the innervation fields of these nerves all depend on them. The constancy of particular standardized cell types is older than whole phyla of animals, while the constancy of ultrastructures, such as the endoplasmic reticulum and the mitochondrion, is older than the whole animal kingdom. Certain giant molecules, like cytochrome c as shown in Fig. 8a–b, are fixated in the whole organic world. And it is immediately obvious that the mutation of a pyrimidine base in the genetic code to a strange type of molecule must always be futile.

c. Standard-part superselection

All selection has to do with environment. But it is only for the whole organism that environment corresponds to what we call the biotope, the external evironment, competition, the market or the advertising trend. It would be absurd to think that the selective value of the pyrimidine bases depended only on wind and weather, on predators or on new trophic niches in the biotope. Indeed it has to do with all these things, but not with them only. In my view this huge span of causal connections is the most astounding part of the matter — so much so that the self-certainty of the reductionists can be understood, at least psychologically. But the viewpoint that I now present assumes totally different dimensions for the 'environment'.

1. *Superselection*. I use this word for the effect of those prescriptions which, over and above those of the external world, are added by the systemic conditions of the organism itself. The higher these mountains of prescription and consequence rise in the internal environment of the organism, the more must the practicable prospects of improvement diminish. Until in the end everything will be monitored and rigidified into a superdimensional system of control.

2. *Self-design by superselection*. In the case of standard parts this is determined by the parameters of size of collective, the grid effect and hierarchical position. It arises with the evolution of phyletic groups and reacts on their evolution. Freedom for one group will be rigidification for another. But freedom got yesterday by the mass employment of replicative features is always paid for today with lack of freedom caused by rigidifying standards.

3. *Canalization by superselection*. This results from this back-payment. It consists of a layered structure of standard parts which, as time passes, increase in number, universality, and rigidity. The result is the ordered system of stratified, constant, and identical individualities and realities which we have defined as standard-part order. It is one of the preconditions for understanding living structure and so describing and predicting it.

4. *Canalization of thought*. This is a further consequence. Standardized order is a reality outside our thought processes. The extensive conjunction between natural patterns and thought patterns can therefore be explained. For normative thought, using standard units, will have been taken up necessarily as a category of thought in the mechanism of our brains. Selection, ruthless in this case also, will have chosen these thought patterns which approach the structural patterns most closely and which make it possible to recognize and predict these structural patterns with the greatest reliability.[7]

3. Norms and standards in civilization

A final glance at the benefits and drawbacks of the norms or standards that we have developed round us in civilization, and which indeed we presuppose, will illustrate the principle of this interplay of freedom and slavery. To a reader who does not hope to penetrate in earnest into the 'natural history of the human spirit',[8] this will be mere analogy. For those who do so hope, it will be a logical consequence.

We must be cautious in shipping 'wisdom' to thought-continents of totally different complexity. Nevertheless, the old question is still with us: How shall we explain the resemblance between organic standards and the norms or standards of civilization? Is it accidental? But we know already that accident is most effective as a designer of order, when it has least freedom. The normative or standard-part aspect of thought is a consequence of the standardized results of evolution. And if so, should we not expect that the normative aspect of civilization results from thinking in terms of standard parts — from the thinking that created that civilization?

In the interplay of civilization we find that the same two mechanisms have led to the development and rigidification of standardized order. The only difference is that they seem self-evident. This insight on its own would scarcely be worth writing an essay about.

a. Success and mass-production

People change their aims between sleep and waking, but always strive 'to possess more than their neighbour, and tomorrow more than today'.[9] This is such an archaic principle of evolution that a comparison with preconscious evolution should be reliable.

One of the tragic features of this second evolution, which has escaped from the slowness of the genetic mechanism, is the experience that 'quantity succeeds'. This has led from the standardization of the stones of the pyramids in the first advanced cultures to that of cars in success-oriented industrial society. Success consists, as with preconscious organic evolution, in attaining a maximum of production (effect, influence, stipulation, $bits_D$ with a minimum of knowledge (instruction, insight, law content, $bits_L$. The results in both cases are classes of identicalities with reduced individuality or none — standardized products and producers. Both, again, are the building blocks of higher structure and functions — the tiles on the cathedral roof, the cars of industrial economies, the classes of political parties, all being parts of the State.

In a selective world, in which, very shortsightedly, systems are 'honoured' with stability according to how they flood the market with their product, such a development is obviously a necessary result. The analogy with preconscious evolution is total. The subject is full of topical relevance. However, as an anatomist, I shall be excused for sticking to my theme.

Think of the importance of standard parts in civilization (Fig. 30a-c), of the problem of incompatability and inspection of standards (DIN, ASA, and the International Organization for Standardization (SO) founded in 1946). Consider also the economies of standardization and rationalization correlated with reduction of product range and series production. The controversy between standards science and positivism is also relevant.[10]

b. Tolerance and the collective

Established standards result in floods of regulations prescribed by governments and bureaux of weights and measures, by taboos and fashion, by Party members, and the standards departments of industries. All these authorities restrict what is tolerated.

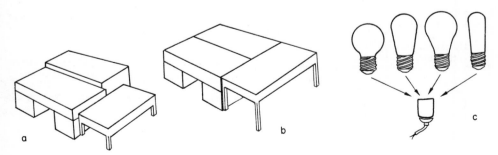

Fig. 30 a-c. Standard parts in civilization. (b) Standardization of previously unstandardized (a) furniture. (c) Standardization of electric-light fitments. (From the *Brockhaus Encyclopaedia.*)

It is instructive to note that these stratified authorities, just as in organic evolution, have taken on a self-legitimizing quality in establishing their regulations. We no longer leave it to the market (to the environment) to police the functionally necessary tolerances for the bolts used by the car industry. Every authority inspects, not because the world will thereby be better, but because the next higher authority requires it. It is astonishing to see how the tolerance permitted again depends on the same parameters — the size of the collective, the grid connection, and the hierarchical position of the part. This is true both of the tolerance permitted to the collective of the products, and that permitted (unbelievably enough) to the collective of producers.

The result is an extraordinary restriction of what is realized as against what is potential. Indeed it is impossible to picture the degree of intolerance which our civilization specifies as standard. Even the concepts which an annoyed reader will use to argue against me are standardized. This goes for language, syntax, the grammar with which he writes to me, if he chooses to write. The letters of the alphabet that he uses, the paper, the stamps, the post office, and my own head, which will seek to understand him.

This was what I wished to show. We are back again to the starting point. Standard-part order is a universal principle.

———————   ———————

On subjective grounds it may be annoying that, using a biological tool kit, I make the collective of standard parts to be a necessary component of natural laws. But in the next chapter I shall discuss the natural law to which the collective is subject. Others might do the converse. I shall therefore return to the reliability of anatomical methods and come to the next pattern of order — the hierarchical pattern.

NOTES

1 Extensive literature was cited as early as Binding (1872). See also Kaufmann (1954) and Lautmann (1969).
2 The minimum duration depends on the precision of observation. It lies at about 10^{-20} seconds, which is the life-time of the eta-mesons. The so-called resonances are two or three orders of magnitude smaller.
3 Doblhofer (1957) with widely understandable examples; epistemological questions dealt with in D. Campbell (1966 b) and also in Popper (1962) and Lorenz (1973).
4 A remarkable study of this theme is that of Erwin Schrödinger (1961) *Über die Nichtvielheit*.
5 The discovery was announced by Roggen *et al.* (1966). Bird (1971) gives the most recent summary.
6 Remane (1936) gives a particularly extensive survey of the condition of vertebrae.
7 This has been most brilliantly confirmed by Konrad Lorenz (1973). The normative aspect of our 'calculating apparatus' must have arisen before we became conscious of logic or concepts. It would have appeared in a 'ratiomorphous' condition (Brunswik, 1934, 1957), as a precursor of reason, enforced by selection which forced 'hypothetical realism' on the organism.
8 Konrad Lorenz communicated these thoughts in a lecture at the University of Vienna in December 1971. Subsequently his work appeared (1973) with the subtitle 'Attempt at a Natural History of Human Knowledge'. It confirms what underlies my own book, which is that we are dealing here with true connections.
9 I came to this conclusion in connection with the problem of energy flow through the biosphere (Riedl, 1973a, b). Compare the quantitative analysis by Odum (1971) and the qualititative one of Hass (1970).
10 Compare, for example, Husserl (1928), Lalande (1948) in connection with standards science and Klein (1970) in connection with industrial standards.

CHAPTER V

THE HIERARCHICAL PATTERN OF ORDER

A. INTRODUCTION AND DEFINITION

The second basic pattern of order in living organisms is the hierarchical pattern. The study of it leads into comparative anatomy, palaeontology, and systematics. It is not easy to present its form and consequences so I shall have to write at greater length. (Compare the contents lists of Chapters IV to VII.)

For three reasons there are additional difficulties in the presentation. First, to discuss hierarchy requires a knowledge of the systematic consequences of morphology. Among present-day biologists such knowledge is less widespread than the knowledge needed to understand the other patterns of order. Second, there is much interplay with the cause of the interdependence pattern (Chapter VI) so that anticipation cannot be avoided. And, third, all hierarchical concepts arise from fields of study which, up till now, have been transcausal in method, so that all the correlations and causal connections need to be established for the first time, without path or guide.

Hierarchical order is characterized by features (or concepts) whose fields of validity do not overlap but are contained within each other, so that several lower concepts of equal rank are usually included in a higher concept. The higher concept specifies the significance of its lower concepts, and the latter specify its contents. This pattern, once it has been explained, makes in itself no further difficulties (see for example, Fig. 36a-c). Its consequences, however, whether functional or conceptual, can be so complex as to reach the limits of conceivability.[1] I shall derive the basic conditions which lead to hierarchy and the limits within which these conditions apply.

The political system of hierarchy (Byzantine Greek — sacred or priestly rule) gives a first conception of the meaning of the word. For we must remember: first, that all organic structures are hierarchically organized; second, that all the communal affairs of men are hierarchically ordered; and third, that we cannot form a concept except that it receives meaning from higher ranking concepts, and content from its lower ranking concepts. The agreement between these three phenomena is so extensive that we cannot avoid thinking that the three are causally connected.

I shall deal, in the first place, almost exclusively with biological morphology. I do this because I have most confidence in that field and because I believe that within it I can show that the formation of hierarchy is a necessity. Having done that, however, I shall then ask how we can understand the conjunction with thought patterns (Section V C3*d*) and with the patterns of civilization (Section V C4).

First, there are already more than ten million structural concepts in anatomy and systematics which only acquire meaning and content from their hierarchical position. Thus a Tiger Swallowtail signifies nothing except as a butterfly, an insect, an arthropod, a representative of the Articulata, the Protostomia, the Bilateria, the Metazoa, and of the animal kingdom. A gene exists only in a chromosome, of a cell, usually of a tissue, of an organ, of an individual, of a population. And 'rhinoceros' only has content if we know whether it refers to a bird (hornbill), a beetle or a pachyderm. The relevant literature fills half of biological libraries. But the question of causes has remained unanswered.

Second, hierarchy as a structural form of civilization is a field where the history and effect of the pattern can be especially studied – its benefits as against the drawbacks of establishing and preserving it. In this very topical field of study[2] we are dealing with authority, dominance and subordination, ranks and classes of social entities, from industries and class structure to the organization of the church (whence the concept originated).

Third, our brains use the hierarchical pattern to put order into inorganic phenomena even including physical ones and the conceptual products of man. A doubt thus arises whether the pattern pre-existed or whether it was introduced by the brain. Even in the definition, groups of features were taken as synonymous with concepts. Even languages are all of hierarchical structure.[3] I shall come back to these questions later and shall be able to resolve the doubts. At present it will be enough to recognize that the hierarchical pattern is so deep-rooted in our imaginations that it would be difficult even to think of a world without hierarchy. In fact, most of what is understood would sink into unintelligibility, as I shall show immediately.

Those hierarchical patterns of anatomical and systematic concepts which are the single homologues (Section II B2*b*) are especially open to analysis since they can be described rather objectively with the coordinates of 'time' and 'morphological distance'. Time here is in the geological dimension of phylogeny; distance in degree of structural similarity. The appropriate higher concepts are characterized by greater range in both these dimensions – by greater persistence in time as by the greater number of sub-features that they include. In short they are more widely valid.

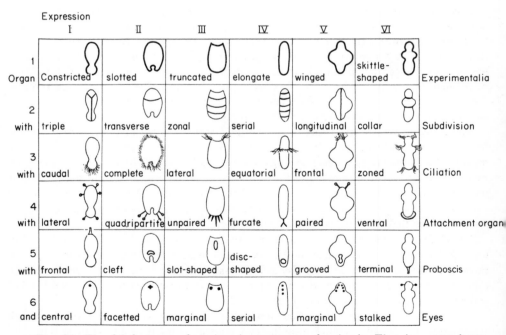

Fig. 31. The 36 features of an imaginary group of animals. The six organs have Arabic numbers and their character-states ('Expressions') have Roman numbers. A diagnosis of a group of organisms, for example, which only showed character-state I, would be as follows: 'grooved experimentalia with tripartition, caudal ciliation, lateral attachment organs, rostral snout, and a central eye'.

In the first instance it will not be necessary to go into the various sorts of systematic method because in this question all schools have come in principle to the same conclusion, from Remane (1952) on the one hand, to Sokal and Sneath (1963) on the other.

a. A fantasy world without hierarchy

I shall now try to form a conception of the ordering consequences of unequal constancy and, at the same time, of the characteristics of a world without hierarchy. I shall do this by assigning equal prospects of change to all the features of a combination, instead of unequal prospects as is usual. Experimentally I start with the imaginary organism in Fig. 31, made up of the features I 1, to I 6. The organs 1 to 6 can occur also with the character-states II to VI if the organisms reach adaptive channels II to VI which favour these character-states. The result is decided by throwing dice. Two throws will symbolize the number of generations in which a mutant can be completely successful. The first throw specifies the channel; the second throw specifies the organ to be adapted. Accident thus decides the sequence of adaptive radiation.

Rules of play, 1. We begin with six organs in character-state I (cf. Fig. 32a, adaptive channel I). Throw twice for each generation-group in each occupied channel. The first throw will decide in channel I whether one of the empty channels shall be occupied, and which one. In the other channels the first throw will decide whether a change will occur in it (whether a mutation will strike). The second throw decides in each case the organ which shall change in the sense of the appropriate channel, if this change has not already happened. If these accidents do not happen with either throw, move on until, in the adaptive channels (II to VI), all organs have become adapted.

The result shows (Fig. 32a-c) that all similarities change in the same way until, after 34 generation-groups, they have completely disappeared. As a rule no more than three 'species' in a channel have four character-states in common. No hierarchical pattern arises.

If in this experiment (Fig. 33a-b) a particular sequence of adaptive changes is prescribed, then there arises at first a broader field of similarities, but by generation-group 34 these have entirely disappeared again.

Rules of play, 2. As with the first rules of play, but with one difference. The second throw again decides on the organ to be adapted. When it affects an organ that can still be changed, however, then in each channel II to VI, organ 6 will be changed and afterwards all others in descending order from 5 to 1. (Fig. 33a can therefore be based on the recorded die scores of Fig. 32a.)

A phylogeny of this sort would never form a hierarchical system. Instead it would produce a world of organisms with features completely equal in value, each feature being realized in only one recent species. Such a world of organisms is unknown and indeed inconceivable. Features restricted to single species are known, to be sure, though not numerous. Systematists recognize them as accessory features – special spots of colour, small structures, or patterns. They are useless for the recognition of relationship and only suitable for confirming the identification within the narrowest limits. They have all remained nameless, because they cannot be used in making comparisons.

A world of organisms without hierarchical order would allow no relationships to be recognized, not even definable groups. The only usable concepts which could be formed in it would be analogies (scales, thorns, pedicles etc.). The concepts would thus take on that type of comparability which must be carefully excluded from all controversies about phyletic relationship (cf. Section II B2a).

b. Preconditions and forms of hierarchy

Starting from this fantasy world, we can take the next step. What additional requirement is needed in our thought-experiment in order to allow hierarchical order to arise?

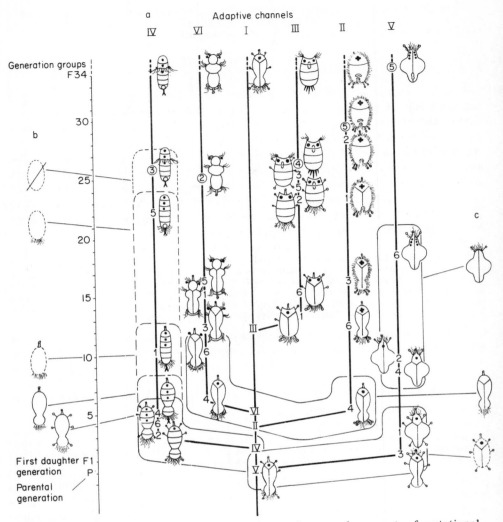

Fig. 32 a-c. Disappearance of phyletic similarity given equal prospects of mutational change in all features. (a) The occupation and adaptation to adaptive channels II to VI as decided by the dice, is reached by accident in 34 generations (see rules of play 1). Decisions causing alterations are marked according to generation and channel. (b) As a result, at each step a sixth of the resemblance to the original form is lost. (c) Four agreeing character-states occur in a chain of three, or at most four, of the closest related mutants. After 34 steps no relationships can be recognized any longer.

This requirement is, in fact, a gradual adding and fixation of features according to the branching of the supposed flow of determinative decisions. Biologically speaking, this means according to the separating paths of genetic connection. Three end results can then be reached, both in experiment and in Nature (see Fig. 36a-c).

Another precondition is the diagnosability of the whole assemblage, i.e. of the overall hierarchical framework of the group of animals. In fact the combinations of features in Figs. 32a and 33a (i.e. the species) had no common characteristics. The various species could not be defined by a diagnosis but only described by listing all the 36 character-states of the six organs. Some fixation of character-states therefore had to be presupposed at the beginning e.g. the character-state I of the body shape.

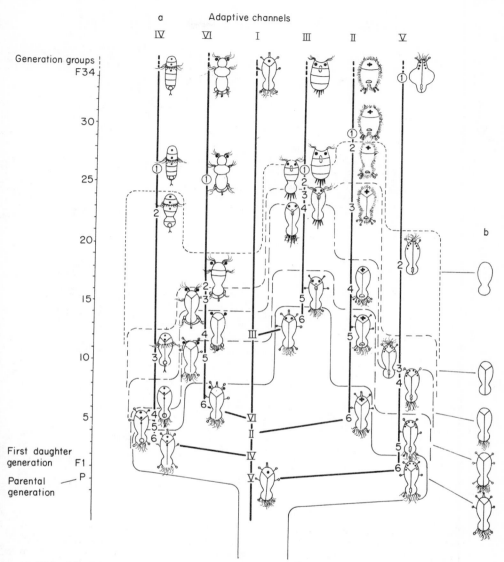

Fig. 33 a-b. Gradual disappearance of similarity resulting from equal-chance mutations whose sequence has been specified beforehand. (a) The occupation of channels II to VI and adaptation within them based on the same random scores as in Fig. 32a, but with rules of play 2. (b) Note the gradual decrease of similarity and, in the end, its total disappearance.

1. The first kind of end result is found when the newly added features are added only in one phyletic branch, which thus becomes the main branch. This gives a hierarchical pattern in which the number of frame concepts equals the number of alternative concepts. Such a 'box-within-box' or *sequential hierarchy* (Figs. 34c and 36c) often occurs in the major parts of natural classification involving the broader degrees of similarity and the longer periods of time. An example is the subdivision of the Chordata in which the number of ranked frame concepts (Craniata, Gnathostomata, Tetrapoda, Amniota, Mammalia, Theria) is equal to the number of alternative concepts of nearly the

121

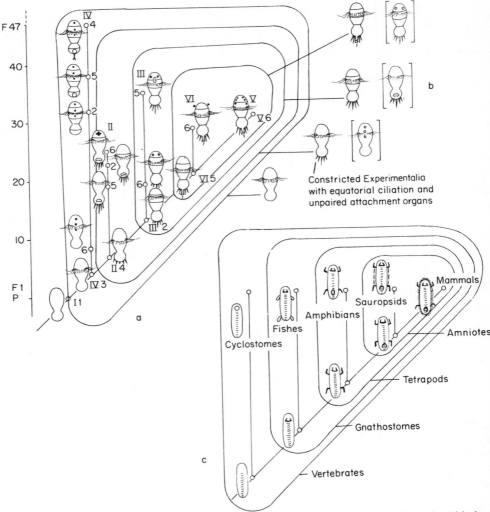

Fig. 34 a-c. The Origin of a sequential hierarchy by the fixation of newly added features (see rules of play 3). (a) Accidental result affecting 'Experimentalia' (Fig. 31), beginning with I1. The obliquity of the axis I-V indicates increase in morphological distances. (b) Content of the higher-ranking concepts. Note the asymmetry of the alternative concepts in brackets [ ]. (c) Sequential hierarchy of the vertebrates, up to the mammals. Not the symmetry of the alternatives.

same rank (Acrania, Cyclostomata, Pisces, Amphibia, Sauropsida, Prototheria). A similar concentration of frame groups is common in the upper levels of human organization and the older structures of society. Biologically speaking they are historical remnants, not totally necessary for the practice of systematics but wholly necessary for the logic of morphology (cf. examples in Fig. 34a-c).

Rules of play, 3. The starting point is a basic feature showing some particular character-state (e.g. Fig. 34a). All other features are added later, and after the separation of the adaptive channels, are permanent. There are three throws of the die per generation group and channel. The first throw decides whether a mutation shall occur in a channel (or whether to pass). The second is repeated until the next previously unoccupied channel is specified. The third is repeated until the next previously unoccupied organ is specified (cf. Fig. 34a).

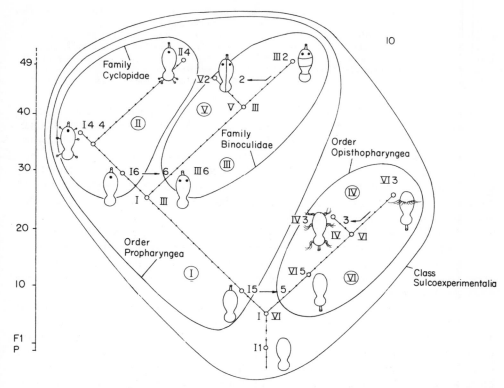

Fig. 35. The origin of a dichotomous hierarchy of alternatives by the fixation of alternative character-states of the same organ after the dichotomous branching of the stream of determinacy (cf. rules of play 4). Points signify unsuccessful throws of the dice, and circles signify successful ones. The arrows indicate when dice decisions have been carried over to the alternative channel. The artificial groups of animals have been given artificial names.

2. The second kind of end result (Fig. 36a) occurs when the branching points and fixation points for character-states can no longer be specified within groups. This gives a *hierarchy of collectives* in which the number of subgroups of the same rank can be a large multiple of the next higher-ranking group. This is characteristic for the smallest and youngest groups of the natural classification, with thousands of examples. In societies it corresponds to the lowest hierarchical levels in mass organizations such as churches, armies, and parties. In biology such hierarchies either represent early states in study, or else are simplifications that allow for present-day ignorance of the relative importance of single features.

3. The third end result (Fig. 36b) is a middle position, the *dichotomous hierarchy* of alternatives. This is characterized by a uniform distribution of the fixation points of character-states over all branches of the determinative connection. As a rule there are two subgroups in each group of higher rank. These two subgroups are usually distinguishable by alternative developments of the same organ. This pattern is widely distributed in the natural classification. It is probably the basic pattern. In the hierarchy of collectives it cannot yet be recognized while in the sequential hierarchy it can be recognized no longer. I shall give a final example (Fig. 35).

Rule of play, 4. The starting point once again, is a basic feature in some particular character-state (see e.g. Fig. 35). The first throw specifies, for each channel and generation, whether a further adaptive canal shall begin (the die number shows the number of the new channel). The second throw specifies the next unoccupied channel, or is repeated until this is specified. The phyletic path therefore divides. The third mutation will affect the first of the branches. The fourth throw, repeated if necessary, decides the next adaptable organ which has not yet been fixed in this particular phyletic line (the die number decides the character-state). It is then assumed that the same set of die decisions applies to the second phyletic branch.

The dichotomous hierarchy of alternatives best corresponds to the logic of classification (e.g. Remane (1971) Fig. 42; also Fig. 36a-c herein) But even this has become a reason for doubting whether the natural classification is natural. I shall later show the consequences of this doubt, and how unjustified it is. In actual fact our conceptual thinking follows the dichotomous hierarchy to such an extent that, even here, it can be seen as a basic pattern. Whenever it is not obvious, either in the classification of organisms or the systematic ordering of thought, this is because the fixated features differ in value in the two branches of a dichotomy, or else the distinction between alternatives has been partly erased. Biologically the dichotomous hierarchy is a necessity because, so far as our knowledge reaches, the splitting of the flow of determinative decisions, or break-up of a gene pool, is always dichotomous. And also because fixation of character-states can occur before any subsequent branching, even when the subsequent branching happens very quickly.

The step-wise fixation of features is therefore crucial for every hierarchy, and indeed the definitive fixation of newly-added features. Many features may be subsequently altered, or even eliminated entirely. But, by definition, diagnostic features cannot be so, neither in systematic thought nor in the classification of organisms. Likewise a fixation cannot once have existed, then to disappear and reappear, nor happen twice in separate phyletic lines. The feature and its mode of fixation are, as will be shown later, unique events with unique determinative connections.

The pattern of hierarchy which has arisen by the playing rules of the thought experiment is still very different from the patterns of the natural classification of organisms. The reason for this difference does not lie in the lack of sufficient complexity, nor in the naivety of the examples, but rather in true lack of structure. In the first place, fixation is not a dice decision, but a gradual, lawful process. Second, we have ignored the other equally fundamental patterns of order in organisms, i.e. interdependence, traditive inheritance, and the standard part.

Having clarified the preconditions of the hierarchical pattern we can now ask how this pattern arises. This involves two steps. First (Section V B) I shall show that the process of fixation is a matter of law, i.e. that it takes place under precisely definable preconditions. Second, I shall develop a theory to explain the process, beginning from known connections.

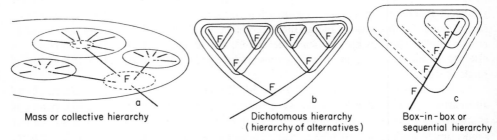

Mass or collective hierarchy   Dichotomous hierarchy   Box-in-box or
                                     (hierarchy of alternatives)   sequential hierarchy

Fig. 36 a-c. The three forms of hierarchy distinguished in the text. (a) The position of branching points and fixation points in uncertain, either through lack of knowledge or insufficient differentiation. (b) The branching is obvious with alternative fixation for each pair of branches. (c) The branches are not equal in length and the alternatives are obscured or unrecognizable.

B. THE MORPHOLOGY OF THE HIERARCHICAL PATTERN

If hierarchy, as I have asserted, depends on the definitive fixation of additional features, than I first need to show that this is a common and regular phenomenon among organisms. I shall do this using the previous example of single homologues, i.e. the vertebral column (cf. Section II B2b). These homologues have the advantage of being known in large numbers, for up to 10^7 have been described, and they therefore offer abundant comparative material. First, I shall present the general properties of such features as are necessary for comparing their degree of fixation. Then I shall describe how these properties are correlated, so as to derive the conformities to law which are hidden in the phenomenon of fixation.

1. Identification of general properties

The question of setting limits to the concept of a feature is particularly difficult when applied to homologous features. We are considering unities with minimal homologues as their lower hierarchical boundary (Section II B2b); these unities are so complex and similar, in any one case and from species to species, that we cannot doubt the presence of basically identical genetical conformities to law. These homologues are indeed abstractions; they are mutually corresponding unities or systems, which we separate from organisms by making comparisons, e.g. a nervous system, a gill slit. But this thought-dissection is both legitimate and necessary, as will be repeatedly confirmed, so long as it follows the rules. I shall therefore touch on these rules (*a*) and their results (*b*) before describing the actual properties (*c-h*).

a. The rule for separating homologues

This rule is that we think in terms of functional systems.[4] We cannot separate the systems from the whole organism by geometrical planes or sections. Rather the scalpel of thought follows the boundaries by which the organizing process has arranged subunits into structures that perform higher-ranking functions. This method is continually being tested whenever we need to understand functional unities, whether in the molecular or the ecological realm. It is especially familiar at the organismic level of the anatomists. The concept of systems began at this level and has become one of the fundamentals of biology — as for example the nervous system. No mistakes can be expected here.

b. The arrangement of systems

This is complicated. It is not a mosaic but a matter of organization. To a layman it may seem inconsistent that the chains of hierarchy of homologous systems are included within each other like boxes within a box, while the limits of neighbouring systems can overlap, or even coincide. This overlap is not accidental, however, but merely a consequence of functional stratification. Within themselves the systems are hierarchically arranged, box within box, when we are thinking of partial functions (supportive system, vertebral column, cervical vertebral column, atlas, odontoid process, ventral articular facet, cf. Fig. 11, Section II B2b). But they overlap when the functions cut across each other (nervous system, musculature, limb, skin, vascular system). Certainly the limits are unequal in clarity, just as the function and structure are unequal in difference. However, if position-structure times number of instances ($L \cdot a$) is enough to put the identicality of the compared systems beyond doubt (Section II B2d), then no great mistakes in drawing the boundaries can be expected. I shall now consider the properties of systems.

We have already met the two properties which need to be correlated (Section IV B2). They are position within the system (Section V B1*e*) and constancy (Section V B1*h*). Each of these is made up of subproperties which are important because they contain quantitative components. These subproperties (Section V B1*c, d* and *f, g*) must be examined in detail because this is a region where mistakes of indefinite size must on all accounts be avoided.

c. Degree of complexity

The systems and subsystems have degrees of complexity which can be estimated with sufficient precision. The single homologue can be used as the unit of quantity. In this connection I recall the two basic rules (cf. Section II B2); these are that the arrangement of the homologues is hierarchical and that, with progressive analysis, counting stops at the minimal homologues. These smallest units should all show a comparable degree of complexity since we can still recognize the presence of identical law content in each one of them. Let us assume for them a complexity constant of 1 ($c = 1$), since they cannot be further divided into subhomologues. The complexity of each cadre homologue (divisible into sub-homologues) will likewise equal 1.

The degree of complexity of a system will then correspond to the sum of its single homologues, this sum being made up of all the minimal homologues plus all the cadre homologues of intermediate rank including the homologue that gives the system its name (cf. Fig. 11, Section II B2*b*).

For example, the odontoid process of a human skeleton contains about 4 minimal homologues (the apex, the ventral articular facet, the dorsal articular facet, and the constriction of the collar of the odontoid process). With one cadre homologue (odontoid process) it therefore has a complexity $c = 5$ (Fig. 11). The centrum of the axis vertebra, including the odontoid process ($c = 5$) and the principle centrum ($c = 4$), includes 10 c. The axis with its centrum (10 c) and arch (29 c) contains 40 c. And so forth.

Naturally in individual cases it can be a question for comparative research whether a possible minimal homologue should still be counted as such, should be treated as a cadre homologue within the hierarchy, or should be neglected. With sufficient knowledge, however, the results of these counts prove to be completely reproducible. Differences as high as 1:2 are unusually great. This is important. It shows that the mistakes connected with a quantitative estimate of **c** are tolerable. For the actual differences in the value of **c** for different organ systems range over more than four orders of magnitude.

A conception of the degree of complexity of the larger systems can quickly be gained by looking at the subject indexes of the atlases of normal human anatomy.[5] These describe the locomotory apparatus, for example, as having 3 to 6×10^3 homologous features, or including right and left, from 6×10^3 to 1.2×10^4 homologous features. The numbers for the nervous system go even higher; 2.1 to 3.6×10^3 are listed. Including right and left, therefore, these are 4 to 7×10^3. Seeing that about one-third of these features are repeated about 30 times each in the spinal nervous system we would get 2.2 to 7.5×10^4. The definitive number would therefore lie between 5×10^4 and 10^5. The smallest functional systems are sufficiently complicated to establish their homology, but the greatest systems exceed them in complexity by a factor of 10 000 or 100 000.

d. The degree of integration

This is a second general characteristic of features. Within it I distinguish degree of coordination and degree of dependence.

1. *The degree of coordination.* This can be described as the precision with which a feature must be fitted into its functional surroundings. It is the degree of tolerance which must not be exceeded in the given functional connection. In a car the dimensional tolerance permitted to the foot mat differs from that of the ignition key. In a similar way the tolerance of biological systems also varies. In the human optical system, for example, the marginal feature 'eyebrow' has a tolerance of centimetres, the fit of the lens has a tolerance of fractions of a millimetre, while that of the retinal cells has a tolerance of micrometres.

2. *The degree of dependence.* This can be described in terms of the direction and importance to life of the complex of connected features. In a car, for example, this can be illustrated by the functional chain: differential, half axle, rim, hub cap. It is not difficult to specify direction and importance here. We would all drive without a hub cap and would not take much notice of a bent rim, but nobody would drive with a bent half axle. In networks of functions the limits are harder to see. The cervical vertebrae would not function without ligaments; the complex of ligaments and vertebrae would be meaningless without muscles; while neither ligaments nor vertebrae nor nerves would function without blood supply or innervation. We have to work out the limits of systems and mentally abstract those systems from the complete mechanism in order to understand the parts (taking the skeleton as the chassis, the ligaments as screw fittings, and the nerves as wiring).

A quantification of degree of coordination and degree of dependence would have advantages. However, it can be neglected except in limiting cases, so I shall not introduce it. I shall be operating in dimensions several orders of magnitude beyond such fine distinctions. Coordination and dependence are so distinct, however, that we are certainly dealing with the integration of two different characteristics.

e. Position within a system (and burden)

The position of a feature within a system is therefore of crucial importance and a quantification of the degree of integration (i) is necessary. By degree of integration or systemic position I again understand the functional situation (whether marginal or central) that a feature occupies in the network of a system. To specify it we must know the limits of complexity of the system quantitatively and the integration network qualitatively.

To work out position within a system we must remember that the minimal homologa are the lowest units of the hierarchical pattern of order while the cadre homologues are intermediate units. The systemic position can be specified, then, by knowing the number of homologues which depend on the feature in question, within the integration network of functions in the system. This implies a quantification of the obvious comparability of homologues within a functional system. The systemic position of cadre and minimal homologues can likewise be specified so long as their difference in complexity is remembered.

The question, therefore, is: How many units will be caused to disintegrate, becoming functionless or disconnected, if a given number of other units are removed.[7] In the overall system the features can be specified as being of varying scope and of central or marginal position (Fig. 37a-e). If, for example, the main artery of the right thumb is removed (Fig. 37e) then only the right thumb would be put out of action with five units (four subarteries as minimal homologues). If the distal end of the right brachial artery is destroyed (Fig. 37c) then the arterial system of the right arm goes out of action, with 80 units. If the right subclavian artery is destroyed, then the whole arm and a large part of

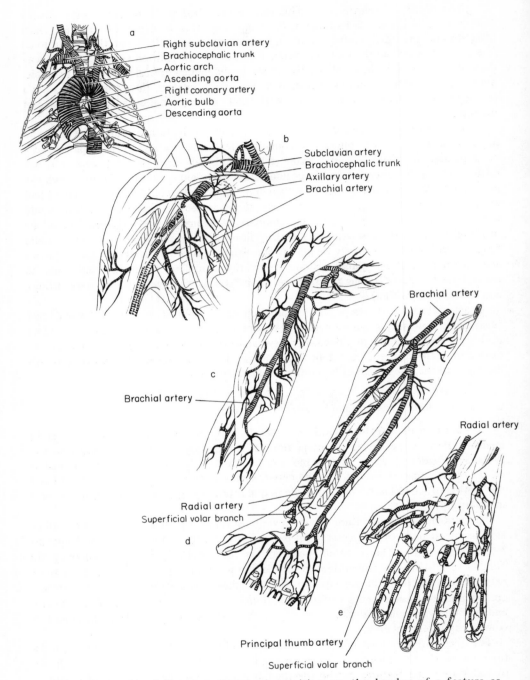

a
— Right subclavian artery
— Brachiocephalic trunk
— Aortic arch
— Ascending aorta
— Right coronary artery
— Aortic bulb
— Descending aorta

b
— Subclavian artery
— Brachiocephalic trunk
— Axillary artery
— Brachial artery

Brachial artery

c

Brachial artery

Radial artery

Radial artery
Superficial volar branch

d

e

Principal thumb artery

Superficial volar branch

Fig. 37 a-e. The influence of systemic position on the burden of a feature as illustrated by the functional chain of the human arterial system. (a) Ascending branch of the aorta; (b) Axillary artery; (c) Brachial artery; (d) Radial artery; (e) Main artery of the thumb. For simplicity only one stage of dissection (usually the superficial stage), and only the inner side of the arm, is shown, After Hochstetter (1940-46).

the thorax of the right side of the body would be put out of action with 150 units. And in the case of the ascending branch of the aorta (Fig. 37a) the whole arterial system would go out of action, except for the lungs and the coronary vessels — thus more than 1500 units and about half of the human vascular system.

Such chains of systemic position can also be recognized within the locomotory apparatus, e.g. the girdles etc. It should be remarked that the hierarchy of the supportive and nervous systems is as independent as that of the vessels, glands or muscles but is expressed in terms of the respective cadre system. The parts of the different systems therefore behave differently in phylogeny.

Burden (B)

This can be illustrated anthropomorphically by thinking of the responsibility with which a feature is burdened. Objectively it corresponds to a systemic position (i) within the complexity of the system under study. Differences in burden thus range over three to four orders of magnitude. Some burdens are a thousand or ten thousand times greater than others. This general characteristic of features is very important for what follows.

Burden does not necessarily depend on the complexity (c) of the feature itself. It is true that a more complex feature carries somewhat more burden than a less complex one of similar systemic position. Thus in the above example, if the aortic bulb is added to the ascending branch of the aorta then the burden of the coronary arteries is also added (cf. Fig. 37a). The increase, however, is by no means proportional. Usually it is possible to choose the limits of features in a chain so that their complexity is the same. Nevertheless the burden will steadily increase, with increasingly central position. To recognize this is useful in confirming the conclusions.

After explaining this concept of systemic position or burden another connection between concepts must be presented. Constancy or fixation can be deduced from similarity and age.

f. Similarity

The task of quantifying similarity (as indicated in Section II B2e) has still not been satisfactorily completed. This remarkable fact can be explained both by the complexity of biological structures and by the complexity of comparative thinking. The latter is obvious from the extent and epistemologically difficult situation of the theory of form (Gestalt theory) which will be dealt with later. At this point I wish briefly to develop a practical approximation which will allow a quantitative estimate of degree of similarity. This is desirable as a support for judgements of similarity. In actual fact a mere estimate will be sufficient, because degree of similarity is much less important than age. This is unexpected but I shall explain it below (Section V B1h2).

Biological similarity is different from geometrical similarity because qualities obtrude into it and have unequal importance.

The necessary quantification is limited, as usual, to homologues, i.e. to identical parts. Purely proportional similarity (i.e. analogy) is useless. Similarity only becomes important within identical parts. Another crucial point is that the special qualities of a minimal homologue can, by definition, no longer be broken up into identical individual qualities. They can only be analysed into proportions, i.e. into quantities, as soon as we leave the realm of singular identicalities (or single homologues). All cadre homologues can be broken up into minimal homologues and these can be analysed by measurements. Consequently every system can be quantitatively described as well as its degree of agreement with corresponding related systems. The units of measurement are derived from the lengths of lines between corresponding points (and the angles enclosed between

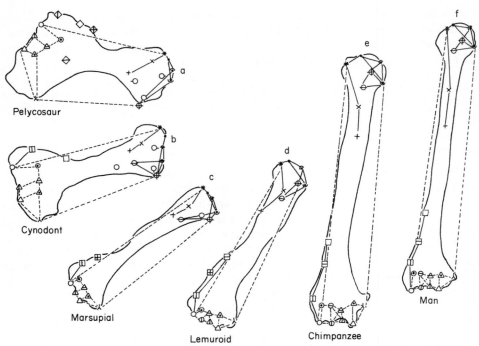

Fig. 38 a-f. The quantification of similarities by comparison of the positional differences of minimal homologues as shown by the humerus from reptiles to man. (a) *Ophiacodon* (Permian); (b) Cynodont (Triassic); (c) *Didelphis,* to represent a marsupial of the Cretaceous; (d) *Notharctus* (Eocene); (e) *Pan;* (f) *Homo* (Recent). Identical minimal homologues are indicated by identical symbols and some of the corresponding positional connections by the straight lines between the symbols. After Gregory (1951), somewhat elaborated.

these lines). The weighting of homologues can be obtained from the quantity of data that they yield (i.e. their extent of determinacy). The similarity can be obtained from that of the comparable individual values (e.g. the smaller as percentage of the larger). Figures 38a-f will illustrate this.

The framework within which representatives should be examined for similarity is easy to define. It corresponds to the systematic group at the base of which the system appeared as a feature, or is deduced to have appeared on grounds of relationships. I shall return to this later.

The maximal morphological distance can easily be defined as the greatest distance between two representatives within the group. As a rule it is desirable to consider individual similarity classes. For the range of variation matters more when it applies to a great mass of species rather than to merely a small minority of extreme forms. In the section on constancy I shall return to this subject.

g. Age

The next general characteristic of a feature is its *age* (a) which likewise needs to be precisely specified. In many groups this is not in fact difficult. If it is documented by fossils then the age can be determined directly, by geological and palaeontological methods. Otherwise it can be established indirectly, by correlation with features that are documented by fossils. The age of a feature can be defined as the span of time that has passed from its origin until its disappearance, or until the present day. We shall be concerned with time spans from 10^5 to 10^8 years, or at most ranging from 10^4 to 10^9 years.

h. Constancy (and fixation)

The constancy or 'steadiness' (s) of a feature is another crucial characteristic. It is made up of constancy in form and in time, and thus of the similarity and the age of a system. The range over which many morphological concepts are realized could be plotted on a graph with 5×10^8 years of fossil history as the ordinate and 0 to 100 per cent similarity as the abscissa. The concept of 'the aortic root' (Fig. 37a) would appear as a vertical band while horns of antelopes and sheep would in contrast form a horizontal band (Fig. 42a-g). Constancy and freedom would thus be clearly distinguished.

1. A composite measuring rod, more biologically useful, is obtained by considering the chances of conservation for features. The average duration of species[6] features can be taken as a unit of measurement. This is the duration of the features which are most susceptible to the mechanism of change in shape. Ignoring a few living fossils, this unit can be roughly estimated at 10^6 years (2×10^5 to 5×10^6). It follows that, after a million years, the accidental prospect of survival, or of constancy, for a species feature is equal to its accidental prospect of alteration. As with heads or tails $P = 0.5$. How large then is the prospect in the subsequent millions of years?

Assuming that the laws remain the same the *prospect of constancy* for the feature (P_s) will fall by a half for each successive *million years (Y)*. Thus:

$$P_s = 2^{-Y} \tag{30}$$

Consequently the prospect of conservation for a Middle Pliocene feature (2×10^7 years, $Y = 20$) would be only 9.5×10^{-7}. That of a feature from the Older Tertiary (10^8 years) would be 7.9×10^{-31} and therefore already virtually impossible. And the prospect of conservation for a Cambrian feature (5×10^8 years) would be 3×10^{-151}, although we know many features of this age among living organisms. I shall return to this subject later (Section V B2b).

Such differences in dimension raise two broad considerations before finishing this discussion of the general features of biosystems.

2. The first consideration is the subordinate importance of degree of similarity. If the decreasing similarity of changing systems is seen as depending on the gradual disintegration and replacement of their subsystems, corresponding to the evolution of complex features, then the probability that one-tenth of the similarity will be conserved in unit time is ten times greater than the probability that the similarity will be conserved completely. A conserved agreement of one-thousandth is 10^3 times more likely, but would be scarcely discernible. But even this increase in probability would be insignificant for an Old Tertiary feature because 10^{-31} or 10^{-28} are in effect equally improbable. The lowest level of similarity that forces us to assume the presence of homology (cf. e.g. Fig. 8a-e, Section II B2a) will oblige us to assume the action of some fixation mechanism.

3. The second consideration applies when we need to be more precise so as to compare sequences of differing constancy (cf. Tables C and D, Section V B3f). We then need to take account of changes in degree of similarity. It will be sufficient here to give an example of the method that I shall apply. It takes degree of constancy (s) as the product of degree of representation (r), of the degree of identicality of single homologues (h), and of their similarity of proportions (p).

The degree of representation (r) is the percentage of species of the group which show the cadre homologue of the feature. The degree of homology (h) is the percentage of subhomologues which are constantly present within **r**. And the degree of proportional similarity (**p**) is the average percentage agreement of lengths and angles within **h**.

Fixation (**F**)

This term can be used for the complex of systems factors which tend to create a high level of constancy. It has anthropomorphic overtones, just like the connection between burden and systemic position (cf. V B1e). I shall show later (section V C) that the causal connection between systemic position and constancy can be understood as that between burden and fixation. However, I shall begin to use the dynamic terminology immediately, as being easier to understand.

4. *The nature of the mechanism.* When similar phenomena, such as those of constancy, differ so completely in dimensions, then it is likely that the causes also vary. Some additional phenomenon, such as fixation, is extremely likely because the range of orders of magnitude for constancy is inconceivably large, even exceeding that of the limits of the whole known Universe. Compared with 3×10^{-151} (which was estimated above as the chance of accidental survival for a Cambrian feature, assuming no fixation mechanism) the greatest range of our space concept is only 37 orders of magnitude (wavelength of gamma rays = 10^{-10} cm, limits of the visible world 2×10^{27} cm). And the greatest range of our time concept is 40 orders of magnitude (the duration of the shortest describable elementary particles is 10^{-16} seconds while the greatest assigned age for the universe is 3×10^{24} seconds).

2. Extreme degrees of freedom and fixation

'Freedom' is such an indefinite and emotionally loaded concept that it might seem good to avoid it in a scientific work, though 'fixation' in the special sense of determinacy is better. 'Freedom' has the most obvious psychological associations among the available alternatives, however, and has often been used in the phylogenetic literature, so I also shall use it. I apply it in a sense of indeterminacy or lack of direction, or better as lack of commitment or predictability. It thus means a large field for the play of accident, and a small one for necessity.

The concept of freedom is understood in this sense in modern biology, systems theory, and natural philosophy.[7] I shall examine it in detail as applied to anatomical features. I do this by studying the extreme types of systems, their general characteristics, and their modes of occurrence.

a. Systems of maximal freedom

These have been repeatedly surveyed in the literature.[8] Rensch calls them 'examples of directionless transpecific evolution'.[9] Textbook examples are the horns of antelopes, goats and sheep and the plumes of birds of paradise. To those I would add the lips of orchid flowers and the dorsal anatomy of the Membranacidae (buffalo tree-hoppers). I could give many other examples.

A modest notion of these extraordinary 'fantasies of Nature' can be got from Fig. 39a-g. They affect striking external features and fully agree in general characteristics; of these I shall take the quantifiable ones first.

1. *Structural components.* All these features of maximal freedom are of low complexity, using the word as I have defined it. The observer should not be deceived about this. Despite the geometrical complication of form and the variation in proportions there is no homologue which changes in principle. In Membranacids and orchids the

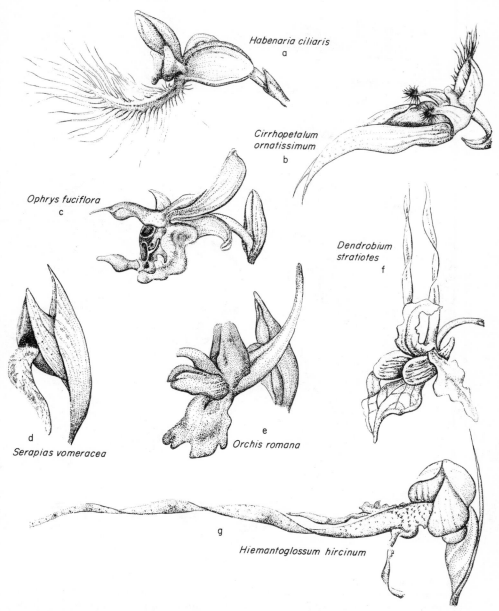

Habenaria ciliaris
a

Cirrhopetalum ornatissimum
b

Ophrys fuciflora
c

Dendrobium stratiotes
f

Serapias vomeracea
d

Orchis romana
e

Hiemantoglossum hircinum
g

Fig. 39. Orchid flowers to illustrate features of maximal freedom. Note the extra-ordinary and almost directionless variation in the petals and especially in the lips. Partly original, partly from Danesch and Danesch (1969).

varying substrate is a very low-ranking cadre homologue. With the horns it is almost a minimal homologue and with birds of paradise the feature need not even be homologous. This agrees with the fact that no two substructures of these varying features can be reliably homologized — neither the cuticular bladders of the tree hoppers, nor the curves of the horns, nor the points of the lips of the petals.

133

Hamma nodosa

Oeda inflata

Bocydium globulare

Lycoderes gaffa

Sphongophorus inflatus

Umbonia spinosa

Sphongophorus ballista

Fig. 40 a-g. Features of maximal freedom as shown by buffalo tree-hoppers (Membranacidae). These are mainly tropical forms varying in length from millimetres to centimetres. They have extremely bizarre thoracic appendages which vary even within one species (Fig. g). Partly original, partly from Haupt (1953) and Heikertinger (1954).

Second, all these features are extremely peripheral in bodily position. If they do belong to a functional chain they are at its extreme outer end for the individual. No other homologue of the individual is directly dependent on them. In fact it is becoming more and more obvious that all of them are signals — they correspond to the outermost part of a radio corporation, its transmission tower. This has recently been shown for antelopes[10] and even the ludicrous dorsal processes of Membranacids can be understood as mimetic.

Thus all these features carry extremely low burden. In the sense used above they are also textbook examples of almost burdenless features.

2. *The time component.* The age of these features is very low. So far as fossils allow it to be dated, the separation of the modern species of goat *(Capra)*, for example, is not older than late Pleistocene.[11] No living species of goat is older than 3 or 4×10^5 years. Even the whole of the Bovidae arose as a group 'only in the Upper Tertiary'[12] which would be only 10^7 years ago (cf. Fig. 44, Section V B3a). Most Bovid species would be not older than 5×10^5 years. Even for mammals, which in general change remarkably quickly,[13] these are strikingly short times. Among the forms not represented as fossils, equally short, or sometimes even shorter, times of differentiation are suggested by the mode of occurrence and by the fact that all are merely species features.

The constancy is also extraordinarily small. Scarcely one of the features extends beyond the limits of the species. Even among the species of *Sphongophorus* (tree hoppers) the generic feature is very indistinct (Fig. 40e and g). In addition sexual dimorphism is often obvious and in single species (cf. Fig. 40g *Sphongophorus ballista*) the variability is excessively high so that: 'scarcely any two animals are totally like each other',[14] Much the same is true of the birds of paradise and even for the shapes of horns (Fig. 42b and c). In the imagined diagram of constancy (Section V B1h) all the features in question would fill the whole range of dissimilarity but would only take up one-thousandth of the scale height (5×10^5 out of 5×10^8 years). Quantitative considerations thus confirm the pre-existing conclusion that all are features of extreme 'freedom'.

3. *The distribution* of such features that combine least burden and least fixation is extraordinarily wide. Those mentioned are merely the best known. For such quixotic developments extend over the whole realm of organisms as a special form of what systematists call simply 'accessory features'.[15] It is crucial to appreciate this and I shall therefore next consider the less known and simpler of such features so as to show that the correlation between least burden and least fixation holds universally.

Accessory features are those which fundamentally do not extend beyond the species limits, which have no parallels even in the most closely related species, and even less in the repeated basic pattern of the genus. They are often colorations, little processes, thorns or superficial structures which characterize the species but are useless for phylogenetic studies. In all species with a large number of features a few such accessory features are known and the number of recent species is about 2×10^6. In Recent populations, therefore, there may be 5×10^6 features of least burden and least constancy.

Great expertise is needed to find hidden, uniting characteristics in, for example, the generally random-looking colour patterns of a low-ranking systematic category.[16]

4. *General qualitative characteristics.* These emerge as soon as the quantitative correlation has been made. All features which unite the quantitative characteristics of 'low burden' and 'low constancy' function in a peripheral bodily position; are prone to substitution; are highly variable within the species (especially at the distal ends of the feature); are represented only in the smallest systematic categories; and often involve the phenomenon of homoiology, i.e. analogy on a homologous base (cf. Section II B2a). These qualitative characteristics are instructive because they show graduations in exactly the same same as the quantitative features.

b. *Systems of maximal fixation*

These are of a totally different type. However, it is striking that, although they are much the more astonishing of the two extremes, they have never been surveyed as such.

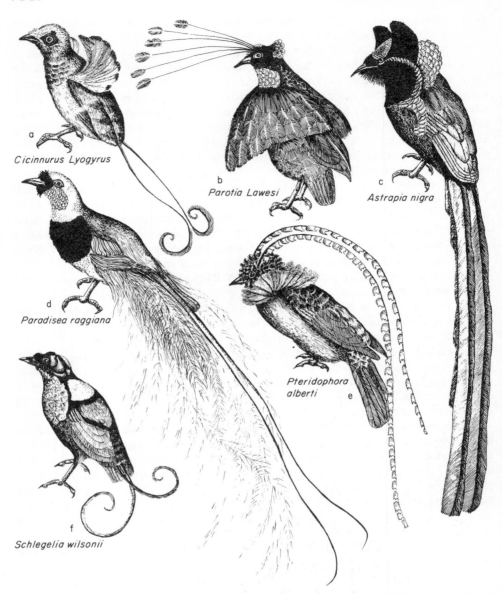

Fig. 41 a-f. Features of maximal freedom as shown by birds of paradise. Note the great and almost directionless variability of plumes going off the head, sides, and tail. After Grzimek and Schultze-Westrum (1970).

'Living fossils' are a well-known concept. 'Fixated system', however, though they are much more constant and must have arisen much earlier, are a new concept.

Overwhelmingly they are features of the internal anatomy, often not particularly noticeable so that they are only seen after deep morphological analysis. A few examples may illustrate this — or, rather, explain it. For it is a general characteristic of these features that they cannot easily be illustrated in a figure. This has nothing to do with lack of definiteness in the way they are expressed. On the contrary, it has to do with the

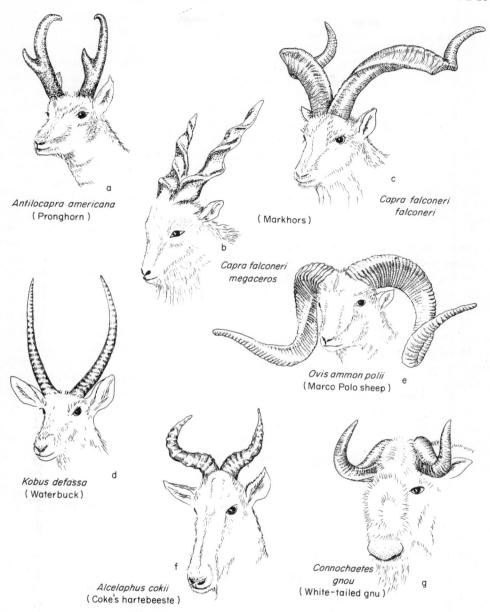

Fig. 42 a-g. Features of maximal freedom as shown by the horns of Bovids. Note the great variety of spirals and whorls in the horns, even as between close relatives as in (b) and (c). After photographs and drawings from several authors.

powerful and numerous transformations which have affected them since their origin and which are totally impossible to show in a single picture.

Examples in the organization of the vertebrates are: the fourth aortic arch (cf. Fig. 69a-c, Section VII B3*a*), the central canal of the dorsal nerve cord, the optic chiasma, the Wolffian duct, the eye with lens and inverted retina, the hypophysis. These are not features which, through lack of importance, have been pushed into peripheral positions. On the contrary, they are parts with important functions which have been preserved despite their obscure anatomical position or adaptive incompleteness or even despite total change in function. If we include in the list homologues with considerable changes in proportions then we find that the extent of change to which they have been subject has even hindered their formulation as single concepts. Think, for example, of the homologue which was called on the one hand the notochord and on the other the nuclei pulposi; or of the homologue known respectively as the hyomandibular or as the stapes (cf. Fig. 7 a-e, Section II B2a). There are many other such examples.

Similar features could be quoted for the coelenterates, the bryozoans, the brachiopods, the echinoderms, the molluscs and the arthropods. That is to say, for all large and old groups well documented by fossils. To list them is neither possible nor necessary. For, first, they are very numerous, and second, they have already been examined, sorted out, and summarized in the diagnoses of the phyla (cf. Section V B2b3). Both these facts are crucially important in what follows.

1. *Structural components.* First, all these features are parts of the most complex systems. Thus in the nervous systems of the phyla — systems which tend to reach the highest levels of complexity in the ground plan of each major group — all the basic features belong here. With reference only to the great functional complexes of the differentiated supportive, excretory, and vascular systems, the burden (**B**) is mostly more than 10^3, and may even reach 10^4 or 10^5 homologues.

Second, the degree of complexity of these features need not be great. Indeed it is important to appreciate that the complexity may be very small, corresponding to a single minimal homologue. Indeed in all the larger features, corresponding to cadre homologues, it is possible to find a minimal homologue which is as old and constant as the whole. Thus we only need to consider a part of the neurohypophysis, a section of the fourth ventricle of the cerebral canal or, in the optic system, only the anterior attachment of the left posterior rectus muscle (Fig. 43 a-b). We are here dealing with homologues of any degree of smallness within larger systems.

Third, the position of these features in the great functional systems is always somewhat or extremely central. Thus a general characteristic of all of them is a high, or

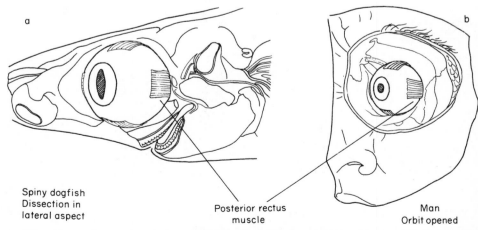

Spiny dogfish
Dissection in
lateral aspect   Posterior rectus   Man
                                            muscle   Orbit opened

Fig. 43 a-b. Example of a minimal homologue of very high constancy. The anterior attachment of the posterior or external rectus eye muscle in shark (a) and man (b). The muscle attachment has been preserved independently and unchanged for 400 million (4×10^8) years. (a) after Marinelli and Strenger (1959); (b) after Pernkopf (1960).

extremely high, burden. The burden carried, for example, by the remaining left fourth aortic arch of a mammal has already been indicated (Section V Bl*e*). Even taking only the proximal part of the descending branch of the aorta as a true homologue in man (i.e. excluding the ascending branch cf. Fig. 37a Section V Bl*e*) then the burden for the arterial system is still about 10^3. The burden of the hypophysis or of a part of the fourth ventricle is even higher.

In addition to this burden, carried in the adult, there is a second burden which must not be overlooked. This can be called ontogenetic burden. It consists of the sum of the homologues of those organs which depend on the anlage of a particular feature, i.e. whose ontogeny is controlled by this anlage. I shall not anticipate further here, but refer to Chapter VII which is devoted entirely to this subject. It will be obvious how many homologues of the adult depend on the anlage of the Wolffian duct or on the notochord – almost the whole excretory system and almost all the dorsal subdivisions of the body respectively.

2. *The time component.* The age of all these features is at least 4 to 5 × 10^8 years, for they are all found in Recent groups whose ancestors separated in the Lower Palaeozoic or before. We have assumed that the prospect of conservation for a usual species feature is 50 per cent per 10^6 years (Section V B1*h*). On this basis the propect of conservation for these features of high constancy would be inconceivably low at 10^{-151}, corresponding to a number with 150 zeros after the decimal point.

Admittedly the degree of similarity varies. It extends from well over 50 per cent, as in the external rectus muscle, to a mere 1 per cent in the case of the hyomandibular (cf. Fig. 43a-b and Fig. 7a and e, Section II B2*a*). However, the similarity in either case is large enough to leave no doubt about the homology of the features. Consequently the improbability of their conservation is reduced because of this variation in similarity by at most 1/10 or 1/100.

We are therefore dealing with features of extremely high fixation, of the most constant orderliness and regularity. They are preferred by systematists when defining this orderliness and regularity.

3. *Distribution within systematic categories.* The distribution of the features of highest burden and fixation within the natural classification can be indicated exactly. Indeed their pattern of distribution has been established by the labour of generations of anatomists and systematists. I say this with relief: first, because this huge field would be impossible to document; and second, because we are thus beginning to see the causes for the reality of natural classification. This classification is the monumental achievement of biological morphology, although its existence in the real world of Nature has been doubted. The features of high burden and fixation are all contained in the diagnoses of the great systematic groups of all higher organisms, as differentially diagnostic single homologues.

This pattern of distribution is so regular that no feature of high burden can be found in the diagnosis of a low systematic category. Nor is there any homologue of low burden which can be used to characterize the great systematic groups.

The establishment of a correlation of such general validity needs to be made more precise, however, particularly as the principles of systematics are not generally known.

First, we are speaking of single homologues. The homonoms, being systems which occur several or many times in the individual, as with cilia or mammalian hair, are not under discussion here. The correlation of such standard parts is just as unequivocal but, as already shown, has different patterns of burden and is differently fixated.

Second, only differentially diagnostic homologues are under consideration. These are represented in all the members of a phyletic group, but not represented in any member of any other group. Features which are selective, in the sense of not being represented in all members without exception, will be dealt with under special condition of representation. Accessory features do not appear in this connection.

Third, only full or maximal diagnoses indicate the full extent of the differentially diagnostic features of groups. Practical systematics, on the other hand, often requires minimal diagnoses consisting of those differentially diagnostic homologues which just suffice to distinguish the group with certainty.

Fourth, for brevity of description, it is usually only the defining concept of the homologous system which is written down (e.g. 'ventral heart' for the vertebrates, 'dorsal heart' for the arthropods). The total of subordinate homologues is assumed to be known.

The number of differentially diagnostic homologues in the maximal diagnoses of the great systematic categories depends partly on the state of research. It will not be an overestimate to assume 10^4 of them for there are at least 20 phyla, 100 classes, and 380 orders with an average of at least 20 differentially diagnostic single homologues. No precision is here necessary. We only need to establish that these systems of maximal fixation are realized abundantly in evolution.

4. *General qualitative characteristics.* These can also be recognized for features of greatest burden and constancy. They are converse to the characteristics for features of least burden and constancy. All have a central position in the functional network of the system; have almost no possibility of being substituted; have a variability which is only expressed over broad phyletic groups and which does not alter the principle of the feature; are represented in the very largest systematic categories; and are never involved in homoiology (unlike features of low burden).

3.   The correlation of burden with fixation

To sum up, I have so far shown that least constancy of features in phylogeny goes with lack of burden, while greatest constancy goes with greatest burden. I now have to show that, in the middle region between the extremes, the constancy of single homologues behaves in general as a mathematical function of burden.

To show this it will probably be less convincing to bring a great number of examples than to explain the principle behind the correlation. Only by knowing this principle can deviations or supposed exceptions be taken into account, so testing the postulate. However, a great range of examples, as in Table A, will show the general connection between quantitative and qualitative characteristics for different amounts of burden and fixation and how these are connected with the pattern of occurrence in the natural classification.

a. *Quantitative characteristics*

In the first place the degree of fixation (**F**) and degree of burden (**B**) are so strongly correlated on average that one can be described as a mathematical function of the other: $\mathbf{F} = f(\mathbf{B})$. As Table A and Fig. 44 indicate, the fixation time for a minimal burden of $\mathbf{B} = 1$ is $\mathbf{F} = 10^6$. We know this value already for we have met it as P_s (cf. equation 30, Section V B1*h*) and also as $P_m \cdot P_e$ (cf. equation 26).[17] Beyond this I now assert that the function, if expressed in log — log form, is a straight line such that a change in the burden of three orders of magnitude gives a change in the fixation of two orders of magnitude. Therefore:

$$\mathbf{F} \approx \mathbf{B}^{2/3} \times 10^6 \tag{31}$$

As evident from Fig. 44, this function is the most cautious possible estimate of the fixation. The power of **B** and the constants may be somewhat bigger. Shortness of duration for a feature is less obvious than the converse, however, and caution is therefore appropriate. The foregoing cautious estimate will therefore suffice in the first instance.

It would be unjustified and unnecessary to insist on precision. First, because the quantitative estimates have the merit of simplicity and second, because a direct connection between burden and fixation is biologically unlikely. It will be enough, for

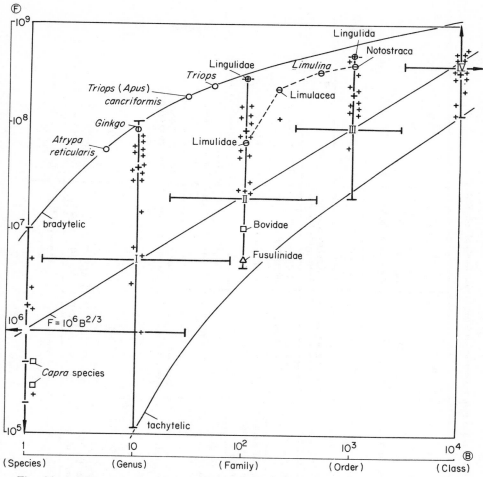

Fig. 44. A summary of the general correlation between burden and constancy. The figure shows the mathematical function (equation 31), the means, and the ranges of levels of complexity I to IV; the values (+) and mean values (+) of features documented by fossils (after Müller (1963, Vol. I, p. 191); the oldest 'living fossils' (and one long-lasting extinct fossil, *Atrypa*) are indicated by circles; examples of extremely brief durations or quick evolution are shown by triangles and rectangles respectively (see also Section V B2a). (After several authors, see text.)

the present, to establish that a connection certainly exists and that, on average, every homologue added to a functional system as additional burden, will add a million years to the fixation.

Variability in the relationship of burden with fixation (Fig. 44) seems at first to be considerable. In the levels of complexity along the abscissa it ranges from about 1/2 to about 1/10, but these values are large so as to err on the side of caution. They ought really to be reduced since each degree of burden is definable. Along the ordinate (fixation time) the range of variation is likewise about 1 order of magnitude for high burden. This wide range, however, is purely the result of 'widely scattered examples'. It would not be expected that the correlation would be quantitatively the same for birds of paradise and orchids, or for the vascular system of vertebrates and the nervous system of snails. When

141

Table A. The general characteristics of functional systems and their expression arranged in four levels of complexity

Levels			I	II	III	IV
Burden Characteristics		**B**				
Correlation characteristics	Complexity (in correlated homologues) =	**c**	Mean 10 Range 1 to 10^2	10^2 20 to 5×10^2	10^3 3×10^2 to 3×10^3	10^4 3×10^3 to $> 10^4$
	Type of integration	**i**	Loose	Obvious	Narrow	Very narrow
Functional characteristics	Functional position		Peripheral	With many inter-connections	A precondition for many other features	Central
	Propect of substitution		Large	Limited to a particular direction	Reduced in all directions	Almost none
	Freedom of variation		At the functional distal ends almost unlimited	One-sided	Very canalised	None as concerns the principle
Systematic characteristics	Taxonomic representation		Mean: Genera Range: (Species — Families)	Families (Genus — Order)	Orders (Families — Classes)	Classes (Orders — Phyla)
	Constancy in the group		Very low (accidental)	Moderate (common)	Large (present as a rule)	Almost absolute (with no exception)
Fixational Characteristics		**F**				
	Age, in years of conservation	**a**	Mean 5×10^6 Range (10^5 to 10^8)	2×10^7 (4×10^6 to 3×10^8)	10^8 2×10^7 to 5×10^8	4×10^8 (1.5×10^8 to $> 5 \times 10^8$)
	Type of change	**s**	Explosive	No longer in all directions	In trends in a few directions	Little or no alteration

Examples	Number of 'known' cases as order of magnitude			
	10^6	10^5	10^3	10^2
20 individual examples (out of more than 100 cases studied) and the systematic categories for which they hold	Plumes in genera of birds of paradise *	Claw shape in families of higher crustaceans	Types of girdle in orders of tetrapods	Circulatory system of the tetrapods
	Horn shapes in genera of ungulates *	Foot shape in families* of ungulates	Abdominal segments in orders of crustaceans	Ambulacral system of the starfishes
	Dorsal processes * in genera of buffalo tree hoppers	Genital apparatus* in families of turbellarian worms	Dorsal armour in sub-orders of higher crustaceans	Cerebro-visceral system of snails
	Rostral teeth in genera of shrimps *	Attachment organs in families of tape worms	Armour of tortoises	Brain of mammals
	The lips of flowers in genera of orchids *	Uropod shape in families* of higher crustaceans	Types of mouth parts in orders of insects	Peripheral nervous system of bony fishes

The tables shows the correlation of burden characteristics (B) with fixational characteristics (F) and gives a number of broadly scattered examples (*indicates systems in which homoiologies are known). The number of 'known' cases is the product of the number of systematic categories known for the level of complexity times their differentially diagnostic cadre homologues. The ages are after Moore (1965) and Müller (1968).

considering the individual groups (Section V B3*e*) I shall be able to eliminate this source of uncertainty.

The correlation can nevertheless be taken as a reality, as concerns the principle. The fields of burden for systems of high and low constancy do not overlap, nor do the fields of constancy for systems of high and low burden.

b. The qualitative characteristics associated with high and low burden

These all show a gradation, exactly corresponding to the two quantitative characteristics (cf. Table A).

1. *Position within a functional system.* In all systems of low burden this is always very peripheral. Indeed comparison shows strikingly the very different ways in which peripheral positions are realized. For these positions are almost always at the extreme adaptational forefront, being the pacemakers of adaptation and diversification. In the middle categories of burden dependences of the features become obvious. And in the higher categories the dependences increase still further so that the position of the feature is a precondition for a growing number of other features. In the highest categories the features seem to have become fundamental preconditions with a basic or fundamental position.

2. *The prospect of substitution.* In the lowest category of fixation and burden this is almost unlimited. A complete replacement of the feature is not surprising even in the closest related of species or even races. With $B = 100$ replacement still happens but, interestingly, only in particular directions or under particular conditions, so to speak. With $B = 1000$ the prospect of substitution has been obviously reduced. A whole series of further structures and functions would need to be substituted before the feature could conceivably be replaced. With $B = 10\ 000$ substitution is almost inconceivable.

3. *The kind of variability.* Besides a quantitative gradation there is also a qualitative alteration. Not merely the extent of variability alters, but its kind changes basically. In category I there is an almost total 'freedom of the ends' — the freedom of 'the lash on a whip'. The next position that the outer ends will take up is totally unpredictable, although the feather, horn or leaf, whose outer ends have this freedom, always remains basically what it was. In categories II and III limitations and canalizations occur. Until in category IV a completely different freedom or variability remains — the freedom of 'harmonic proportions'. By this I mean the extremely regular 'metamorphoses' or transformations of a feature which is not alterable in principle — transformations that are expressed harmonically and geometrically by Cartesian transformations (Fig. 53a-h, Section VI B1*d*).

4. *Representation within the natural classification.* In passing from one category to another, not merely the number of representatives changes (as shown in Section V B3*c* below) but also the type of representation. In passing from category I to IV, the occurrence of a feature within its field of representation changes from accidental, to common, to almost universal, to totally without exception.

5. *The tendency to homoiology.* This property, marked * in Table A, shows a similar change. The gradation is very steep, however, for homoiology is a 'marginal feature' of homology. In category I all the homologues in the examples are suspect of being of homologues — analogies on a homologous base. In category II homoiologies still sometimes happen. in category III they are at most very rare, for I know of no example. And in category IV they never occur.

Our scale of the characteristics linked with burden and fixation is therefore based not merely on a few quantifiable values but also on a harmonious sequence of many qualitative characteristics.

c. The pattern of distribution within the natural classification

The categories show a purely hierarchical pattern of distribution which entirely agrees with the systematic framework. These two facts are important. The hierarchical distribution allows us to recognize the consequences of the levels of burden and fixation and how these levels relate to the orderly gradation of characteristics which has been noted above. The agreement with the systematic framework connects the results of this quantitative study with the structure erected by systematists on a qualitative basis. It confirms this structure and is confirmed by it.

There is an increase in mean size of the groups to which the characteristics of the different categories apply. Categories I to IV contain the differentially diagnostic characteristics for genera, families, orders, and classes respectively. As burden and fixation increase, the size of the phyletic groups to which the characteristics apply also increases, and these fields are arranged inside each other, box within box, as the hierarchical pattern requires. The definition of hierarchy, however precise, does not state the number of subgroups that a group may contain. Consequently the groups to which the same burden-fixation characteristics apply differ in the number of contained species. For example, the number of species in a phylum ranges from 8 (Priapulida) to 838 000 (Arthropoda) that of a class from 1 (Somasteroidea) to 750 000 (Insecta).[18] The systematic rank of a group to which a particular kind of burden-fixation characteristics applies can only be defined by considering the relation of groups to each other and thus by the number of hierarchical levels which are lower or higher in rank.

The variability of ± 1 hierarchical level for each burden-fixation category is partly due to caution since the examples have been compiled at random. To this extent it has the same origin as the range in fixation (Section V B3*b*). It depends on 'widely scattered examples'. For nobody would expect that the criteria for priapulid genera need agree quantitatively with those of insects. The hierarchical rank of a systematic category has flexible limits in absolute terms, but the position of a systematic category in its own context is precise. This seeming discrepancy corresponds perfectly to the rules of hierarchy and has been unjustly criticized by non-systematists. It is sufficiently striking that, with absolute measurements, the correlation of the burden-fixation categories with systematic hierarchical ranks should appear at all, even with a scatter of about 40 per cent of the scale breadth. In considering the individual hierarchical sequences of the groups this scatter will disappear (Section V B3*e*).

Before proving this, however, I shall deal with the limiting cases of the burden-constancy correlation.

d. Living fossils

This term is applied to species with extremely high constancy, if they are still living. Biologists have justifiably given them much attention.[19] They will indicate the maximal values for constancy when the burden is low.

One of the most extreme examples is the crustacean *Triops (Apus) cancriformis*. Even the species-specific characteristics seem to have existed already in the Triassic (1.8×10^8 years ago).[20] Recent Notostraca are not numerous and fossil documentation is limited. Consequently species characteristics are rather coarse and include probably 30 homologues (though in fact there is also talk of a subspecies *T.c. minor*). This puts the species precisely at the upper edge of the band of scatter for the burden-constancy relation (cf. Fig. 44). The same is true for the generic features with about 100 homologues (!) – for recent members, knowing the soft parts, these would be equivalent to family features. The appropriate constancy is 2.2×10^8 years. Some other forms recognized as extreme likewise fit exactly into the uppermost part of the scatter in the diagram if the size of group is taken into account. This is true of the *Limulus* and the *Lingula* groups. All others are far behind these.

Thus in Fig. 44, the order Notostraca is about 4×10^8 years old and the suborder Limulina at most 3×10^8, the superfamily Limulacea 2.4×10^8 and the family Lingulidae 7×10^7.[21] For the order Lingulida the age is 5×10^8 years and for the family Lingulidae about 3×10^8. In the lower categories caution is appropriate: 'The lower stratigraphical range of the family is not precisely known. Many Ordovician species have been loosely referred to *Lingula,* but the internal structure of the valves is unknown and in these circumstances even the family assignment is doubtful.'[22] And moreover '. . . no existing species of *Lingula* goes very far back, contrary to current belief.'[23]

Among fossil species of longest duration the brachiopod *Atrypa reticularis* is foremost.[24] The species characteristics change very little through 6×10^7 years. Even so (Fig. 44) with at least 5 homologues the coordinates lie in the expected range of scatter.

'Living fossils', and the longest lasting of extinct fossils, are still astounding since they exceed the average age and minimum age of their category by one of two orders of magnitude. Nevertheless within the limits of the correlation they do not constitute exceptions.

e. Correlation of burden and constancy with the hierarchy of representation within a phyletic group

The scatter in the relation between burden and constancy, shown when comparing 'broadly scattered examples' (cf. Section V B3*a*, Table A, and Fig. 44), disappears when we concentrate on single paths of evolution and single systems (e.g. the Limulid sequence in Fig. 44). For this eliminates the component of scatter due to random mixture of slowly and quickly evolving lines. Such lines are often distinguished from each other as typogenetic and typostatic phases. When both kinds of evolution occur together in one species, we have mosaic evolution. By concentrating on single systems the difference between quickly and slowly changing groups of features falls away.

Even including the scatter, the correlation shows (Fig. 44) that the most constant of reliable species-specific features is scarcely as constant as the the shortest lasting class feature. However, the principle which *requires* burden and constancy to be correlated is still not evident. It will be obvious as soon as the functional connection of the features is considered.

I shall consider this connection from two aspects. First of all I shall show that the correlation of burden with constancy depends on the pattern of representation within a phyletic group (Section V B3*f*).

Table B shows the differentially diagnostic features of four organ systems for six systematic groups hierarchically arranged within each other, i.e. the sequential hierarchy above the mammals.

This short sequence, leading from phylum to class, ranges in representation from 43 000 to 3700 species and in constancy from 5×10^8 to 2×10^8 years. Even so, within individual functional systems, the functional dependence of lower ranking features on higher ranking features is recognizable. On grounds of space I shall only discuss the supportive system (skeleton).

The differentially diagnostic features of the post-cranial skeleton of the subphylum (Vertebrata) are the vertebral column and the branchial skeleton. These are built upon corresponding features of the phylum (Chordata), i.e. on the notochord and the branchial apparatus. The jaw and gill arches of Gnathostomata and the vertebral centra presuppose the vertebral column and branchial skeleton of Vertebrata. The girdles and skeletal parts of the paired limbs of Gnathostomata presuppose that the paired limbs should somehow have been prepared beforehand.[25]

In Tetrapoda the junction of the pelvic girdle with the vertebral column, the differentiation of this column into regions and the development of an elbow or knee joint between upper and lower limbs are special differentiations of the pre-existing vertebral column, pelvic girdle, and paired limbs of gnathostomes. In Amniota the disappearance in the adult of the remains of the notochord, the differentiation of atlas and axis and the annexation to the sacrum of a second vertebra can only be understood as developments based on broader features of tetrapods. In mammals the characteristic fixation of the number of cervical vertebrae (with a few exceptions, i.e. sea cows and sloths) is a special case which in essentials must have been prepared within the amniotes and then realized by further processes.[26]

Equally these connections continue into the class Mammalia. The coracoid process of the scapula, diagnostic of the subclass Theria, assumes the pre-existence in principle of a coracoid bone. The pro-ad supination of the hand, diagnostic of the Primata, could only have evolved out of a pre-existing position of the two bones of the forearm (ulna and radius).

The evolutionary process thus described as 'building upon', 'presupposing' or 'special differentiation' of the preceding layer, is characteristic of hierarchy which in logical principle is simple, but in concrete detail is bewilderingly complicated. For in any group in a hierarchy the diagnostic features of the next higher group apply; no feature of the group is possible without the sum of features of all higher ranking groups, nor can the feature otherwise function or be understood.

At the same time it is clear that, in ascending the classificatory hierarchy of groups, the features of each next larger group will carry the next larger burden. This is partly because several groups of lower concepts usually depend on each higher concept and partly because a great quantity of still narrower groups will have been built on it. This is confirmed in the other examples of Table B.

We thus see how the burden and constancy of the same single homologue increase with phylogenetic development. Among the 41,700 vertebrates, we know of none where the heart lacks an auricle and a ventricle. Among the 43,000 chordates, however, there are more than a thousand without this differentiation; indeed, there are even some that have no heart (as with *Kowalevskaia* among the appendicularians). In fact the burden of the heart in appendicularians only amounts to a few homologues, whereas in the mammals it has a burden of thousands of homologues. Again with the notochord, its loss would be lethal for all species of vertebrates while there are about a thousand chordates among the Ascidiacea and Thaliacea where the notochord is lost after the larval period or even absent throughout the life history. The 'path into fixation' will be discussed in detail later (Section V B4).

The burden of a feature within a functional system must therefore increase with hierarchical position in the actual classification to the extent that features of subordinate hierarchical position are built upon it.

f. *Correlation of burden and constancy with the hierarchy of position*

The functional connection of features within a functional system is easy to see. It is only necessary to select features which form one of the many functional chains. A single example will show this and illustrate the correlation of burden and constancy with position in the physiological system. At the same time it will throw light on the connection of burden and constancy with the representation of the features within a systematic group.

1. As an example I choose the arterial system within the vertebrates and in particular the chain of vessels which stretches from the human thumb (one of the many peripheral features) to the centre. It is summarized in Fig. 37a-e, Section V B1e and in Table C.

Table B. Dependent series of differentially diagnostic features. The functional dependence of differentially diagnostic features, as illustrated by four functional systems in six sequentially hierarchical levels down to the mammals.

	Supportive system	Vascular system	Heart	Excretory system	
Chordata (phylum) are bilaterians with ...	dorsal notochord situated ventral to the central nervous system	antero-ventral principal vessel and paired gills	medio-ventral contractile portion of the principal vessel	when present myomeric subnotochordal kidneys	43,000 species beginning 5 × 10^8 years ago (Cambrian)
Vertebrata (subphylum) are chordates with ...	Jointed vertebral column and branchial skeleton	six paired, ventro-lateral, primary aortic arches	heart divided into auricle and ventricle	paired Wolffian ducts with postero-ventral openings and proximal tubule portions	41,700 species beginning 4 × 10^8 years ago (Ordovician)
Gnathostomata (superclass) are vertebrates with ...	jaws (visceral arches) and two pairs of limb anlagen			pronephros now only functioning in the larva	41,650 species beginning 3.5 × 10^8 years ago (Silurian)
Tetrapoda (division) are gnathostomes with ...	primarily 5 vertebral regions, 2 girdles, and stylopodium and zeugopodium	differentiation of the pulmonary (6) carotid (3) and aortic (4) arches	pulmonary portion of heart and formation of an interauricular septum	regularly developed glomeruli	21,100 species beginning 2.8 × 10^8 years ago (Devonian)
Amniota (group of classes) are tetrapods with ...	the centrum of the atlas vertebra mostly fused with the axis, with atrophied remnants of the notochord	reduction of the ductus Botalli and disappearance of the 5th aortic arch	formation of the intraventricular septum and reduction of the truncus arteriosus	complete loss of the nephrostomes	18,600 species beginning 2.4 × 10^8 years ago (Lower Carboniferous = Mississippian)
Mammalia (class) are amniotes with ...	primarily a fixation of seven cervical vertebrae	conservation of only the left aortic arch and loss of the renal portal vein	complete separation of the chambers of the heart (as also with the birds)	formation of the renal pelvis; complete development of Henle's loop, renal medulla, and cortex	3,700 species beginning 2 × 10^8 years ago (Trias)

Note that in each functional system the features of lower systematic value presuppose the features above them in the column. In the right column note the decrease in age and number of species. Groups after Romer (1959); age of groups after Müller (1963); numbers of representatives after Mayr (1969); age of groups after Müller (1963)

Table C. The connection of burden and fixation in a chain of subsystems of unequal age. Illustrated by the vascular system of man compared with the other species of vertebrates

Single examples of four subsystems in a functional chain

		Principle thumb artery subsystem	Left aorta — subclavian artery subsystem	Free aortic arch subsystem	Aorta — ventricle — auricle subsystem
B Burden:	Functional position in the arterial system	terminal	intermediate	near the centre	central
	Complexity (in homologues) = **c**	5	10	12	20
	Integration (in dependent homologues) = **i**	5	80	400	1,000
F Fixation:	Representation	Order: 200 species	Class: 3,700 species	Division: 21,000 species	Subphylum: 41,700 species
	Maximum of relative representation = **r**	0.47 %	8.5 %	48 %	100 %
	Minimum of constant homologues = **h**	25 %	30 %	60 %	70%
	Minimum of identical proportions = **p**	10 %	20 %	30 %	80 %
	Constancy (in °/oo) (= $\mathbf{r \cdot h \cdot p}/10^3$) = **s**	0.118 ‰	5.1 ‰	86 ‰	560 ‰
	Age (in years of conservation) = **a**	2.5×10^7	6.5×10^7	1.9×10^8	3×10^8

Note the correlation of integration **i** and and constancy **s**. **r** is defined in vertebrate species that have the relevant subsystem. **h** is a quantitative estimate of the number of constantly represented homologues within **r**. **p** is the extent of conserved geomtrical proportions within **h**. (Sources as in Tables A and B.)

149

Table D. The connection of burden and fixation in a chain of subsystems of the same age. As shown by the skeleton of the posterior limbs of man compared with the other species of tetrapods

The four subsystems of the functional chain

		Phalanges subsystem (toes)	Metatarsal—tarsal subsystem (middle foot)	Stylopodium—zeugopodium subsystem (upper and lower limb excluding foot)	Pelvis subsystem
B Burden:	Functional position	terminal	subterminal	intermediate	central
	Complexity (in homologues) = c	39	75	60	32
	Integration (in dependent homologues) = i	39	114	174	206
F Fixation:	Representation = r	93%	93%	94.5%	95%
	Minimum of constant homologues = h	5%	15%	40%	50%
	Minimum of similar proportions = p	2%	5%	15%	20%
	Constancy = s ($r \cdot h \cdot p / 10^3$)	0.93‰	7‰	56.7‰	95‰

Note here, as in Table C, the correlation of integration i and constancy s. The age of these subsystems is about equal and corresponds to that of the tetrapods at 3×10^8 years. The representation is based on comparing the 21,100 Recent tetrapods. (Further explanation in Table C, sources in Tables A and B, p, 142 and 148.)

In the four selected features of the chain (Table C) the burden increases towards the centre by more than two orders of magnitude (from 5 to about 1000) and the constancy by several orders of magnitude. The age increases more than tenfold (from 2.5×10^7 to 3×10^8 years). The representation increases by more than two orders of magnitude (from 200 to 43 000 species). And the conservation of resemblance within the vertebrates, based on an index involving representation, conservation of homologues and proportion, increases by more than three orders of magnitude (from 0.12 to 560 per mille). The functional connection between the position in the functional system and constancy explains why none of the four values for the selected features shows an inverse correlation in the sequences for representation, for minimum of constant homologues or for minimum of similar proportions. Only such a correlation would contradict the proposition.

The necessity of this connection between position within a functional system and constancy is easy to see by looking at extreme marginal positions. Outwards from the principal thumb artery two or three analytical steps are still within the homologues of the arterial system. After that we reach the lower limit for minimal homologues and meet vessels which can no longer be compared as corresponding individualities. Indeed they do not correspond even between the right and left thumb of the same person.

Consequently, as already shown, it does not matter that the quantitative data for degrees of similarity are only estimates, for the truly incomprehensible degree of complication and variation near the homologue-homonom limit would require deeper and deeper comparative study.[27]

2. We should expect the same principle to apply for features of the same age. To document that this is true consider the bones of the pelvis and the legs within the tetrapods. They all existed in the labyrinthodonts (2.8×10^8 years) and even in the crossopterygians (3.2×10^8 years) though in these they can no longer be homologized with certainty. A summary is given in Table D (see also Fig. 48a-d and 49a-e, Section V, B4*b*).

The subsystems in this example are of equal age and almost equal representation. Nevertheless along the chain of features there is an increase of almost an order of magnitude in the burden while the index of conserved similarity (s) increases by almost two orders of magnitude.

The necessity of the connection between position within a functional system and constancy is again confirmed by looking at extreme peripheral positions. Think for example of the burden and constancy of the caudal vertebrae of mammals; the number varies within wide limits and has repeatedly been reduced (cf. Figs. 26a and 73a, Section IV B3*b* and VII B4*a* respectively). The representation of a last caudal 'vertebra' is in every case minimal while that of the first caudal vertebra is 100 per cent. For the total number need not agree even between siblings, as is shown by the variation in ossification, individuality, and fusion of the last caudal vertebra (no. 5) in man. The functional connection between burden and constancy may not seem obvious at first. However, the principle is simple, or even compelling. Indeed it is almost a triviality that each inner member of a functional chain will carry all the outer ones.

This presentation of the connection between burden and constancy for single homologues has been rather extended. This was necessary to prove with certainty that the ruling principle exists, before considering the phenomenon from a dynamic viewpoint.

4. The path towards burden and fixation

There is no reason to think that fixated features of high burden and free features of low burden first appear as such in phylogeny. On the contrary everything suggests the opposite, especially in the fields of fossil history, species formation, and genetics. We should expect that the building-up of burden and constancy, as well as the conservation of unburdened freedom, would both result from phylogeny.

If this is true then it is required that the initial state of a feature and the path towards specialization would agree with the observed patterns, and that some principle can be found to explain the process.

a. The zero instant for a feature

This can be defined generally as the moment when the feature first appears as a permanency in the phylogeny of a phyletic group. Evidently it will then be at most a species feature or more probably a feature characteristic only of a single population. Species features, as shown already, have the lowest burden and lowest constancy in our scale (Section V B2a). We should remember also that successful changes in species characteristics are very small (Mayr, 1967) and in the first place often are purely physiological. To establish a feature as being even the smallest homologue, however, it must be sufficiently fixated to be recognized in a certain number of descendant species. Consequently at its zero instant a feature must be orders of magnitude less complex then when it first can be recognized as a minimal homologue. It is therefore certain that most features at the zero instant must have had the lowest conceivable burden and constancy.

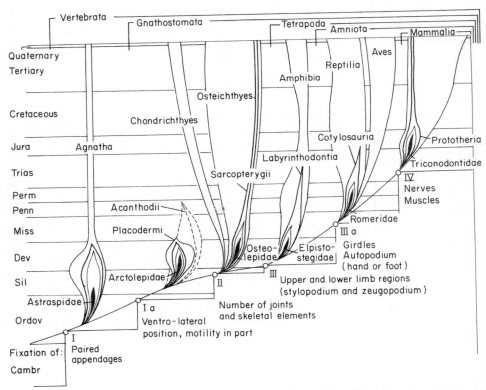

Fig. 45. Taxonomic survey of the precursors of the mammalian limbs according to time and estimated morphological distance. Along the supposed ancestral line I have shown the sequence of consecutive groups, with the families in black and three higher hierarchical groups up to the class. Above the diagram the systematic categories of the sequential hierarchy are shown. Beneath it are inserted the features becoming fixated in the fixation stages I-IV. Compare Table E and Figs. 46-49. (After several authors; original.)

Goldschmidt's hypothesis (1940, 1952) should be briefly mentioned. He held that large mutations (similar to the bithorax mutant of *Drosophila* cf. Fig. 55b, Section VI B2*b*) might play a role in phylogeny. Such would be, from the very beginning, heavily burdened new features. Nearly all large mutations, however, are lethal (Hadorn, 1961). The probability is almost zero that they could ever have been successful (Mayr, 1970). I shall corroborate this later (Section V C and Chapter VI).

But if features at the zero instant are extremely low in constancy and burden, then every case of higher burden and constancy that we establish must result from later processes. I shall now show that this is true.

b. An example of a fixation path

Thousands of fixation paths can be deduced from the data of systematics, palaeontology, and functional anatomy. Unfortunately, all of them are difficult to explain because of the specialized vocabulary required. For this reason, and also on grounds of space, I shall describe only a few examples from a single chain of fixation.

The processes involved in fixating the paired anterior appendages of vertebrates, from the origin of the appendages up to the mammals, can be presented as five stages. Figure 45 shows the systematic groups concerned and their stratigraphical ages and Table E presents the logical connections.[28]

1. *Agnatha* (Fig. 46a-h). The oldest vertebrates known up to the present are Heterostracans from the Middle Ordovician, *Astraspis* being the oldest known form. In

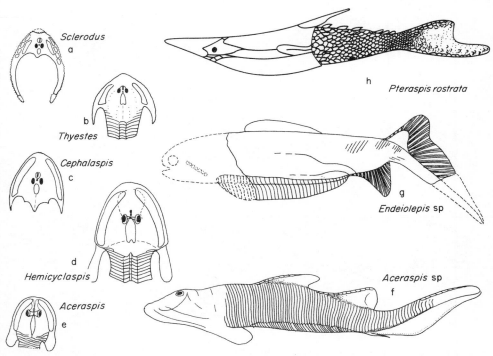

Fig. 46 a-h. Representatives of anterior paired appendages in early Agnatha. (a-f) Superorder Osteostraci, Order Oligobranchiata (Upper Silurian to Lower Devonian); (g) Superorder Anaspida (Lower Silurian to Upper Devonian); (h) Superorder Heterostraci, Order Pteraspida (Upper Silurian to Middle Devonian) as an example of an agnathan with no trace of paired appendages. Note the variability of paired processes at or behind the posterior margin of the head. After Gregory (1951) and Müller (1966).

these earliest forms paired fins are unknown cf. for example, *Pteraspis* Fig. 46h. In the Anaspida, which range in age from Lower Silurian to Devonian, there are ventrolateral rows of scales which finally, after 60 million years, come to resemble fin folds, as in *Rhyncholepis* and *Endeiolepis* (Fig. 46g). Finally, true 'fin lobes' occur in the Osteostraci from the Upper Silurian and Lower Devonian — both in the Orthobranchiates (like *Hirella*) and a few Oligobranchiates such as *Aceraspis* (Fig. 46f). These lobes are plated projections from the head; they had no known internal skeleton, acted as stabilization surfaces and it is doubtful whether they were actively mobile. The representation of these lobes within the group is small and many of these Agnatha carried horns instead of fins, or thorns, or showed no trace of paired appendages.

The burden was therefore small, the prospect of substitution was still high and the position within the functional system was still peripheral. In 10^8 years of experiment the

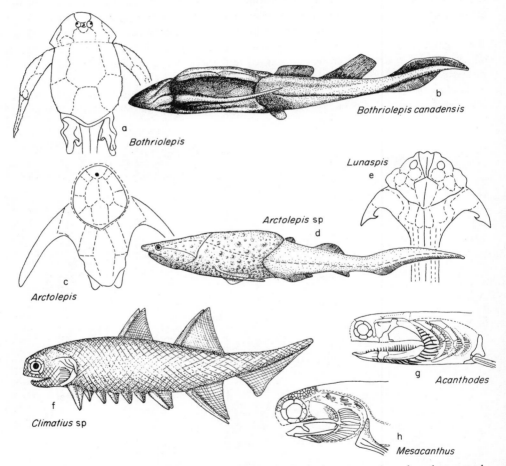

Fig. 47 a-h. Representatives of primitive fossil gnathostomes, i.e. placoderms and acanthodians. (a-e) Placoderms. (a-b) Superorder Antiarchi (Middle to Upper Devonian). (c-e) Superorder Arthrodiri, Order Coccosteiformes (Upper Silurian to Upper Devonian). Note the partial articulation of the spines to the body armour and the occurrence of belly fins. (f-h) Acanthodians (Upper Silurian to lower Devonian); in more recent reconstructions the membranes behind the spines are not believed to exist. Note the ventrolateral rows of movable spiny fins. After Gregory (1951), Müller (1966), Romer (1966).

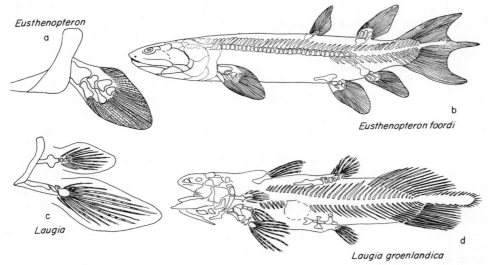

Fig. 48 a-d. Early crossopterygian (Sarcopterygii, lobe-finned fishes). (a-b) Order Osteolepiformes (Middle Devonian to Lower Devonian). (c-d) Order Coelacanthiforms (Middle Devonian to Recent). Note the two pairs of articulated fins which nevertheless have dissimilar bony axes. In *Laugia* the second pair has moved far forward. Literature as in Fig. 47.

fossil Agnatha had not fixated the basic pattern nor even the mere occurrence of paired appendages.

2. *Primitive gnathostomes* (Fig. 47a-h). The oldest gnathostomes appear in the Upper Silurian as the acanthodians, placoderms, and Rhenanida. In them paired anterior appendages are represented throughout. In the few placoderms that carry hollow spines (the arthrodires such as *Arctolepis* is *Lunaspis* (Fig. 47c-e)) it is thought that the structures truly agree with fins, and indicate active mobility in the whole group. This is connected with a further attainment — the occurrence of an internal skeleton and joints, together forming an arthropterygium. The form of the skeleton, however, varies very freely, corresponding to an adaptive experimental stage. In acanthodians (Fig. 47f-h) the fin is formed of a spine. The spine rests on two bones which are analogous to what, after fixation, will be called the scapula and coracoid, or which indeed may be homoiologous with these, e.g. *Mesacanthus, Acanthodes*. The appendages of the placoderms are totally different, for among these a two-jointed dermal armour predominates covering a two-segmented arm, as in *Pterichthys, Bothriolepis*, (Fig. 47a-b). In some placoderms, however, there has been extensive reduction of the head armour, leaving a girdle-like ring for the insertion of a soft pectoral fin supported by three bones. The Rhenanids are different again, for they have big fins like those of a ray or skate, supported by a few or very many elements arranged into a fan mosaic.[29] At the same time the representation of paired pelvic appendages is uncertain; indeed in acanthodians the number of pairs of spines varies from three to seven (Fig. 47f). Paired occurrence thus seems to have been fixated and already always carries the burden of supporting girdle elements, though these in their turn are in the phase of adaptive freedom. Next we must follow the main line leading to the tetrapods, after the sharks and the greater part of the bony fishes had branched off from it (see Fig. 45).

3. *Sarcopterygii* (Fig. 48a-d). At the junction of the Silurian and Devonian the paired pelvic appendages also become fixated and the representation of all four paired fins, in

about 21,000 known species of cartilaginous and bony fishes, is by now complete. In principle the internal girdle was constant in the Sarcopterygii and supporting elements were always present inside the limbs. On the other hand, the arrangement of the supporting elements inside the limbs is extremely variable over 3×10^8 years up to the present day, so that homologization of the elements is to a large extent uncertain. The recognition of homologues with the tetrapods first becomes possible with the sarcopterygians of the Middle Devonian (crossopterygians) and the establishment of a bony axis for the limbs (*Laugia, Eusthenopteron*). The single-jointed fin becomes a many-jointed fin (Fig. 48a and c) and six elements of this bony axis as well as three in the girdle gradually become recognizable individualities which can be homologized (cf. Fig. 49c-e). There is considerable increase in complication as adaptation to greater limb movement, in passing from a hydrodynamic stabilization surface to a 'walking fin'. There is no fossil evidence concerning the musculature, innervation or control by the brain, but comparative anatomy and embryology indicate that these also would have become more complex.

In evidence of this, consider the separation of the dorsal and ventral musculature of the fin, or of the limb bud in a tetrapod limb, and the complexity of the extensors and flexors. Likewise the spinal nerves, which in cartilaginous fishes are disposed regularly, become reticulated in tetrapods to form the complex brachial plexus, the sensory and motor components become sorted out in the spinal roots while the relevant nuclei are differentiated in the rhombencephalon.

A number of supportive elements inside the fin and the girdle elements become fixated and carry the burden of individualized systems of mobile elements as well as of an autopodium which increasingly moves downwards from the body.

Within the appendage the stylopodium (i.e. the upper limb, with the humerus in the arm) is the most fixated portion. The zeugopodium (ie. the lower limb, with the ulna and radius in the arm) is less definite. The autopodium (hand or foot) has the carpal bones very variable while the grouping of phalanges into fingers homologous with those of tetrapods is not yet discernible.

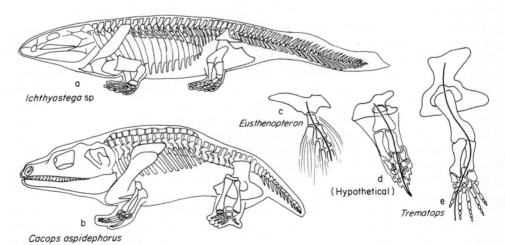

Fig. 49 a-e. Early and primitive amphibians (Labyrinthodonts) and the theory of the evolution of limbs. (a) A representative of the Ichthyostegalia from the Upper Devonian. (b) A temnospondyl (Lower Carboniferous to Upper Trias). (c-e) Phylogenetic transition from the pelvic fin to the posterior pentadactyl limb. (c) Fin of an osteolepiform. (e) Posterior limb of a temnospondyl. The probably homologues bones of the limb axis are connected by a line. After Gregory (1951) and Müller (1966).

4. *Labyrinthodontia.* These are very near to the root stock of the tetrapods (Fig. 49a-e). Even in the oldest order Ichthyostegalia, from the Upper Devonian, the zeugopodium or lower arm is well defined out to and including the articular surface for the carpals. It remains well defined in 88 per·cei.. of all of the 21 000 known tetrapod species. In the Ichthyostegalia the zeugopodium carried a hand or autopodium with five differentiated fingers. Comparison with the most closely related Recent forms among the Anuromorpha suggests that the bones of this hand would be burdened furthermore with the attachments of the forearm and hand muscles, which would now have been differentiated.

The zone of adaptive freedom has now moved out into the autopodium. The degree of freedom can be judged by comparing extreme autopodia like those of plesiosaurs or ichthyosaurs (Fig. 27, Section IV C1b) which came after the amphibians.

Table E. The fixation path of a group of features as illustrated by the mammalian limb. The table illustrates the building up, step by step, of the fixation layers I to IV in passing through fixation stages 0 to 4 of certain limb features, within the systematic groups closest to the ancestors of present-day mammals. In the sequence of these groups the class or subclass names are better known and have been used in the table instead of the family names. Thus the Agnatha correspond to the Astraspidae; the Placoderms to the Arctolepidae; the Sarcopterygians to the Osteolepidae; the Labyrinthodonts to the Elpistostegidae; the Cotylosaurs to the Romeridae; and the Prototheres to the Triconodontidae. Compare also Table A and Figs. 45 and 50, Sections V B4b and B5b respectively

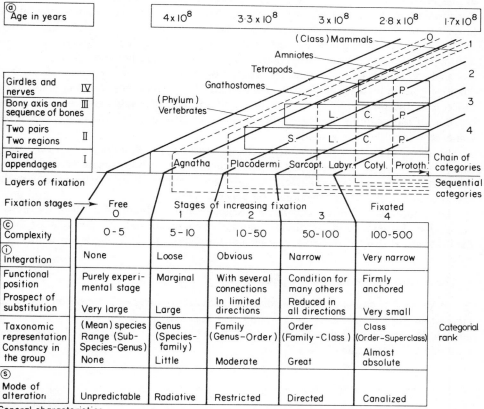

5. *The triconodonts.* After the branching-off of these reptiles, the triconodonts continue the mammalian line as the first primitive mammals. The number of fingers is fixated (apart from reduction) and there is extensive fixation of the carpals. Even in aquatic mammals there is no increase in the number of fingers (Fig. 27, whale). Even the number of spinal nerves that take part in the brachial plexus is limited, which again is connected with the high fixation of the number of vertebrae. Only the distal end of the autopodium remains free, as expressed by the fact that the number of phalanges may increase.

In summary, therefore, the constancy of the features increases step by step in the course of evolution. According to the burden carried, this constancy spreads outwards peripherally from a central position. Deeper lying features leave the zone of free adaptability as they are burdened by new features, in a state of free experiment, which build on them.

c. The principle of fixation

This principle becomes evident by comparing the increase in burden and constancy in different systems. Fossil evidence is not indispensable in this connection since an inspection of the phyletic connections and representation of recent features, together with their functional anatomy and developmental physiology, will provide evidence for an enormous number of different fixation paths. An example is the increase in the burden of the heart in passing from tunicates to vertebrates, or the high burden of the notochord in the process of ontogeny of vertebrates, whereas in more primitive forms it can be dispensed with.[30] Another example is the build-up in the nerve and blood supply of the lungs, together with the associated moving parts.

The principle of fixation is connected with an increase in the 'responsibility' or 'presupposedness' laid on one functional stage as soon as another stage is erected on it. An analogy would be the successive storeys of a factory which is continuously and adaptively being built upon. Shifting the bottom of a staircase becomes more and more difficult as the number of storeys increases to which it is functionally connected. The position of a lift-well in the lower storeys cannot be altered if the lift has to go up to the higher storeys without interruption.

The substitution of a feature, or its change in principle, presupposes that everything that depends on it in the functioning system of the whole organism can also be substituted or changed in principle. Obviously the prospect of this will decrease with the extent of preconditions, connections, and dependences.

These questions of prospect will be discussed later in detail (Section V C). Before doing that, however, I must discuss another phenomenon which, like that of 'living fossils' seems at first to contradict any general correlation of burden with constancy.

5. The rhythm of free and fixated phases

I have tried to show, up till now, that differentiation processes in evolution impose an increase in burden on most features, and that this increase in burden will produce an increase in constancy. If this assertion is true then the phylogenetic paths of all organisms must end in fixation and rigidity.

a. A hyperbolic course for cladogenesis

The individual evolutionary paths, when plotted against the coordinates of 'time' and 'structural change', would be expected to show a hyperbolic course for cladogenesis, because of the correlation of burden and constancy. Such a course is, in fact, generally observed. It is a general rule that the individual branches of the phylogenetic tree begin with a typogenetic phase in which they traverse a great morphological distance in a short time. This is followed by a typostatic phase which leads to a continuous increase in constancy and to fixation for long periods.[31] The individual paths are repeatedly seen at first to follow the morphological coordinate, then to swing up parallel to the time coordinate. However, and this is the crucial point, how can we then understand a new change from a typostatic back to a typogenetic course? How can new freedom follow on old fixation?

There is no doubt of the fact that a typogenetic phase can follow a typostatic one. Indeed this fact is the cause for the second basic component of the phylogenetic pattern of order, as I shall later show. For when typogenesis does not happen, or when it fails, then those peculiarities occur which are called 'living fossils'. Or else 'typolysis' happens with the appearance of extreme forms and signs of dissolution,[32] such as often precede the extinction of a group.

Sometimes, however, it seems that lines which have become fixated can achieve new freedom. If this is so then the correlation 'constancy = burden' would no longer hold in general. In fact, however, even these renewals would be expected, for I shall now show that the apparent contradiction depends on a logical confusion — a confusion between the collective results of evolution and the fate of individual features. Indeed I can show that these new freedoms provide yet further evidence for the correlation of burden with constancy and indeed that the special dynamics of evolution depends on them. They are the only component of this dynamics in which, so far as I can see, pure accident reigns.

b. New freedom by means of new features

New typogenetic phases depend on the adaptability of new features which are relatively unburdened. This insight is the key to the problem.

Systematists know that, within a single phyletic group, the differentially diagnostic characteristics of subgroups of like rank tend to affect the same feature. On the other hand, those of systematic groups of different rank affect other features. Non-systematists are aware that in systematic identification keys and in diagnoses, taxonomic groups of like rank are distinguished by different character-states of one group of features, while subgroups within these groups are largely distinguished by different character-states of another group of features.[33] Table F gives two examples among the countless sequences in the natural classification.

1. *Stratification of features.* The features whose character-states allow closely related groups of the same rank to be defined, are closed entities or qualities, in the sense that they never recur in Nature. The concept of the particular feature may recur in groups of next higher or next lower rank but the relevant character-state will then be on a different level. These levels are referred to by systematists as general or special expressions of the feature (see 2, below). Usually, however, the groups of next lower rank are distinguished by character-states of quite other features.

Thus the classes of vertebrates are distinguished fundamentally by the structure of the respiratory organs, and differentiations of the heart, principal vessels, body covering and embryonic membranes (amnion). The subclasses of the class Mammalia, however, are distinguished by the structure of the girdles (coracoid, marsupial bone), provision for the embryo or the young (pouch, teats, umbilical cord), dentition, and tooth succession. The orders of the subclass Placentalia are distinguished by specialization of the teeth and limbs. The suborders of the order Primates are distinguished by the closure of the incisors and the orbits — and so on down to the familes, genera, and species.

2. *The process of generalization in phylogeny.* 'Specialized' features are young features. The later they appeared in phylogeny, the smaller is their representation and the

Table F. Change in the features adapted in evolution, as shown by the character-states distinguishing groups of equal rank in two sequences of systematic groups

Example 1. Stages in the sequential hierarchy leading to *Maia squinado* (the spiny spider crab)

Example 2. Stages in the sequential hierarchy leading to *Homo sapiens*

Number and rank of the respective equal-ranking sub-groups	The most important differentially diagnostic features	(Example 2 features)	(Example 2 number and rank)
13 phyla of the Protostomia	The basic features of the body cavities segmentation and the structure of the body covering.	Basic forms of the body cavities, symmetries, and skeleton	6 phyla of the Deuterostomia
9 classes of the Arthropoda	Subdivisions of the body; basic features of head structure and of head appendages	Respiratory organs, appendages body covering, heart, kidneys, and vertebral column	8 classes of the Vertebrata.
10 subclasses of Crustacea	Type of body regionation and the distribution of the appendages.	Jaw covering, tooth replacement, food supply to embryo and young	7 subclasses and infraclasses of the Mammalia
13 orders of the Malacostraca	The extents of the regions and the basic forms of the appendages	Teeth, lumbar vertebrae, shoulder girdle, hands and feet and their accesories (hooves, claws etc.)	17 orders of the Eutheria
7 tribes of the Decapoda	Pincer formation on the walking limbs; mouth and mouth parts	Form of the orbits, sternum, incisors and cheek teeth	8 sub- and infra-orders of the Primates
31 families of the Brachyura	Development of the rostrum; armour, abdomen and types of appendage	Proportion of the lower jaw and of the face skeleton and calvarium and of appendages	8 families of the Catarrhina
140 genera of the Maiidae	Eye sockets, dorsal processes, shape of the marginal teeth and of the second antennae	Inclination of the alveoli, of the foramen magnum, chin development, stance	2 genera of the Hominidae
15 species of the genus *Maia*	Special forms of spininess, colour, size and proportion	Proportions of the facial angle supraorbital ridges and canine teeth	2 species of the genus *Homo*

In descending sequence, corresponding to the temporal axis of evolution, different features undergo adaptation and become features of systematic, differentially diagnostic importance. (Sources Claus, Grobben, and Kühn (1932), Balss *et al* (1940-61), Fiedler (1956), Claus (1960), Riedl (1970) and Remer (1950), greater their specialization in an adaptive functional connection. 'General' features behave conversely. However, they did not arise adaptively as general features, but as highly specialized character-states. They became 'general' later. Their fixation path, as already discussed (Section V B4) is also responsible for the generalization of their characteristics. Their passage out of the adaptive phase will be discussed again later (Chapter VII).

For example, the basic types of supportive system (skeletons etc.) referred to in the diagnosis of the deuterostomes (Table F) must, as Recent chordates show, go back to the Lower Cambrian. At that time they would have been extremely modern features in the phase of immediate adaptation and selection. The basic forms of jaw were in the same condition in the root stock of the gnathostomes in the Silurian. The basic forms of dentition and other jaw characteristics of the Mammalia were in this phase among eutherians at the beginning of the Eocene etc.

3. *The sequence of features.* This shows itself to be chronological in passing from groups of higher to lower rank, since the individual features functionally presuppose each other. This holds for features and for their character-states and it is important because, even without fossil evidence, the chain of events is not in doubt (cf. Table B, Section V B3*e*, with Table F).

Thus the basic subdivisions of the class Crustacea into subclasses depends on the type and extent of regionization of metameres. This is a precondition for the division of the subclass Malacostraca (higher crustaceans) into orders according to the mode of closure (carapace, fusion of segments) of the thorax. This is a precondition for the division of the suborder Brachyura (crabs) into superfamilies according to the type of anterior end of the carapace (rostrum etc.). And this is the precondition for the subdivision of the superfamily Oxyrhyncha according to the form of the rostrum. And this is a precondition for the division of the subfamily Inachinae (spider crabs) into genera according to the length of the rostrum etc.

4. *New diversification.* The traversing of new morphological distance, and the origin of new diversity, will depend on adaptive modifications of the youngest features added, these being least loaded with burden. The hyperbola-like evolutionary paths, appropriately shown in many phylogenetic diagrams as swinging horizontally outwards from the ancestral path, reflect the sum of all alterations. It would be totally mistaken to suppose that old features had gained new freedoms. The total morphological distance is constituted by many separate features. But if we sort out only a few of these (cf. Figs. 50 and 45) then it becomes obvious that the more constant a feature remains, the older and more burdened it is, and that the new morphological distance is gained by the youngest and therefore freest features.

Thus in Table F and Fig. 50 the evolutionary path of the notochord at the time of its transformation into the bony vertebral column was already climbing steeply, almost parallel to the time axis, and after the notochord had been subdivided into the nuclei pulposi nothing else in it changed. To take another example, the path of the lung system was likewise settled when the mammals arose in the Jurassic. In principle there have been no further changes in the pulmonary vessels despite the origin of the running, climbing, flying, and marine mammals. Similarly, the basic form of the mammalian dentition, with teeth limited to the edges of the jaws and inserted in alveoli, has remained the same although they have evolved extraordinarily diverse special forms (e.g. Fig. 23a-j, Section IV B1*c*). Even the basic pattern of the autopodium (hand or foot) has rigidified, although the most amazing transformations are used in flying, swimming, and running.

5. *Vigour and niches.* It has long been proven that new periods of vigour and new diversifications coincide with the conquest, by means of these new features, of new ecological niches, i.e. of new possibilities offered by the environment. For example, the numerous fin experiments, from agnathans to crossopterygians, were necessarily connected with locomotion in water. The experimental division of the limb into stylopodium, zeugopodium, and autopodium, from crossopterygians to reptiles, necessarily depended on adaptation to the land. And the differentiation of the autopodium, which began with the amphibians, necessarily coincided with the occupation of niches corresponding to the various locomotory possibilities on land. All these were necessities. Only the way in which the adaptability of all these features coincided with the opening of new adaptive niches was accidental.

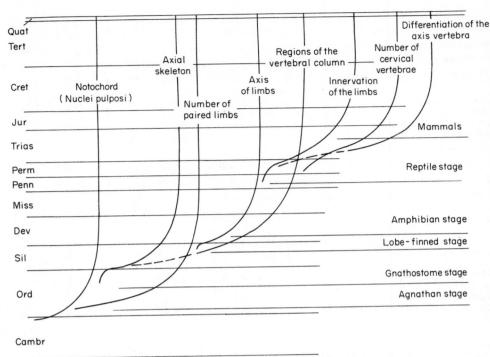

Fig. 50. The building-up of typogenetic features upon typostatic ones, as shown by certain characteristics of the axial skeleton and limbs in mammals and their ancestors. The figure shows time and morphological distance on the same scales as in Fig. 45 so as to facilitate comparison with the sequence of ancestors of the mammals and of the overall morphological distance. To the left the time intervals are shown and to the right the sequence of systematic groups.

c. The hierarchy of the fixation patterns

This hierarchy is the necessary total consequence of all the phenomena so far described. Each new diversification is represented by the systematic categories of next lower rank which systematists use to express the phyletic relationships.

This connection of diversification with systematic grouping is so close that even the extent of newly acquired features per evolutionary stage is expressed in terms of systematic value. Thus limited changes are expressed as intermediate categories (e.g. Gnathostomata, Amniota, Theria, Eutheria among the vertebrates), while more striking transformations, in which changes in several features coincide, characterize the main groups (Vertebrata, Mammalia, Cetacea).

It is impossible not to admire the systematists, especially those of the nineteenth century. For they created a system which, despite gaps in the fossil record, and despite the huge abundance of forms and vagueness in epistemological concepts, has in principle withstood all modern re-examinations. Even in the middle of the nineteenth century this system was sufficiently correct to be the basis of Darwin's discovery.

No less amazing is the extent to which the structure of our comparative thought can reflect such complex natural phenomena. Merely on the basis of conjunction of features, similarity, and representation, the systematists foreshadowed all those connections which Darwin began to explain causally, and which are not yet completely explained.[34] It is even possible to understand the critics who pretend to see the natural classification as a projection by the systematists, consider the systematic categories as artificial, and regard morphology, on which systematics depends, as an art form and not a science. I shall come back to these epistemological questions (Chapter VIII).

6. A summary and an anticipation

Looking back, I have shown that position and constancy within homologous functional systems are indubitably correlated. What is gained by this? This question must be answered before continuing since to understand the hierarchical pattern makes great demands on knowledge (as I said in the Introduction and as the reader will now agree). In answering it I shall only emphasize the consequences necessary for the next step. A complete summary will be given in Chapter VIII.

a. The reality of the hierarchical condition

As already stated (Sections II C3 and 4) some have doubted the reality of hierarchical order. It has been held to be, if anything, a thought-projection. As against this I have been able to show that the general characteristics of homologues can be demonstrated objectively and indeed quantitatively. Moreover, their constancy reaches and exceeds hundreds of millions of years, so their reality is not truly possible to doubt. Rather do species appear to be the very temporary carriers of these much more important realities.

But if homologues are realities, so also are morphotype and groundplan, being made up of homologues. And so also are systematic groups and the hierarchy of natural classification, for these can only be an expression of the same phenomenon.

b. The origin of hierarchy

The hierarchical pattern arises, as already shown, by the fixation of added features after genetic paths have become separate. I have also shown that paths of fixation are connected with the building-up of burden, while this increase in burden depends on local differentiation processes within the organism. But when the die shall fall and what feature it chooses (as in the model in Section V A*b*) depends solely on the coincidence of functional structures with ecological niches. We cannot predict this and can therefore call it 'accident'.

However, as soon as a fixation happens, by the addition of new features, then it decides not merely the possibilities for these features but also its own fate, as a reaction to these new features. What began as accident finishes as necessity. The preconditions for the reign of necessity as opposed to accident are by definition satisfied. The hierarchy of natural classification is a necessary consequence.

We are confronted with a process of self-design, or self conception. The form of the pattern of order is a necessity although the origin of its contents is an accident. Necessity here implies determinable regularity, a superdeterminacy, a second profound limitation of the freedom of evolution.

c. The causes

I have already explained the cause of fixation, homology, and hierarchy in the molecular realm (Section III C1*c-d*) but have not yet discussed it for the morphological realm. In this, systemic conditions and functional considerations must likewise cause each other mutually. Two things, however, are already probable.

1. Fixation must be the result of increasing burden, rather than the converse. But how can the systemic position of a feature change by fixation? Increase in burden must have a further cause. For why should the amount of burden of itself change so consistently?

2. Increase in burden must be connected with differentiation. The degree of hierarchical burden depends on complexity, while the latter depends on the process of differentiation.

If fixation results from burden, and burden results from differentiation which is a production of evolution, then the process of fixation must be a consequence of evolutionary mechanisms. Will it be a consequence of mutability or of selectivity? What can produce superdeterminacy of such extraordinary dimensions? Obviously determinative decisions in the molecular code could by mutation become loaded with burden, as already shown. In the last analysis, however, it must be selection which finally decides.

In the morphological realm the cause must be a superselection as with standard-part order (Section IV C). It must again be a unverisal principle that we search for.

C.   THE SELECTION OF RANKS

By now the reader, if he has patiently followed me this far, will be able to foresee the mechanism which enforces the hierarchical pattern of order in the morphological realm. It must be a question of laws of selection, of a form of superdeterminacy by superselection. Indeed we must be dealing with a mechanism just as universal and rigorous as standard-part selection (Section IV C). It will depend in the same way on the systemic conditions in the 'internal environment' of the organism. But in the present case it will be the systemic conditions of the hierarchical pattern actuating a mechanism of hierarchical superselection. We might suspect that the principle of 'profits today for losses tomorrow' will now reappear in the hierarchical pattern. For it is obvious that the huge hierarchical burden carried by some features (Section V B3) implies a constancy totally beyond the average.

The selective advantages and disadvantages, i.e. the prospects of accomplishment and success, will again depend on the relationship between the determinative requirements of the events of features on the one hand, and, on the other hand, the probability of accidental origin for the decisions that carry these events.

If, as with standard-part selection (IV C), I started by assuming that decisions were independent, then it would be easy to work out this relationship. In actual fact, however, the decisions in the genome are not independent. On the contrary, they are highly dependent on each other, being that network of dependences called the epigenetic system.

Before considering selection further I shall therefore estimate numerically not merely the relationship between events and decisions, but also the degree of dependence of these events and decisions.

1. Organization of events and decisions

A comparison will specify the relationship between degree of organization of events and of decisions. This comparison is between the degree of organization of a phene system (i.e. the number of purposively intercoordinated homologues in it) and the degree of organization which the associated determination complex, as defined in the next paragraph, is purposively able to alter. It is a comparison of degrees of complexity. The degree of organization of phene systems has already been dealt with (cf. Section V B1c).

That of the determinative decisions will be investigated more fully in Chapter VI. Here I shall limit myself to a definition.

A determination complex can be described as a number of homologues which can be changed by altering a single determinant unit in a genetic code — for instance a single mutation of a cistron. This is a question of the effects on the events or features which result from changing a single overranking decision. The relevant examples are supplied by genetics and developmental physiology.

a. The degree of organization of determination complexes

The degree of organization of such a complex of determinative decisions can be specified by stating the limits of the effects of the complex and measuring these effects on a suitable scale.

1. *The limits* of purposively organized change can be recognized by comparison with the organizational characters of the phyletic group in question. (Even the most purposive ophiuroid arm would be purposeless as a replacement for a beetle's leg.)

In the three-toed mutant of a horse, as an example of spontaneous atavism, we can recognize the additional toes of the horse's ancestors (*Miohippus, Protohippus*) as well as those of recent relatives such as the rhinoceros or the hind leg of a tapir (odd-toed ungulates or perissodactyls). The homology theorem shows that we certainly have here the inner lateral toes of mammals and the extent of the change can be measured by comparison with *Equus*.

Similarly the effect of the bithorax mutation of *Drosophila* (cf. Fig. 55a-c, Section VI B2*b*) is a doubling of many features of the thoracic region of the species. In other cases there is repetition of a feature in an unexpected position. Thus the tetraptera mutant has wings instead of halteres, tetraltera has halteres instead of wings, proboscipedia has legs instead of mouth palps and aristopedia has legs instead of antennae (cf. Fig. 56a-n, Section VI B2*b*).

2. *The scale* for measuring alteration is again that of homologues, as proposed in Section II B2 and applied in general features of phene systems in Section V B1. Within the limits just defined the complexity (**c**) of the organic change is given by the sum of minimal and cadre homologues which are changed.

As already mentioned (Section V B2*a*) changes in the phenotype that are crucial for evolution are mostly too small to be measured on the homologue scale. But here I am attempting to measure the greatest possible changes so as to recognize the degree of organization of genetic determination complexes. For changes of these dimensions we can ignore the similarity problem (as in Section V B1*f*).

3. *The number* of homologues comprised in a determination complex lies, for most of the known examples, between 40 and 100 **c**, where **c** is the complexity in homologues. For the bithorax mutant of *Drosophila* or the three-toed mutant of *Equus* it is probably between 100 and 500 **c**.

This shows that accident can be totally excluded as a cause of these changes, since about 100 single features coincide meaningfully. The complexity of the organized determinative decisions involved in such changes is astounding. But it does not approach the complexity needed for faultless functioning of such great phene systems. There is a gap. That is the decisive point.

b. The gap in organization

This organizational gap between determination complexes, on the one hand, and the respective functioning phene systems, on the other, can now be estimated. Knowing the limits of a determination complex we can quantify the *complexity of an organized change* due to the mutation of a single gene (this c can be called **cg**). We can compare this with the complexity needed for the functioning of the respective system of phenes. (**cp**).

The degree of organization of genetic changes such as tetraptera, aristopedia or proboscipedia is 50 to 100 **cg**. But the complexity of the respective phene systems is about 200 to 500 **cp**. For the musculature of the mutated systems is grossly deficient, the innervation is lacking or confused, not to speak of those homologous paths which must be assumed for the coordinated switching of the activity of the system. What then are the **cp** values for still higher values of **cg** as in the three-toed horse mutant or bithorax? With these there is the same difference; **cp** is twice as great as **cg**, or even exceeds it by an order of magnitude. Once again there are gaps and mistakes in the internal organization of the mutated part (as already pointed out by Waddington (1957) and cf. Fig. 55c, Section VI B2*b*).

Wherever there is evidence, we find that **cp** > **cg**. This is particularly obvious on remembering (Section V B2*b*) that the complexity of the organization of the larger phene systems easily reaches 10^4 to 10^5 **cp**, although genetic information complexes attain in this connection at most 10^3 **cg** to 5×10^3 **cg** and this is only for these coarse mutations, all of which are unsuccessful.

c. The prospects of success

It has long been a fundamental of the synthetic theory of evolution that the most extensive mutational changes have the least prospect of success, while the least extensive have the greatest prospect.[35] Simpson (1955, p. 93) states: 'As an over-all tendency, subject to exceptions and irregularities in some particular cases, the importance of mutations in all sorts of evolutionary processes tends to be inversely proportionate to their size.'

1. *The complexity of successful mutational changes.* As already mentioned (Section V B1-3) quantitative study shows that hereditary changes with the greatest prospect of success are of very low complexity. The changes of species features (V B2*a*, V B4*a*) may even be too small to be measured on the homologue scale, i.e. **cg** < 1.

Genetic determinative decisions are, no doubt, highly organized, but the complexity of organization of individual complexes is always less than that of phene systems (**cg** < **cp**). Moreover, those organized information complexes with a prospect of successful change must be considerably smaller than the average phene system that is being adapted. All this agrees with evolutionary theory as now accepted and we are now close to explaining the cause of the hierarchical pattern.

2. *The difference in complexity.* Biologists have, of course, already noticed the arithmetical difference between the complexity of organization of mutational changes, especially successful ones, and the complexity of the corresponding phene systems, i.e. **cp-cg**. But the question of understanding large phylogenetic changes, when faced with this discrepancy, has become a point of contention, separating schools and trends from each other.

As is well known, the synthetic theory of Neodarwinism came to the opinion that only tiny accidental changes decide the course of evolution. The critics of Neodarwinism, on the other hand, consider that pure accident would explain neither the ground plans of the great systematic groups, nor evolutionary trends (literature in Section II C2, 3). I shall now show that both viewpoints can be confirmed at one and the same time.

3. *The problem.* Translated into our terminology the problem can be stated thus: Can the gap in complexity be bridged by the known mechanisms of intraspecific or micro-evolution, and, if so, how? We are dealing here with the modifiability of the more

extensive phenes, of those features characterized as complex, highly integrated, of central position, and of high burden, close similarity and great age (Section V B1-2). The adaptive modifiability of simple functional systems is explained by the synthetic theory in an unexceptable manner. But how can large, functionally indivisible complexes of features be altered, when changes that promise success always affect only part of such complexes? Formulated in this fashion the question is near to being answered.

d. Three hypothetical solutions

In principle there are three conceivable hypotheses and all three have been developed in the literature. According to the mechanism suggested, I shall call them the hypotheses of simultaneity, accidental coincidence, and storage.

1. *The simultaneity hypothesis.* This begins from the unequal sizes of mutational changes which range from being unnoticeable to the doubling of a whole body region. Goldschmidt (1940) argued that large mutations might once have been successful, especially considering the enormous time available. Perhaps a fundamentally new systematic type could arise, like the bithorax mutant, at a single stroke. This would be, so to speak, a 'hopeful monster'.

All subsequent work argues against this possibility and Goldschmidt's theory has everywhere been rejected. Indeed, as just shown, the organizational gas between cp and cg increases with the extent of the change. Lack of coordination increases by several orders of magnitude, and every single error in coordination can be lethal. Large mutations will always produce 'hopeless monstrosities' however much time there may be. Reciprocal gene effects certainly exist but their prospects of success diminish as the size of the mutation increases.

2. *The coincidence hypothesis.* This emphasizes the possibility that accident would permit the coincidental occurrence of two or several mutations. The long periods of time available would supposedly permit the accidental conjunction of mutually complementary mutations so as to close the organizational gap.

Simpson (1955, p.96) worked out, however, that such a coincidence, even under the most favourable conditions, would be so extraordinarily rare as to be negligible. Assuming a mutation rate of 10^{-4} and assuming that one mutation would double the chance of another occurring in the same nucleus, then the prospect of five mutations coinciding is 10^{-22}. Given a population of 10^8 individuals and a generation length of only one day, such an event would occur only once in every 274×10^9 years. This is about one hundred times as long as the period during which life has existed on earth. Coincidence, therefore, can only be effective to a very slight degree.

3. *The storage hypothesis.* This is at present the current opinion. It begins from the fact that mutational changes, even if not beneficial, could be stored in the genome so long as they were not disadvantageous. As soon as all the changes necessary for the improvement of a complex functional system have been stored they will all act together and produce the improvement by collaboration. This assumes that all these changes are so slight that they do not disturb the coordination of the system in question.

The difficulty is that a mutated allele will need to increase in frequency so as to increase its chance of occurring together with the next necessary mutant, but it can only do this if it has some advantage. However, as the functional system in which a gene acts becomes more complicated, the portion of it influenced by the gene decreases. And consequently the advantage attainable by a single mutation will also decrease. Here again, therefore, difficulties must increase with the number of gene loci that are needed for a successful change in the system.

The importance of storage of adaptively advantageous alleles in the genomes must not be ignored. But neither can we ignore the fact that successful changes will be limited to very small, harmonious modifications, even though they affect extensive systems. In no case can they overstep functional burdens.[36]

e. The consequence of the organizational gap

This consequence is now evident. The organizational gap increases with the number of gene loci which need to be altered simultaneously. The prospects of success for a change in a feature must, therefore, decrease with the number of phenes dependent on that feature.

Even a slight difference between **cp** and **cg** will make a system enormously more difficult to alter. A middle-sized difference will make alteration virtually impossible. As shown by equation 24 (Section III C2a) the prospect of successful change P_e decreases as the power of the number of individual events that need changing (E); i.e. P_e^E. The prospects of success therefore decrease as $P_e^{(cp-cg)}$, i.e. exponentially with the size of the organizational gap. *For single features the organizational gap increases with the burden, so burden will be inversely proportional to the prospect of successful change.* The probability of decrease in success as a result of change ($A_{e(neg)}$) will be:

$$A_{e(neg)} = P_e{}^{(cp-cg)} \tag{32}$$

This is the theoretical part of the link between the hierarchical pattern of order and the organizational gap in genetic determination. The rest of the link consists of facts and their necessary consequences. Even the theoretical part of the link can be experimentally verified, as will already be obvious to the biological reader.

In actual fact the existence of a link between the organizational gap and hierarchy has already been anticipated by the synthetic theory . Thus Mayr (1967) has maintained that the greater the number of subfeatures involved in determining the phenotype of a major feature, the less probable is its alteration by selection. Kosswig (1959) stated that the nexus of a gene with its background causes a narrowing of evolutionary possibilities. And Mayr (1970, p.367) says that the greater the number of genes that contribute to a feature, the less easily can it be modified by natural selection.

I shall add only three things to this recognized concept:
(1) The condition which causes a decrease in alterability can be specified quantitatively; (2) this condition can be seen as a necessary product of evolution; (3) the reduced alterability has specifiable effects on the nature of organic order. I shall take these three points in turn.

1. *The condition* tending to make dependent phenes hard to alter can be specified quantitatively as the burden (V B1-2). This ranges from < 1 to 1 000 homologues and thus over more than three orders of magnitude. Constancy is a mathematical function of burden, for it increases approximately with the square of the burden (cf. Fig. 44). Burdens of 10^2 to 10^3 homologues involve constancies of 2×10^7 to 10^8 years (cf. Fig. 44 Section V). They thus seem to be fixated even on a geological time scale.

With burdens of such dimensions it becomes difficult to conceive how any alterations can be possible at all. However, even features of the highest burden, such as the truncus arteriosus, the medulla oblongata of the brain and the atlas vertebra, do show modifications, though these are almost unnoticeable and only within fixed limits. I shall later analyse in detail the factors that thus oppose fixation (Section V C2c, VI C1c, VII C1c).

2. *The process* that leads to high burden has already been considered (Section V B4). It is connected with the increase in the differentiation and integration of features (V B6c). It seems highly probable that prospects of success for a mutational change are causally connected with the size of the organizational gap. The causal sequence involved in the connection can also be postulated, therefore, with the same high probability.

It seems to be as follows: evolution leads to an increase in differentiation and integration and thus to local increases in burden within the organism. These result in an increase in the organizational gap, which decreases the prospect of successful adaptive change. This decrease can be described as the phenomenon of superdeterminacy or constancy, or as the phenomenon of fixated features.

This then is the causal connection of the process. It answers to all the demands that must be made on a theory – it corresponds to all the data and operates with convincingly demonstrated causal connections.

3. *The result* is that the determinative decisions which evolve and continually change in phylogenetic lines become fixated in ranks one above another. The primary result is the hierarchical pattern of order. The fixated complexes of features, however, enforce this pattern in a still more potent manner which I shall now consider in more detail.

2. The advantages of ranked decisions

As already pointed out, given the high complexity and burden reached by many events (or systems of features), it is not surprising that they are fixated. What is more surprising is that they can change. In other words, some features burdened with hundreds of homologues and with the single decisions required for these homologues, show continual modification in the course of phylogeny. How can this be explained when the accidental conjunction of so few as five relevant single decisions is virtually impossible (V C1*d*)? This paradox brings to mind the possibility of hidden redundancy and its cause.

a. *The origin of redundancy*

We can now take a significant step forward and describe the origin of decision redundancy in the genome using the example of the hierarchical pattern. The mode of origin was not totally obvious in discussing the standard-part pattern of order.

Taking the simplest conceivable case, such as the transmission of events I II III and IV, (cf. Sections I B2*d* and B4*b*) there will be eight decisions required and predecisions 2 and 4 will be redundant. As concerns biological events (or features), however, we must expect adaptive changes resulting from accident. This means, in the first place, that redundancy disappears if we suppose that each permutation of the four events will have the same prospects of success. But, in the second place, if it turns out that I and II are functionally dependent so that they could gain new adaptive prospects only by adopting the form III and IV, then predecision 2 would be redundant in a dynamic sense also. *Thus a decision becomes redundant if selection will not tolerate the change in the event made possible by changing the decision.* We have already shown that such selective conditions accumulate with the advance of differentiation and burden.

b. *The possibility of dismantling hidden redundancy*

In principle a redundant decision can be dismantled if the deciphering mechanism can 'remember' the command issued by the relevant overranking preliminary decision until this command is countermanded or cancelled (cf. Section III C1*c*). This mechanism, i.e. 'apply until further notice', is realized in the structure of the genome by the so-called operon system. Enormous selective advantages can thus be gained by dismantling the redundant decisions of a hierarchical pattern.

We gained some measure of these selective advantages (A) by showing that every decision dismantled (in $bits_R$) affects the negative power of the probability of accomplishment or realization (P_m), i.e. $A_a = P_m^{-R}$. For the special case of the hierarchical redundancy pattern, I have described these advantages as an exponential function of the events involved (E) (equations 22 and 23, Section III C1c). This means that, in a phene system having only two hierarchically dependent events (i.e. a redundancy content $R = 1$), there can be a ten thousandfold selective advantage which will increase exponentially with every increase in the complexity of the system.

I shall now show that the features (events) of functional phene systems take up a hierarchical pattern of dependence as evolution advances. This will provide yet another causal connection. As already discussed we found the formation of hierarchical patterns of function (Section V B) to be a universal result of differentiation.

c. *The necessity of ranking*

There must necessarily be a ranking of determinative decisions in the genome. For whenever the process of differentiation can be followed, several adaptive single features are built on the presupposition of a preceding feature, the functional chains branch outwards and every cadre homologue is burdened with several sub-homologues.

The hierarchy of successive layers of adaptation can be illustrated by the evolution of the mammalian limb (Table E, Section V B4b). The hierarchy of outward branching functional chains can be illustrated by an arterial system (Fig. 37a-e, Section V B1e). And the hierarchy of homologues can be illustrated by the structure of the human vertebral column (Fig. 11, Section II B2b).

Moreover, the functional dependences of events or phenes exactly correspond to the hierarchical pattern of the operon systems of gene decisions and to that of hidden redundancy. We should therefore expect that evolution would have organized the epigenetic system in large measure hierarchically, so as to make use of the selective advantages of dismantling redundant decisions.

Indeed, even in systems of very low complexity the advantages are extraordinarily large and with moderate complexity they are enormous. Saving a single decision gives a ten thousandfold advantage. Saving two gives a hundred millionfold advantage. The epigenetic system must be hierarchically differentiated, and to a high degree. I propose this here on stochastic grounds. The proof will be given in the next chapter (VI B).

d. *The imitative hierarchy of decisions*

Where are we then? In this section (V C2) I intend to show the selective advantages, and the necessity, of establishing the hierarchical principle in the realm of decisions. This realm of decisions, however, is that of the genome. The proof that decisions necessarily establish systemic conditions implies a prediction about the structure of the epigenetic system. This is important.

For the epigenetic system — which is the structure of reciprocal gene effects — is the system which poses the biggest problem in biology. Also it includes one-half of the mechanism which enforces order in living organisms. It will be the starting point for any experimental proof of the theory of systemic conditions that is proposed in this book.

We can state at the beginning that the epigenetic system must be hierarchically organized and in high degree. At the same time we can go considerably further by asking why the exact hierarchical pattern should be expected. The answer is twofold. First, because such a pattern is possible from the molecular genetic viewpoint. And second, because the pattern best corresponds to the hierarchical functional pattern built up by advancing differentiation. 'To correspond best' means here 'to have corresponded to the greatest selective advantages'.

Given the ability to arrange decisions in ranks, the true cause of the hierarchical pattern consists in the advantage of being able to copy the pattern of dependence of phene-functions. The more complete the correspondence, the more complete will be the possible selective advantage. This connection is even more important because it has even more general consequences. For it means that selection, by offering unequal adaptational advantages, will always require not merely a copying of the pattern in principle, but also of its special qualities. The epigenetic system will strive to simulate the particular functional pattern of dependences of the phene system which it determines.

Consider, to use our old example, the system of the vertebral column (Fig. 11, Section II B2b). The command 'increase in length' must offer a very great advantage seeing that it precedes and outranks all other commands. A single mutation, of the type we shall meet later, can thus produce a comprehensive modification without disturbing the harmony of the parts. On the other hand, it would be totally impossible that all the hundreds of single features should accidentally and at one and the same time receive identical and single commands to increase in length. Even a lengthening of a single neck vertebra would be absurd if there were not a simultaneous lengthening of the dorsal nerve cord, oesophagus, windpipe, carotids, and musculature.

Indeed only an imitative epigenetic system will explain the occurrence of purposive and therefore successful transformations in the most complex system or in highly burdened single features. These transformations are not random but follow definite laws.

3. Canalization and fixation

This increase in the prospects of adaptation depends in the last analysis on a reduction of random possibilities and tends towards some current adaptive aims. This reduction is a directed restriction in the range of random combinations. However, the increased prospect that one particular combination will occur implies a decrease in the prospects of the others.

a. The turning point

The organism can thus achieve freedom and the advantages of adaptability in a particular direction. This continues until a turning point at which the required direction of adaptation diverges from the switching pattern imitatively produced in the genome. We should expect these divergences because of the stage-by-stage building-up of additional features and because of changes in environmental conditions and consequent changes in function in the evolutionary history.

Consider, for example, the change from a lateral spine to a fin (Figs. 46-49, Section V B4b) to the grasping organ which is the human hand. Or the functional replacement of the notochord by cartilage and finally by a bony vertebral column. Or the origin of the human auditory ossicles from the jaws of the most primitive fishes (Fig. 7, Section II B2a).

Thus it is not merely the functional burden which decreases the prospect that a change in a feature will be tolerated by selection. For underneath all this burden, the imitatively systemized genome, and the switching pattern which evokes it, have themselves been selected to give selective advantages in particular directions. Moreover the advantages thus gained in particular directions must be paid for as regards other directions of selection.

At an early stage I showed that the ratio of accident and necessity in a system cannot be swindled, that the sum of determinacy plus indeterminacy in a system remains constant (equation 17, Section I B3c). This will be confirmed for the other patterns of order (VI C2 and VII C2). Besides this, selection is extraordinarily severe. Functional burdens specify the systemization of the genome according to the breadth of the organisational gap and therefore help specify the severity of selection. But this severity is in addition specified by commitment to a particular switching pattern.

171

I shall next consider the effectiveness (V C3*b*) and the hierarchical extent (V C3*c*) of this superselection.

b. The effectiveness of superselection

This can easily be recognized by considering the individuals excluded by evolution even before the 'struggle for **existence**' in the external environment. This exclusion depends on deficiencies of structure or of functional coordination and corresponds to what Hadorn (1955) summarized as lethal factors.

These are: 'Mendelizing units that cause the death of an individual before it reaches the reproductive phase.'[37] They act mostly in the larval, pupal or embryonic stage, or even on the haploid gamete. Selection is thus testing viability in a totally different sense. It is acting on the coordination of the building regulations of the organism, rather than according to the environmental regulations.

There is extensive material to show the extraordinary effectiveness of this selective mechanism. It is coordinative selection which acts by eliminating organizational deficiencies and mistakes. Its effectiveness can be estimated from three absolute parameters and also relative to the selection dictated mainly by external conditions.

1. *The proportion* of lethal mutations, out of the total mutability, is between 92 and 97 per cent — this compares the number of mutations that cause death with the total number of 'visible' mutations.[38] But this figure takes account only of what has been recognized. In addition there is a considerable contingent of sublethal factors where the mutants are fully formed but hopelessly damaged. Viable and semi-viable mutants are probably even rarer. 'The proportion of gene changes which are not destructive can probably be estimated at only a few per cent of the total mutability.'[39] I believe that it is only about 1 per cent — compare the value for P_e (Section III C2*a*).

The proportion of heritable changes used by evolution to drive the mechanism of adaptive change is vanishingly small compared with the number of such changes that is squandered in preserving the prescribed patterns of order.

If man ever reaches the moral level where the principle of the 'survival of the fittest' no longer applies and the environment no longer decides between life and death, then the whole tribute of lethal mutants will, in evolutionary terms, be sacrificed to preserving the pattern of order.

2. *The commonness* of lethal mutants in the total reproduction is likewise high, i.e. as a proportion of potential total offspring. It corresponds to almost the entire mutation rate and can be estimated from the latter. In humans, even assuming only 15 000 genes, it reaches 'the high overall rate of 30 per cent.'[40] The real values very probably are considerably higher.[41]

In wild populations, egg and larval mortalities of 70 per cent have been established, for example in the beetle *Adalia*.[42] Out of ten organisms reproduced, therefore, three to seven are sacrificed to the preservation of order.

In mankind there are, of course, two additional problems. Firstly modern medicine preserves an increasing number of babies and children carrying subvital features. The number bearing genes that reduce vitality is therefore increasing, as well as the number of people involved in their care. Secondly, increase in radioactivity is raising the mutation rate.[43]

3. *The effectiveness* of lethal factors is high. Most of them leave no chance of escaping death by luck or accident. The number of 'survivors' is small. For most mutants no environment could be conceived in which they could find a survival niche.

4. Lethal factors commonly have *priority in time*, for most act early or very early in embryonic development. They are decisive when the organism is still far from reaching the external environment in which it ought to live.

In summary the effectiveness of this coordinating selection must be extraordinarily high, for it excludes 30 to 90 per cent of all mutants before the suitability of the remainder can even be tested in the external environment. *If about 50 per cent of all mutants are sufficient for phylogeny to produce the miracle of adaptation, then the other 50 per cent must suffice to enforce order.*

This also verifies the postulated connection experimentally since I asserted only that the prospects of changing a feature disappear as systemization and the organizational gap increase.

c. Counterselection and superselection

These are a form of selection which is much stricter than prescribed by the environmental alone. It is not necessary, in this connection, to distinguish between 'internal' and 'external' selection, for every test result involves the confrontation of a test object with a test requirement. Nor is it necessary to deny the action of the external environment in 'internal' selection. For such an action can be found, on demand, even when considering the reproductive prospects of a hen with no vagina. The essential point is only that the result of selection, both in extent and in causal connections, goes far beyond what would be expected from mutability ($P_m \approx 10^{-4}$) and external conditions alone.

It would be wrong to suppose that the synthetic theory had overlooked the difference between the anlagen of organs and the potentialities of these anlagen. Mayr[44] says that what differences in selection pressure cannot explain 'can be traced back to the developmental and evolutionary limits which are set for the organism by its genotype and its epigenetic system'. Kosswig asserts that this amplitude, to use Mayr's term, lies in 'the nexus connecting each gene with its background, i.e. with the rest of the genotype'[45] and thus in connections which for the moment 'cannot be analysed in detail'. However, what escapes observation in the realm of genes can be deduced by way of the realm of phenes.

1. *The extent* of this superselection has already been indicated in the chapter on the standard part (IV B3c); In considering the hierarchical pattern we have encountered comparably huge fixation times and I have established that the observed fixation-time can exceed what is expected by a factor of 10^{12}. I shall return later to the improbability that accident could explain such a fixation. This holds in like manner for all four patterns of order.

2. *The cause* in the first place is the burden of the organizational gap, as already described. Additional to this is the effect of systemization in the genome.

Suppose the switching pattern corresponds to the required direction of adaptation and a number of $bits_D$ are converted to $bits_R$. The prospects of realization (A_a) will then increase considerably, to P_m^{-R}. However, suppose also that a later environment demands a direction of adaptation different from that of the systemization pattern. In this case the whole previously attained advantage (P_m^{-R}) will be lost but also will actually be inverted ($P_m{}^R$). The decisions that have been dismantled must be replaced by means of uncertain accident, with a probability of mutation of $P_m \approx 10^{-4}$. This shows that, even if only a few decisions need replacement, the evolutionary mechanism will have no hope of re-establishing the previous configuration.

d. Accident and necessity

The sum of accident and necessity in a system is constant, as already shown. But the limit between the two is in twofold movement, both real and apparent. The real movement depends on gaining determinacy at the expense of indeterminacy. The apparent movement depends on gaining insight at the expense of uncertainty.

1. *Necessity* in evolutionary phenomena is greater, even with the hierarchical pattern, than biologists have hitherto held to be likely. Causal regularity penetrates transpecific evolution just as it does intraspecific evolution — for the latter, this has repeatedly been confirmed since Darwin's time. The arranging of features in hierarchical order in the course of geological time is a necessity.

In the gross course of evolution only one thing needs to be ascribed to accident. This is the encounter of functional structure (with its necessary abilities) with environmental niches (with their necessary conditions).

For example, it was unpredictable that the Devonian limbs of the lobe-finned fishes should later encounter the steppe environment of horses. Just as it was impossible to foresee the fall of the roofer's hammer in the life of the doctor who was hit by it.[46] Similarly in the Upper Devonian, when five was established as the number of digits among the primitive tetrapods, it was impossible to predict that man would have to confront the piano with only five fingers, where seven would make it easier to play (for me at least). Nevertheless the total patterns of hierarchically ranked homologues in the horse's foot and man's hand are regular necessities.

2. *Evidence for an accidental residue* exists only in adaptively convergent structures, the so-called 'Lebensformen'.[47] These sorts of structure are pure analogies, and so must meticulously be excluded from phyletic studies at the outset (II B2*a*), as the principles of morphology require.

The dice can decide which feature of our Experimentalia encounter which adaptive channels (V A*b*) or even which features shall be fixated. To produce hierarchical order our minimal assumptions are sufficient — there must be a fixation after each split in the diverging stream of determinacy (Fig. 35, Section V A*b*). The results of this simple principle, which embraces all life, are truly surprising.

3. *Reality and thought*. No doubt can now remain that the hierarchical order of fixated component parts is real. There are two consequences. Biologically we can deduce the reality of homologues, ground plans, systematic groups, and the hierarchical system of classification and relationship for all organisms (see also Section VIII B2). Epistemologically we must ask the meaning of the agreement with the hierarchical pattern of human thought, for this pattern is just as univeral and completely irreplaceable.

I return to the point where this chapter began by asserting that our thought cannot grasp a concept unless the latter acquires content from its subconcepts and meaning from its overranking concepts. The limits of the human power of thinking coincide with those of the hierarchy of concepts; the subconcepts begin to fail beyond the 'point', the 'zero', and 'motion'; the higher ranking concepts begin to fail beyond 'space', 'time', and 'infinity'.

We have progressed beyond the starting point, however, by discovering what to think about reality and accident. The reality of the hierarchical pattern in Nature shows that this pattern is not a thought projection. And the agreement between hierarchies of thought and real hierarchies is too extensive to be accidental. There must be a causal connection, but the thought pattern cannot be the cause of the natural pattern. Therefore, the hierarchical pattern of organic structures must be the cause of the hierarchical pattern of thought.

Expressed so briefly this sounds like a fantasy since the mechanism of this wide connection is not at once visible. However, the mechanism does become obvious when we consider the preconditions for the origin of thought and the objects that thought was originally concerned with.

First, among all possible forms of memory and comparison, evolution will always preferentially select those which best correspond subjectively to the objective connections in Nature.[48] Second, seeing the costs incurred by the insertion, storage, and calling up of every *bit*, the most economical system must equally have been preferred. Both lead necessarily to canalization into the orderly pattern of hierarchy.[49]

4. The hierarchical pattern of civilization

If the hierarchy of thought is a consequence of the hierarchical pattern of organic nature, how ought we to understand the conjunction of it with the hierarchical pattern evident in civilization? In discussing the standard part we already met this problem in principle (IV C3). Is it analogy, projection or a product of accident?

Once again we cannot trust to accident. In considering the highly probable causal connections, however, the pattern of thought again has prior place. The hierarchies into which all human concepts and organizations are divided can be traced back to a projection, or more precisely to a necessity, of our thought mechanism. Or perhaps the things are still more closely related as instances of the same law, or in the last analysis as one and the same thing.

a. The success of ranking

This comes from the economy with which information, commands, and competence can be transferred; from the avoidance of mistakes and of too much redundacy; and from the clear definition of fields of authority. **Determinacy** is connected with prospects of realization over a wide range which begins with human organizations and passes by way of human thought, human phyletic relationships, and human anatomical structures down to the laws in man's molecular genetic code. There is much order with little law, and much certainty with a minimum of thought.

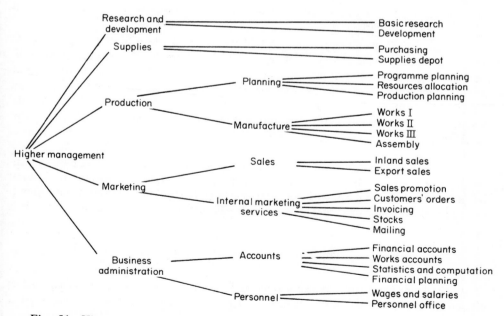

Fig. 51. Hierarchy in civilization as shown by a typical business organization. Divisions, departments, and subdepartments are connected by lines of authority. (After the *Brockhaus Encyclopaedia*, somewhat simplified.)

175

This will be mere analogy for those who believe that the 'natural history of the human spirit' cannot be studied (cf. IV C3). Other people will see it as an analysable mechanism.

In actual fact the increased chances of success of the hierarchical flow of determinative commands have never been seriously doubted — whatever the group, business (Fig. 51), gang, church or state. They are obvious everywhere, whether in the risings of peasants, or the emperor's army on the opposite side.

The breakdown of ranks, when these come to be questioned, always causes the same type of chaos and confusion, with the sole result that the ranks are re-established, oblique to the previous ones (illustrating traditive laws). Indeed ranks are even necessary to persecute the previous rank-holders efficiently.

What happens to hierarchies is even more familiar. As in living organisms they first arise as mass hierarchies of undifferentiated classes of men, as in armies, sports clubs, and savings clubs (which always remain in this juvenile stage). But if they survive long in the turbulence of progress they pass into the senile phase of the box-in-box hierarchy with gentlemen-in-waiting and Princes of the Church. Hierarchy is a necessity. Each attempt to replace it merely replaces order with chaos, until it is itself replaced.

b. Classes and tolerance

In the hierarchies of structure it is clear that the adaptative advantages of yesterday become the adaptational limitations of today, because of the opposed selective requirements of evolutionary mechanisms. The canalization of patterns can lead to complete fixation and thus to the extinction of rigidified systems when these are no longer able to meet the adaptive requirements of the external environment.

Everyone who has read history will recall countless examples. It might be suitable to finish here, but I must add one thing further.

I do this for those who cannot help suspecting that the mechanisms producing hierarchy in evolution and in thought are basically identical. Those who are convinced that the agreement is accidental will gain nothing useful.

It is again one of the features of our second evolution (i.e. the evolution of civilization, freed from the slowness of genetic change) that the differences that necessarily arise in the field of tension between rank and tolerance are not dealt with by the weapons of reason, but always by extirpation. The mechanism of biological evolution explains why the selective conditions of the internal environment (i.e. of the system itself) make any deviation from the established pattern difficult. The same mechanism also explains why rigidified systems are no longer tolerated by the selective conditions of the external environment.

This second evolution, however, has become too rapid for any effective regulating mechanism to be inserted. The hope is, therefore, that it will be able to rationalize the natural necessity of this antagonism with the vital necessity of flexibility. This rationalization is indispensable for humaneness and also for the preservation of patterns of order — it is tolerance.

It may seem subjectively strange that hierarchy is thus explained as being a general law of Nature. If so the natural laws explained in the previous chapter, or the next chapter, may seem more reasonable, so I shall move on.

NOTES

1. The associated problems have been discussed by Simon (1965), Koestler (1968, especially Appendix I), Weiss (1969, 1970a, 1971) and Pattee (1973).
2 For summaries see Weippert (1930), Dahrendorf (1957), Scharp (1958).
3 Polanyi (1968), for example, has shown this convincingly. Compare also Chomsky (1970), Porzig (1971), Höpp (1972), Lorenz (1973).
4 This has been convincingly and repeatedly shown, particularly by Weiss (1971). Compare also Bertalanffy and the remarks at the head of this chapter (VA).
5 Compare, for example, the works of: Adachi (1928-1933), Clara (1942), Pernkopf (1952), Hochstetter (1940-1946).
6 Müller (1963, 4 vol. I, p. 188), Mayr (1967, p. 453).
7 Lorenz (1971), Weiss (1970a), Strombach (1968).
8 Most exhaustively by Osche (1966, 1972).
9 Rensch (1954, p. 67).
10 For antelopes this was shown by Geist (1966) and for the buffalo tree hoppers by Haupt (1953).
11 Thenius and Hofer (1960).
12 Details in Bohlken (1958), Pilgrim (1947), Sokolov (1954).
13 Compare Kurtén (1958), Simpson (1955), Zenner (1946).
14 Haupt (1953, p. 36).
15 In systematics this expression has its own very particular meaning (cf. Riedl 1970, p. 18).
16 See e.g. Wickler (1965).
17 P_s corresponds to 10^6 as the probability of constancy in years; P_m corresponds to 10^{-4} as the probability of mutation; P_e corresponds to 10^{-2} on average as the probability that a mutation will be successful. Only in cases where it was particularly necessary not to underestimate the value have I taken it as 10^{-1}.
18 The most recent survey of the figures for living species is in Mayr (1969, p. 12).
19 The most recent survey is by Thenius (1965).
20 Müller (1963), Tasch (1969), originally Trusheim (1931, 1938).
21 Literature in Størmer (1955).
22 Rowell (1965, p. 262).
23 Hyman (1959, p. 577).
24 See e.g. Müller (1963).
25 Among Agnatha the Orthobranchiata spring to mind and also other Osteostraci such as *Hemicyclaspis* (cf. Fig. 46a-h).
26 This also applies to the development of the brachial plexus.
27 Compare, for example, the arteries of the mammalian foot in Fig. 1150 to 1159 of Zietschmann *et al* (1943).
28 A survey is given by Müller (1966, 1968, 1970) whose bibliography will provide an entry to the literature.
29 Compare *Stensiöella* and *Gmuendina* e.g. in Müller 1966 III (1) p. 91.
30 This was already pointed out by Spemann (1936).
31 Mayr (1942) referred to this connection as the 'hollow curve' of the taxonomists. Westoll (1949) attempted to investigate it quantitatively using the lung fishes.
32 Kaiser (1970) surveyed 'the abnormal in evolution' with extensive documentation.
33 Perhaps one of the clearest examples is to be found in the *Fauna und Flora der Adria*, edited by myself (Riedl, 1970).
34 As Lorenz showed (1973) it was the ratiomorphous achievements of our preconscious 'calculating apparatus' which gave us this early insight into the reality of the natural classification, even in the matter of working out hierarchies (cf. notes 7 and 8, Chapter IV).
35 Compare the viewpoints summarized by Dobzhansky (1951), Ghiselin (1969), Huxley (1942), Mayr (1970), Rensch (1954), and Simpson (1964a).
36 This again agrees with synthetic Neodarwinist theory (for authors see note 35).
37 Hadorn (1955, p. 13).
38 These figures come from Muller (1928) and from Auerbach and Robson (1947).
39 Hadorn (1955, p.47). In recent years the estimated frequency of lethal mutations among all spontaneous mutations has decreased. This is because better methods have made mutations visible which formerly would have gone undetected. Current estimates therefore range between 30 and 90 per cent. I shall use an average of 50 per cent.
40 Muller (1954), Hadorn (1955, p. 41).
41 It is now considered that the haploid genome of a mammal contains about 10^9 pairs of DNA (cf. Britten and Davidson, 1969).
42 Lus (1947).
43 Muller (1950a), Danfurth (1923), Sturtevant (1954).
44 Mayr (1967, p. 475).
45 Kosswig (1959, p. 214 and 215).

46 This was chosen by Monod (1971) to illustrate accident.

47 The concept of the 'Lebensformtypus' was introduced into ecology by Remane (1943) and Kühnelt (1953).

48 It will be remembered that the evolution of thought was a very long process, much longer than the evolution of concepts. This is evident from the organization of preconceptual thought in the animal kingdom e.g. Köhler (1952).

49 I am very pleased to be able to add that this result, which I worked out independently from the facts of anatomy, has now been fully confirmed by Konrad Lorenz (1973) starting from the facts of ethology. Man's thought is the selection product that best corresponds to the real pattern of order in living organisms. Compare also Brunswick (1934, 1957) and note 34. See also notes 7 and 8 to Chapter IV.

CHAPTER VI

THE INTERDEPENDENT PATTERN OF ORDER

A. Introduction and Definition

The third basic pattern of biological order can be called 'interdependence'. It interdigitates with hierarchy, for interdependence makes hierarchy possible, while hierarchy presupposes interdependence. In this sense interdependence is the more basic of the two and the easier to present. The evidence for it comes mainly from mutational and regenerational processes.

The interdependent pattern of order is characterized by the fact that features or concepts (both being events) are only valid and only significant through being connected with other particular features or concepts of equal rank. 'Rank' means hierarchical rank in this regard, and the definition indicates the essential difference from the hierarchical pattern of order. The word 'hierarchy' describes order in which features are fixated one above another or one inside another. But 'interdependence' describes order in which features are fixated one beside another.

As with the other patterns of order it is very difficult to think without using interdependences, although the definition seems so simple. Once again this is because conceptual thought is full of them, or indeed determined by them. Concepts and features are always composite, which means divisible. And most of the subconcepts or subfeatures are impossible to leave out without dissolving the whole into the non-conceptual or unnoticeable, for here already concepts begin to fail.

Thus a human eye is never cubic, never has the pupil at the edge nor the iris at the centre, is never unpaired, nor present in hundreds like pores, is never honeycomb-like nor covered with fur, is never situated on the finger tips, nor found on a worm, is never made of enamel nor of a secretion, never pulsates like a heart, nor is momentary like a cry and never occurs singly. We recognize the phantom quality of such transformations, not their real possibility. However, the colour of the iris varies as does the width of the pupil or the direction of a glance.

Concepts, in the same way as features, are characterized by the firm connection of some of their characteristics, while others vary within defined limits. Interdependences, being characteristics firmly bound together, have chief place among the innumerable facts which confirm comparative experience. Indeed they constitute the thing which is called a definition — the definition of the concept of a natural object however abstract or complex the concept may be. Variables, on the other hand, determine the range of expression of the concept. Once again, in describing living nature, it is the concepts within systematics and morphology which are arranged according to the pattern of interdependence. There are more than ten million such concepts (cf. introduction to Chapter V). All other concepts of human thought, however, are likewise arranged in the same pattern.

Fig. 52. Examples of the breakdown of interdependent order. Hieronymus Bosch combined known structures to produce unknown, absurd, impossible forms. Redrawn from Baldass (1943).

Once again, therefore, we must ask whether thought agrees with phenomenon because an orderliness of thinking has been projected into Nature, or because objective order is repeated in the structure of thought.

a. A fantasy world without interdependence

This in itself is not conceivable, though a world with gaps or mistakes in interdependence *can* be conceived. The uncoupling of characteristics gives the absurdities of dreams, of the paintings of Hieronymus Bosch (Fig. 52), of surrealism or fantasizing realism, of fairy tales or early palaeontological reconstructions.

All these fantastic shapes depend on the breakdown of real individual correlations and the making of unreal ones. According to the mental state of their creators this can be due to lack of control, to intention and fantasy, or simply to ignorance of the facts. The greater the number of interdependent features that are removed from the pattern of reality, the more the product becomes absurd. Finally it becomes unknowable (i.e. indescribable) because even analogy offers no comparison.

It is important to appreciate that a world without the interdependent pattern of order can no longer be described — neither in words nor in symbols. For even the simplest symbols, such as the straight line or the point, contain at least a small number of defined features. Such a world can therefore not even be thought about. Interdependence is a universal principle.

b. The preconditions and forms of interdependence

A notion of the preconditions and forms of interdependence can be got by starting from the indescribable. With regard to any object, we can enumerate characteristics which cannot be removed, neither from the object itself nor from the corresponding concept, if object or concept are to remain the same. Even in the simplest living structures interdependences are numerous. And among higher organisms they reach enormous numbers.

In phylogeny again it is a matter of fixations. Their persistence can be read from the time spans over which the definition of the correlate of a feature applies. Unlike hierarchy, however, we are concerned not with successive fixations, but with relatively simultaneous ones — those which occur beside one another, so to speak.

The differentially diagnostic features which stand alongside each other in the diagnosis of a systematic group are always interdependences of this sort. There are usually ten to several thousand in about 2×10^6 examples observed.

Many interdependences indicate a functional connection, like the gill bars and aortic arches of fishes, the myomeres and spinal nerves of vertebrates in general, or, within limits, the upper and lower teeth of mammals. However, the chains of functional connection have sometimes become long, reticulate, and modified. It is no surprise that some such connections can no longer be recognized, while some, no doubt, cannot be recognized yet.

We can probably always assume that a causal background exists for any interdependence. No one can quote functional connections that have not yet been recognized, but there are several that were not recognized until very recently. Examples are the connection between the growth form of the colony and the position of the polyps in horn corals,[1] between the number of shell muscles and the ribbing of the shell in bivalve molluscs,[2] between the body cavities and the locomotory characteristics of worms[3] etc. Indeed, so many such discoveries have recently been made that a huge number can be expected in the future.[4]

Functional connections that perhaps belong to the past are, of course, just as numerous. Among the differentially diagnostic features of mammals we find associated together: the left aortic arch, seven cervical vertebrae, milk glands on the ventral surface and hair on the skin. The necessity for this combination is unknown. It may depend on still unstudied chains of functions or on coupling by pleiotropic genes, or the functional connection may have been lost, so far as phenes are concerned.

Organisms therefore show two forms of interdependence — functional and transfunctional. Presumably all other forms of interdependence, in civilization and in abstract thought, have an analogous structure. Functional forms are found in the family and transfunctional ones in office organization; functional forms occur in the economy and transfunctional ones in etiquette. As I shall show later the transfunctional form of interdependence always arises from functions, and finishes in traditive inheritance (cf. Chapter VII).

For the present investigation it does not matter in the first instance whether the cause of a combination is visible or not. What does matter is that a particular interdependence is very unlikely to break down. For example, the probability that the features 'neural canal' and 'notochordal anlage' will not occur together in a vertebrate is the reciprocal of the number of individuals in which, up till now, the features have always been combined. This is at least 4×10^4 (species) $\times 5 \times 10^6$ (individuals) $\times 5 \times 10^7$ (generations) and equals 10^{-19} for every vertebrate expected on the earth in future (or 0.000,000,000,000,000,000,1). It is as unlikely as the birth of a griffin, a nasobeme, or a rhinograde,[5] or the world of Hieronymus Bosch.

The precondition for the origin of interdependent patterns, therefore, is that features of equal rank are coupled to each other with totally improbable constancy. The question of how these patterns arise I shall again answer in two stages which, for methodological reasons, must be kept clean apart. First, the correlation between the coupling of features and their constancy must be established as a fact. Second, the two phenomena must be connected together causally by means of a theory.

B. The Morphology of Interdependence

Explaining the interdependent pattern of order depends on correlating the interdependence patterns of phene systems with those of the gene effects on which the phenes are built. A biologist will recognize the basic problems known as co-adaptation (or synorganization) and as homoeosis. The explanation will come later (Section VI C). Here I shall only compare the patterns of order of phene and gene.

1. Interdependence in the phene system

Interdependence in the realm of phenes is such a familiar phenomenon that I can be brief. Apart from introducing the necessary methodological precision I merely need to sort out what is known.

a. The recognition of interdependence

This depends on three phenomena. First, on the certainty of recognizing a similar feature in another individual. Second, on recognizing that two separate similar features are constantly correlated, and third, on establishing how many correlated features there are within a frame of comparison. In this regard, as just shown, the functional connection between correlated features does not matter. It may be unrecognized or a thing of the past. The objective measurement comes from establishing the mere correlation. I shall consider these three phenomena in sequence.

1. *The certainty* of comparing like with like is given by the quantitative theorem of homology (Section II B2d). This theorem permits a quantification of probability that approaches certainty. The comparison involves homologues; individual and mass homologues are compared in like manner, taking the representation and degree of similarity into account (cf. Section II B2e). As already stated, 'representation' is the proportion of representatives within a systematic group that possess the feature.

2. *The degree of correlation* can be quantified in the same way that the representation of two features can be established and compared. The result of the comparison, perhaps expressed as a percentage, indicates among the species that show either of the compared features A or B, how many contain A and B together.

3. We estimate the *number of correlations* that hold within a systematic group by: (a) estimating the number of cadre and minimal homologues; (b) establishing the percentage of correlated homologues among these; and (c) by reckoning the degree of correlation that they reach.

For example take the correlation 'ventral heart: dorsal notochord' within the vertebrates. As just calculated, there have so far been about 10^{19} individuals within this group and representation and correlation are both 100 percent. The probability that this interdependence of two groups of features will break down in the next vertebrate born is therefore 10^{-19}. With respect to known species it is almost 10^{-5} and with respect to the species that have probably existed up till now, it is perhaps 10^{-7} or 10^{-8}.

b. The average degree of interdependence

It is easy to estimate the average degree of interdependence among the features of a group. For, if the group is well known all the important features will have been presorted according to their representation and therefore according to their degree of interdependence. This, again, is the important result of several generations of work in comparative anatomy and systematics. I have already considered this circumstance in discussing the degree of freedom of homologies (Section V B2).

Within a phyletic group its differentially diagnostic features will be completely interdependent. But within this same group, the differentially diagnostic features of subordinate groups will not be completely correlated, and the degree of correlation will decrease as the subgroups become more and more subordinate.

Thus within the class Mammalia the differentially diagnostic features of the mammals will be completely interdependent, as will be those of amniotes, tetrapods, vertebrates, chordates, and deuterostomes. (To take one feature of each, these could be: hair, amnion, two girdles, vertebrae, notochordal anlage and secondary mouth.) But the differentially diagnostic features of the orders, families, genera, and species would be less and less completely interdependent.

c. Synorganization or coadaptation

Whenever we gain insight into the functional connections of an interdependence we speak of organization. Biologists rightly expect it in a functioning machine. Sometimes, however, we do not have this insight and then accident will be invoked, i.e. the reign of a non-causal or transcausal principle agreeing with the interpretation of accident in Sections II B2f and V C3d. Sometimes the functional connection is obvious, but not the causal mechanism that produces and develops it. We then meet the well-known evolutionary problem of synorganization or coadaptation. This can be summarized with the question: originally separate features sometimes develop very precisely coordinated mutually harmonious adaptations; how can the origin and later development of these harmonious adaptations be understood?

Examples are the plumage patterns of birds, extending over several feathers; the stridulating organs of insects for producing sound; the coupling together of front and hind wings in insects; or the way in which the upper and lower teeth of mammals fit together[6].

The explanation of synorganization is still an open question. Critics of the synthetic theory assert that the latter provides no explanatory model, e.g. Remane (1971). For how can external environmental conditions select one of the coordinated parts of the mechanism when the other has not yet been selected? Osche supposes nevertheless (1966, p.889) that such complexes of characteristics 'may offer selective advantages at the instant when they are combined'. Strangely enough, in explaining the cause of synorganization I shall confirm both of these opinions.

At the moment I only need to show that the phenomenon of synorganization is merely part of the interdependence problem — a small part, but specially illustrative. For strictly speaking the problem is just as remarkable when we think we know the cause — as for example the correlation of increase in length with increase in thickness in a bone as it is when we see no functional connection, e.g. the correlation of the right aortic arch with feathers on the skin.

d. The ubiquity of the interdependence phenomenon

This can best be understood if, starting from the phenomenon of synorganization, we introduce two new concepts (later these will again become unnecessary).

1. *'Synhomology'*. This term highlights the fact that every comparable structure, or every cadre homologue, consists of a number of subhomologues whose simultaneous presence is a precondition for recognizing the structure. Synhomology does not make it any less astonishing, for example, that three cartilages in the jaw of a fish can be recognized in the inner ear of a mammal (cf. Fig. 7a-e). Interdependent crossconnections can be invoked whenever function on its own cannot explain the constancy and degree of conjunction of two features of equal rank. Synhomology is the Godfather of all functional connections, as for example in forming articulations between parts that were previously independent or when vessels or nerves find their end organs.

2. *'Synformation'*. I use this term for interdependence of quantities. It is that quantitative control which affects the proportions of single parts, coordinates the proportions of neighbouring parts, and leads to the finely balanced changes in proportion that dominate the evolution of whole groups of organisms. This is called harmony or harmonious transformation.

Such directed changes are often called trends when it is thought that environmental selection will sufficiently explain them.[7] They are called orthogenesis, when it is thought that unknown internal mechanisms must also be in control.[8] And they are called Cartesian transformations[9] when the harmonious change can be described with simple quantitative parameters.

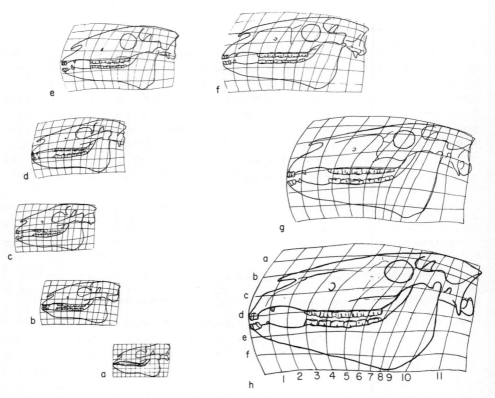

Fig. 53. Cartesian transformation illustrated by the skulls of fossil and recent horses. (a) *Hyracotherium;* (b) and (c) reconstructions near to *Mesohippus* (Oligocene); (d) near to *Parahippus;* (e) and (f) near to *Merychippus* or *Protohippus* (Miocene); (h) *Equus* (Recent). Coordinate transformations on the same scale, after several authors, from Thompson (1942).

Indeed it is hard to see why the skull of the ancestors of the living horse changed in the same direction in the course of 6×10^7 years (Fig. 53a-h). I mentioned this problem of the directedness of orthogenesis at the beginning (Section II B3c and II C2) and I shall explain its causes in the final summing up (Section VIII B5b).

In short, interdependence is not a special case, but a universal phenomenon in the organization and harmony of phenes. In itself this is a trivial result. More important are the causes. The next question, therefore, is whether interdependence can be shown among gene effects.

2. The organization of determination complexes

The question is: Do the effects of determinative decisions show organization, i.e. purposive interdependence? To estimate the purposiveness of genetic interdependence we can ask whether the effects agree with the functional interdependence patterns of the phenes that they determine. Such agreement would imply that a single mutation with complicated results could produce an organized (i.e. functionally meaningful) change.

In considering hierarchical selection (Section V C 1 and 2) I showed that this question must be answered 'Yes'. Indeed, I showed the genome to be highly organized in this respect when I discussed the gap between the organization of genes and of phenes. I must now produce the evidence. This involves studying the complex effects which mutated single decisions can produce.

a. Pleiotropy and polygeny

The first condition for the existence of genetic determinative decisions, in the sense defined, is the fact that a single decision can influence several phenes, i.e. features or events. In fact this is a common phenomenon among mutations and is known as pleiotropy. The study of pleiotropy explains not only the intracellular effects of the mutated decision, but also the intercellular effects, i.e. the mechanisms by which the induction and hormone effects of a cell system can influence other cell systems. Direct and indirect effects can be distinguished (autochthonous and allochthonous effects), i.e. those that act direct on an event and those that act by way of two or three intermediate events (autophenes and allophenes). It is also possible to distinguish between mosaic pleiotropy and relational pleiotropy.[10]

1. *Relational pleiotropy* is always expressed more or less weakly, in pleiotropic gene effects. It provides evidence in ontogenetic development for the second precondition, that single decisions influence each other reciprocally. This reticulation of the effects of different decisions has already been predicted on the basis of the regulator-repressor systems of the molecular-genetic realm (cf. Sections III C3b and III D1c). However, relational pleiotropy tells us one thing else. Hadorn[11] states that: 'connections between autophenes and allophenes become visible because of their relations in time and space.' He asserts further that: 'it is thus possible to arrange features on a "family tree" that expresses the hierarchy of phenes.'

2. This *hierarchy of phenes* shows the way in which decisions and effects are interconnected. With regard to the interdependence problem there is even more convincing evidence that single decisions are interconnected. Their sequence and the parallel 'hierarchy' of effects will be discussed in more detail later under 'traditional inheritance' (Chapter VII).

The more spectacular forms of pleiotropy leave scarcely a single anatomical feature of the mutant unaffected. I could equally say 'undamaged' for almost all of them result in death. The effects can be measured in terms of the number of affected homologues (cf.

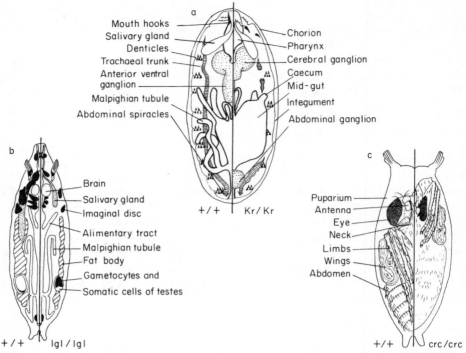

Fig. 54 a-c. Pleiotropic gene effects as shown by three mutants of *Drosophila* affecting the embryo (a), larva (b), and pupa (c). In each figure the normal or wild type is shown left of the line (+/+), while the pattern of damage is shown on the right. Corresponding organs are indicated by the same shading or thickness of line. The abbreviations refer to the name of the mutant. From Hadorn (1955).

Section V C1*a*) and are then shown to be considerable. As regards the degree of organization of the change, however, no purposiveness is visible.

As already mentioned, the degree of organization can be specified by the number of purposively altered homologues within the limits of the ground plan (Section V C1*a*). The examples in Figs. 54-59 will easily demonstrate the purposelessness and disorganization of the change. The degree of order remains large since the crippled organs are still in comparable position and show many identical special qualities.

3. *Polygeny*. This term is used for a common phenomenon of the connection between gene and phene, when a feature (phene) depends on several decisions (genes), rather than a gene influencing several features. For example, more than 40 genes are involved in the eye of *Drosophila*.[1][2] To this extent pleiotropy and polygeny are opposites.

However, when we ask how pleiotropy can be recognized, we meet a more interesting connection between the two phenomena. For how is it known that the altered features of a mutant involve several phenes, rather than one only? The answer is that in other mutants only one of the features is affected. Other genes therefore act on the same feature. This in itself shows that pleiotropy is linked with polygeny. The limitation of gene effects to interdependent complexes can therefore also be measured by means of the polygeny effect, though more indirectly.

It still remains to prove the third precondition for organized interdependence in genetic determination. This is that the composite and reciprocally dependent determination complexes are purposively organized, i.e. are switched on in a fashion that produces meaningfully organized complexes of phenes. The evidence comes from complex but functionally complete phene systems. The respective mutations used to be called major or systemic mutations. Nowadays, however, they are called homoeotic mutations.

b. Homoeotic mutations

These can be called doublings, replacements or forms of spontaneous atavism depending on the changes produced. In each case there is a coherent alteration of numerous single phenes because of a mutation in a single decision, i.e. in a single gene or cistron.

1. *Doublings* are well known — for example the famous bithorax mutants of *Drosophila*[13] in which the metathoracic segment of the thoracic region is formed as if it were a second mesothorax. Figure 55a-c gives some notion of the great number of features which may spontaneously be doubled.

It is not merely the gross form of the mesothorax which is repeated but also its covering of bristles, and indeed each single bristle. The scutellum is repeated. The halteres are transformed into wings showing many details of the veins and bristles. Even the third

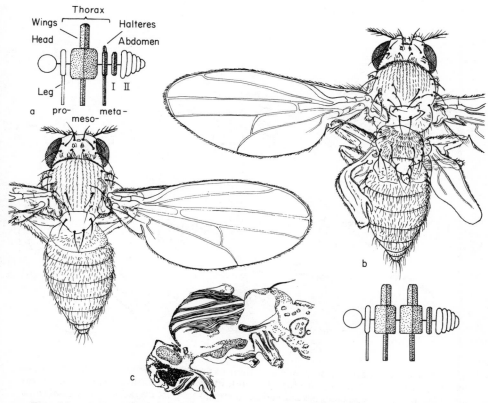

Fig. 55 a-c. The bithorax mutant of *Drosophila*. (a) The normal or wild type. (b) The mutant (both in dorsal aspect, with the diagrams in lateral aspect). (c) Longitudinal section through the head and the thoracic region as affected by the doubling. Note the deficiencies, as for example the lack of flight muscles in the mutated metathorax. (b) and (c) from Waddington (1957) and Lewis (1964).

187

pair of legs takes on features of the second pair. When well developed the supernumerary part is so like the original that, if isolated, it would certainly be identified as a part of the species *Drosophila melanogaster* by an experienced systematist. This is most convincing proof of the extent to which mutually coordinated positional and structural features are doubled.

In the internal organization, however, there are very many mistakes. Parts of the flight musculature are lacking, and there are deficiencies in innervation and coordination and, in any case, in the ability to function. I have already discussed this phenomenon as the organizational gap in determination complexes (cf. Section V C1). The mere existence in a determination complex of deficiencies or of a gap emphasizes the coordination attained when many genes collaborate, all of them depending, as if in a bundle, on a single gene decision.

2. *Replacements* are perhaps even more impressive. In them a complex of features

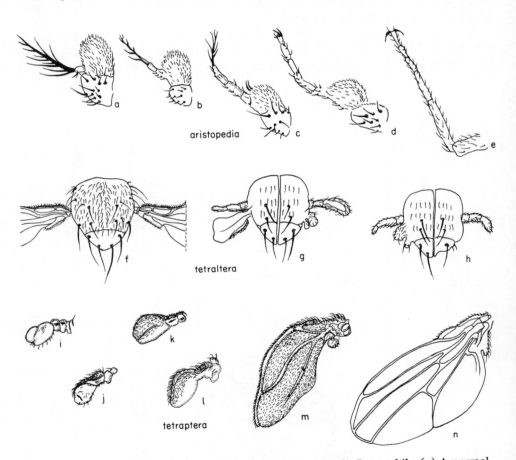

Fig. 56 a-n. Homoeotic mutations and their expression in *Drosophila*. (a) A normal autenna; (b-d) a series of aristopedia mutants with increasing expression of the antennal foot, i.e. leg segments instead of the terminal portion (arista) of the antenna; (e) normal tarsus of the left hind leg; (f-h) tetraltera mutants arranged in a series of increasing expression; the normal wing (as in n) is gradually transformed into a balancer or haltere – a normal balancer is shown in (i); (j-m) tetraptera forms – the halteres (i) gradually become more and more like wings (normal wing in n). (a-e) and (i-m) from Kühn (1955); (f-h) from Goldschmidt (1961).

which in itself is well coordinated occurs in a totally unsuitable part of the body. It replaces, so to speak, some other organ which would have been expected there.

Most of the recorded replacements involve appendages, probably because these are easy to observe. An appendage is replaced by another appendage inappropriate to the position on the body. In *Drosophila*, for example, we find wings instead of halteres, legs or antennal structures on the proboscis, leg structures on the eye or instead of the antennae etc.[14] The replacement of the antennae by leg structures is known as aristopedia and Fig. 56 b-d give some idea of it. A great number of leg features are produced with the correct structure and relative position.[15]

Of course the deficiencies are once again undeniable. The organizational gap is obvious. To appreciate this it is necessary to consider the number of functional impossibilities, i.e. the requirement as concerns innervations, musculature, and brain, for the foot and leg to function in an antennal position. But it is also undeniable that the commands for producing a great number of leg features have been meaningfully organized, that they form a closed complex of determinative decisions, which, by changing a single command, are obeyed or realized in an unsuitable place in considerable detail.

3. *Spontaneous atavism*. Doublings and replacements repeat complexes of structures which exist in the phene pattern of the organism concerned, though in a different bodily position. There are other mutational changes, however, where this condition does not hold. When such a spontaneously occurring pattern is known to have been realized in the ancestors of the organism in question it is called spontaneous atavism.

The most striking example is the three-toed mutant of the horse (already mentioned in Section V C1a). The additional side toes (cf. Fig. 62a-g, Section VII B1a) possess the number of bones, the joints, the insertion points for tendons and the proportions appropriate to their ancient function. In addition the appropriate muscles show in large part a functional arrangement. The meaningful conjunction of these features cannot be explained through accident. It must be due to the re-emergence of an old pattern that is still preserved. I shall discuss the phenomenon in more detail later (Section VII B1).

c. All successful change is organized

To biologists none of this is new. On the contrary, the universality of pleiotropy and polygeny caused the hypothesis of 'one gene, one feature' to be given up long ago, as I have already said. Homoeotic, purposively adjusted phenomena are beginning to be evident in the formative process of all sorts of phenes. Where many commands work together towards a complex realization, they must always be coordinated with each other, and mutually influence, regulate or dominate each other. The more obvious homoeotic mutations are merely those that hit the observer's eye. They are, so to speak, the most macroscopic events in a universal homoeosis of gene effects.

This assertion has now been confirmed everywhere and the phenomenon of homoeosis has become an accepted extension of general theories of development.[16] The aim of these theories is to gain some idea of the interaction of genetically anchored commands, especially in the course of embryonic development. In developmental physiology and genetics this is the problem of the epigenetic system, which we too shall need to investigate.

With random changes, which mutations always are, a large amount of chaos must be produced, to be sorted out later by selection. However, the abundance of homoeotic phenomena proves that the epigenetic system is highly organized. This was all that we needed to show in the first instance.

3. The organization of the stream of determination

I return to the question that I asked at the beginning: Are the determinative decisions purposively organized, i.e. do they show an interdependence corresponding to the pattern of functional dependence which exists among phenes? We have established, first, that the abundance of pleiotropy and polygeny indicates that gene effects are universally interwoven (Section VI B2a1). Second, reciprocal dependence must likewise be a ubiquitous principle (Section VI B2a2). Third, this interdependence shows a pattern in the homoeosis phenomenon which corresponds to the phene patterns in a purposive way. Each 'bundle' of interdependent decisions, dependent on a single general decision, is in itself purposively organized. The proof is the occurrence of an organized condition at an unsuitable place — of the right thing in the wrong position.

There are analogies among the mistakes of civilization. Take, for example, a military exercise in which the attack is carried out according to organized instructions in a completely purposive manner. The advance guard and rearguard maintain the correct distance; the infantry protects the armour; the artillery, staff, and field kitchen are all in the right place. Only the enemy is absent, because someone in the office put the wrong date. Again, consider a complicated equation, which we solve correctly, but with a wrong result because, at the beginning, a decimal point was wrongly placed.

The conclusion already seems certain, although up till now I only demonstrated organization by examples based on mistakes made at the beginning, i.e. mistakes in the original genetic code which we therefore see as heritable mistakes. The result can be confirmed by considering mistakes in the execution, in the flow of organized determinative happenings.

These errors, so to speak, are the results of mistakes made during the exercise by the staff, or by the calculator himself in working out an equation.

The result of mistakes in the determination-flow must also reveal organization. Indeed, the evidence here is much more extensive — it is the material showing transdetermination in the widest sense.

a. Phenocopies

The phenetic result of disturbing a developmental process, by poison, say, or by a climatic stress, will often copy particular spontaneous mutants. Experiment has interfered here with the way in which reciprocal gene effects produce a structure; it has tampered with the process of calculating the equation. It has revealed the sensitive phase in which a wrong decision in the original genetic coded statement (like many initial quantities in a calculation) would have acted. The way such an error would strike has been replaced by the experimental disturbance.

Such phenocopies are known in great numbers. Most of them can be accurately reproduced, for after all, the same mistake at the same point of the same calculation based on the same initial quantities will lead to the same result. They are a chief method of studying the connection and sequence of the epigenetic process, i.e. the system of reciprocal gene effects.[17] Spontaneous mutants can be copied, even to the most complex changes like the bithorax phenotype of *Drosophila*. Moreover, they can be divided into subpatterns and subordinate effects. The same convincing organizational pattern of interdependent gene effects is revealed.

b. Heteromorphosis

The evidence is even more extensive and better known when we consider the 'mistakes' in the course of development which Nature continually produces, quite apart

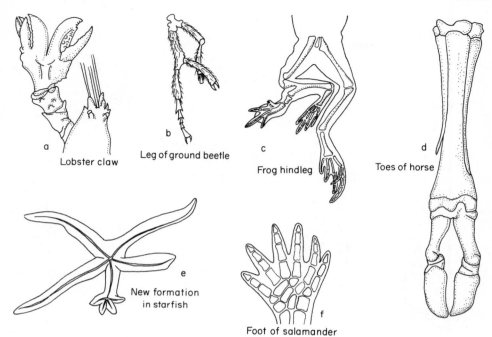

Fig. 57 a-f. Doublings of complex bodily appendages. (a) Claw doubling in *Homarus americanus*. (b) Tripling in *Carabus nemoralis* with extra femur and coxa. (c) Super-regeneration of a foot on the lower limb of *Rana temporaria*. (d) Doubling of the middle toe in the forefoot of *Equus*. (e) Production of a new individual at the end of the arm in *Linckia multiflora*. (f) Supernumary toes (polydactyly) in the foot of a salamander. After several authors, from Korschelt (1927).

from those produced by experiment. As regards such atypical regeneration[18] we usually know the time when the mistake happens but not its cause. This cause, however, must certainly depend on a mistake in the transmission of information. Indeed the agreement with the forms of homoeotic mutations is so great that we can classify them similarly (cf. Section VI B2b).

1. *Doublings*. These, or indeed monsters where the repetition is more than a doubling, are known in the limbs and body appendages from the coelenterates up to the mammals (Fig. 57a-f). Supernumerary crabs' claws, insect legs, and vertebrate fingers are particularly common.

However, doublings and more-than-doublings of the main axis are also not rare, e.g. affecting heads and tails as in Fig. 58a-f. They are known from flat worms, where they can easily be produced by incisions, up to man. It is important to appreciate what a complexity of phenes is 'meaningfully' coordinated when, for example, the anterior end of a human foetus is doubled or tripled.

2. *Replacements*. Organs which in themselves are ordered, but which appear in the wrong place, once more prove the organizational completeness of determination complexes (cf. examples in Fig. 59a-f). They show the grouping-together and interdependence of a great number of genetic decisions whose numerous phenes form functional interconnections. Once again, a single mistake in the installation process shows that an entity of bunched determinative decisions is meaningfully organized. For the entity is coordinated within itself although in totally the wrong place.

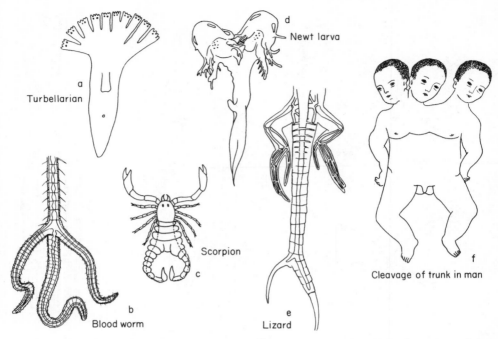

Fig. 58 a-f. Doublings along the main longitudinal axis. (a) *Dendrocoelum lacteum* — ten-headed form produced by repeated incisions. (b) *Tubifex rivulorum* with three posterior ends to the body. (c) *Euscorpius germanus* with doubled abdomen. (d) *Triton taeniatus* with a doubling of the anterior part of the body and head as a result of constriction. (e) *Lacerta muralis* with a forked tail (after an X-ray photograph). (f) Tripling in a boy which partly extends into the trunk. After several authors, from Korschelt (1927).

Organization which is similarly meaningless through isolation, but nevertheless according to plan, is again known in civilization. Consider the planned execution of a search operation when the search ought to be somewhere else entirely; or a foreign mission which does not know that the firm at home has gone out of business; or scattered soldiers carrying on a war without knowing that it stopped long ago.

3. *Atavisms*. These are also known among hetermorphoses and again indicate the action of still-conserved installation instructions (cf. Section VI B2b). In many single cases it can be disputed[19] whether an atypical regeneration corresponds by cause or by accident to an archaic structure. Nevertheless there are many indubitable cases, such as the five-fingered hand in amphibians, the development of a hind leg in whales, the primitive scales on the regenerated tail of a lizard, the primitive regenerated antennae in annelids, or, in a broad sense, the occurrence of people with fur or tails. I shall discuss this in more detail in Chapter VII (Figs. 62, 63 and 64, Section VII B1).

c. Regeneration and propagation

I have discussed the results of mistakes in the effects of genes, whether these mistakes concern the starting conditions or the course of development. I did this because, strange to say, the mistakes in the functioning of a mechanism reveal how its functioning is organized. However, it would be absurd to suppose that this organization is only present when mistakes occur. On the contrary, the astounding thing is that it still exists despite the mistakes.

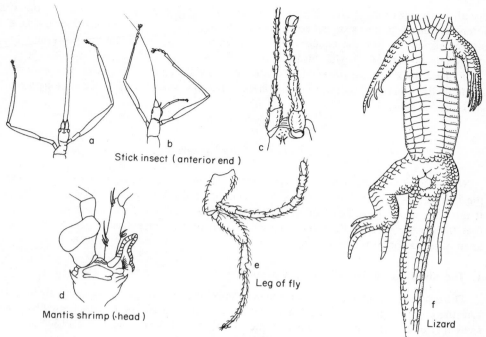

Fig. 59 a-f. Regenerative replacements of appendages. (a-c) Imago of *Dixippus morosus.* (a) Normal anterior end; (b) An atypical leg replacing a lost right antenna; (c) Detail of the same. (d) *Squilla pallida*, with the right eye regenerated as an antenna-like appendage. (e) *Dilophus tibialis* — right foreleg with an antenna arising from the coxa. (f) *Lacerta muralis,* mutilated left hindleg replaced by a tail-like appendage. After several authors, from Korschelt (1927).

We must not forget that successful, error-free regeneration is much commoner. Indeed, in all organisms which propagate asexually by budding and division, successful regeneration has become a reproductive principle. Consider the extent of autotomy, for example, and the organization required for its adjustment. Autotomy has become a form of asexual reproduction dependent on the successful regeneration of parts of the body that separate themselves off. A similar example is budding where small pieces of tissue, or even groups of cells, are able to build up complete organisms. Again homozygotic twins show that the cells formed by the first divisions of the egg have the power, even in very highly organized organisms, of adjusting themselves so as to replace the lost part completely.

Perhaps I am stating the obvious when I assert that regulation and homoeosis are a universal principle of living matter, for everyone has long thought so. The problem lies, not in specifying the limits of this universality, but rather in deciding what mechanism could produce such a purposive power of adjustment in the reciprocal effects of genes. This is the homoeosis problem as presented by every single interdependence among determinative decisions. As a total problem, it is merely epistemological. It is a basic problem whose apparent insolubility gave rise to vitalism and which was considered more scientifically by Hartmann with his 'organic nexus'.[20]

Three questions are hidden inside the homoeosis problem. To what extent, how, and why is the epigenetic system purposively organized? As to the first question the epigenetic system is organized throughout. The second question can now be made more specific: organization of the epigenetic system obviously corresponds to the functional interdependence patterns of the corresponding phenes, except for the organizational gaps already discussed. For the phenes are the measuring rod by which the purposiveness of the system can be judged. The third question will now be dealt with.

If there is a mechanism which causes the interdependence patterns of genes to resemble the functional patterns among phenes, then this would confirm that the patterns are indeed similar.

C. Selection of Interdependence

What I have said about the order patterns of standard part and hierarchy will suggest the sort of mechanism that produces and maintains the interdependence of gene effects. It must again be a form of selection. Indeed, in discussing hierarchical selection, I showed that the advantages must lie in the selection of a purposive connection of gene effects, i.e. in an organization of the epigenetic system.

1. The advantages of imitative interdependence

The selective advantages of the interdependence of decisions are particularly obvious. I can therefore begin to discuss the mechanism of, and necessity for, the synchronous switching of determinative decisions at the point where I previously left off (Section III C3*a* and *b*).

The matter may look simple but I am trying to deduce a causal model of the way in which the epigenetic system acquires its structure. This calls for caution since it has basic consequences for understanding the possibilities of living organisms and their genesis. In the model I shall therefore distinguish quantitative, qualitative, and dynamic aspects.

a. Narrowing the play of accident

This is the quantitative aspect that has been mentioned already (Section III C3*a*). The prospect of the appearance of an event equals the product of the prospects of the independent preconditions for the event. In biological terms this means that the prospect of an adaptive change equals the product of the prospects of the mutations required for it. The mean probability (P) for the occurrence of a mutation (P_m) we have generously assumed to be 10^{-4}. If by synchronous switching one less determinative decision is needed for the event, then the selective advantage (A) in accomplishing the necessary change (A_a) is already ten thousandfold: $A_a = P_m^{-1} = 10^4$.

Moreover, of the single mutations which do occur, only a small percentage are successful, and also alterations must be coordinated (equation 26, Section III C2*a*). For these reasons the selective advantages of a single synchronization will increase to a still higher power.

Polygeny, as just shown above, is a ubiquitous phenomenon. Most phene systems are determined not by two genes only, but by several or many — up to 40. On the other hand, pleiotropy (Section VI B2*a*) is equally widespread and shows how interdigitated the gene effects are. Furthermore, homoeotic mutations (Section VI B2*b*), taken together, show what a large number of single decisions can be regulated by another single decision.

Besides this, the molecular mechanism of synchronous switching can be explained as corresponding to the ubiquitous repressor system. Consequently the interdependence of genes is not merely necessary, on the basis of enormous selective advantages of from 4 to 30 orders of magnitude, but can be taken as fully proven.

194

This is not the essential point, however. Of course, it is pleasing to find complete agreement with the now-accepted theory of epigenesis, though we have contributed nothing more than a notion of the selective advantage of a single interdependent switching event. The next questions are how the synchronous switching is achieved, and whether we can foresee the pattern of this switching.

These also are not totally new thoughts. Indeed a great quantity of previous work points in the same direction. In particular Baltzer's conceptions (1952) have anticipated what follows to a considerable extent, and the same is even more true of Waddington's 'archigenotype' (1957). I shall give proper attention to this earlier work when dealing with the consequences of the theory (Section VIII B4).

b. The imitation of phene patterns by gene patterns

What is the selective advantage of synchronizing two decisions that are anchored in the genome? This must depend on whether, in the process of adaptation demanded by the environment, it is always an advantage to change the individual resulting phenes in the same sense, or not. This will be decided by the functional relationship of the two phenes in the respective environmental conditions.

1. Three possibilities must be considered. For there are three conceivable types of functional relationship between genetically independent phenes — advantageous, indifferent, and disadvantageous. Naturally these three types are not equally common.

Indifferent relationships might at first sight seem numerous. For there seem to be many features which functionally have nothing to do with each other, such as eye colour and wing venation, or toe length and curliness of fur. Indeed at any one moment it may truly be indifferent whether the features are coupled together or not. However, if there is no selection pressure towards such a synchronization, then there is no reason to think that a mutation producing it will succeed. Only if there is some advantage, however slight, attached to the synchronization will it quickly spread through the population. On the other hand, if there is a disadvantage, so that a profitable thickening of the fur involved a totally inadaptive lengthening of the toes, then the novelty will soon be suppressed so long as its burden is small enough to allow suppression.

But ecological niches change and so do their functional requirements. Indeed the required direction of change may be reversed. We should therefore expect that an indifferent functional relationship would not long persist. Because of the sensitivity of selection there will soon be a predominance of advantageous or disadvantageous relationships. Indifferent relationships will not be the commonest sort.

2. As regards the conditions that promote imitation, the requirement that the genome should imitate functional phene dependences will vary in intensity. Its strength will depend on the importance of the functional relationship between the phenes, on the required precision of the relationship, and on how long it lasts.

Thus the *importance* of the fit between the atlas and axis vertebrae is considerably greater than that between the last two tail vertebrae, although the *precision* required is comparable. As I shall show later (Section VII B, Fig. 64a-d) the epigenetic interdependence of these two cervical vertebrae is so great, that they can exchange whole portions with each other but nevertheless produce functional articulation facets between themselves.[21] On the other hand tail vertebrae can be completely lost without any observable compensation.

I therefore suggest that the degree to which the epigenetic system imitates the functional relationships of phenes will increase with the vital importance of these relationships. This will be confirmed in discussing the so-called pattern of induction (Section VII B).

The *precision* of fit will play its own role. The tear duct, which keeps the cornea wet and transparent, is as important for human sight as the lens. But the tolerance in fitting the lens is much narrower. It is therefore no surprise that very special adjustments have been built into the epigenetic system to ensure a proper fit in producing the lens (cf. Section VII B and Figs. 66 and 67). Nothing of the same degree would be expected in producing the tear duct.

It is scarcely necessary to assert that most precision fits are imitated by the genome for this is self-evident. Thus it has long been proven that growing nerves and vessels secure their own connection. It is also well known and self-evident that bones that come to fit together specify each other mutually in epigenesis, while the associated vessels are freer. Likewise adjacent parts alter more similarly than distant parts (cf. Fig. 53a-h).

The *duration* of a functional relationship is seen to be important only when we remember that functions change easily, while the imitative process in the epigenetic system can only change slowly. Functional relationships that change from niche to niche, and from species to species, would probably have no prospect of being imitated by the pattern or reciprocal gene connections. But whatever retains its function over considerable geological periods will be recapitulated to an astonishing degree in the working patterns of the epigenetic system. Vertebrate examples are the notochord, the dorsal nerve cord, the eye, the gill bars, and the jaws (cf. Section V B2b).

We should expect that imitation by the genome would be most obvious with the morphotypic features of the old and large systematic categories, as recorded in the differential diagnoses of these categories. The pattern of reciprocal gene effects must correspond to the major ground plans. And within these ground plans of the large systematic groups they must be identical and homodynamic, i.e. functionally homologous.

All this asserts a great deal. Accordingly I shall discuss it again when dealing with the genotype concept and will justify my assertions when considering the concepts of induction pattern and homodynamy (cf. Section VII B2).

The abstract image of the ground plan or morphotype is not easy to discern in a network of biochemical reactions which often have not yet been clarified. After all, most modern biologists see the ground plan as a vague and distant abstraction and doubt its reality.

c. An imitative epigenotype

The controversial concept of the morphotype, made up of the constant correlation of particular structural features within every phyletic group of organisms, must be recognized as a reality. It is not the result of projection by ordering thought, but an objective natural fact. It is only accessible to us by a large measure of abstraction in thought, but nevertheless it certainly existed before and outside of thought. *The morphotype is the pattern of freedom and fixations which is formed by the collective of features of a phyletic group.* It is a necessity which specifies the direction of every evolutionary path − both the necessary features of its history and the limits of its future possibilities, of its 'hopes'.

I must not anticipate further. We met the problem of the morphotype at the beginning (Section II C3) and again in connection with hierarchy (Section V B6). It will be solved only at the end of the book in presenting the conclusions (Section VIII B 2b).

The selection of interdependence therefore consistently increases the genetic interconnection of those features (or phenes) which have a basic, immediate, and long-lasting functional connection. Consequently the system of reciprocal gene effects will imitate the essential aspects of the phene system — the high-ranking and constant ones. This is imitation of the morphotype of phene patterns. Selective conditions in the inner environment or organisms are extraordinarily effective (Sections V C 3b and c). We should therefore expect that the epigenotype of a phyletic group would long ago have copied its morphotype. This naturally means not the phenotype or collective of features but the morphotype or pattern of fixations. This is constituted by the standard-part, hierarchical, interdependent, and traditive patterns of order and of determination.

These patterns of determination occur in the natural classification, hierarchically arranged for every group of organisms. Their phenetic aspect has been worked out extensively by comparative anatomy in the differential diagnoses of these groups. Systematics has defined thousands of such patterns unequivocally.

It is not easy to show the rather abstract morphotype in a picture (cf. Fig. 78a-b, Section VIII B2b). It is even harder to illustrate the epigenotype. However, its unpicturable features, which are the sequence of determinative reciprocal gene effects, are ordered in a space-time pattern. This does make it possible to show the epigenotype in a picture (Section VII B 2c, Fig. 66, Section VII B2b).

2. The canalization of interdependent patterns

Great selective advantages thus demand that genetic synchronization will adapt the genome to the functional relationships of the phenes. This, however, does not explain why the resulting switching pattern is conserved. Such a fixation, however, is a precondition for the origin of constant epigenotypes. It is necessary to consider this since we must assume that, as functional demands alter, selective forces will tend to disrupt established interdependences.

a. Disruption and alteration

Synchronization and interdependence can be traced back to the regulator-repressor genes. Like structural genes and all others, these are anchored in the triplets of the DNA sequences of the genetic code (cf. Section III C3b). The same mutational conditions should therefore affect them and, what matters most here, their effects can disappear by mutation, just as they can appear. Selection would also decide, by reference to the environmental requirements, whether the coupling of the adaptability of two phenes, which had originally been produced by selection, should be conserved or not.

Features of peripheral position and some adaptive freedom (cf. Section V B1e and 2a) would not remain coupled once evolution, with change in ecological requirements, ceased to select for coupling. If the coupling became an evolutionary obstacle causing negative selection pressure it would persist for an even shorter time.

Thus the coordination of length and breadth of the caudal fin must once have been important in our Devonian ancestors. But we should not expect any trace of it in the human epigenetic system, nor do we find any. If we did, the genetic system would be swamped with unnecessary coordinations. As shown later, the swamping is large enough as things are.

b. Burden and fixation

The freedom to disrupt established interdependences must disappear with increasing burden. The problem of burden has already been encountered in dealing with standard-part and hierarchical order (Sections IV B3a and V B1e) and the effects of burden were considered under standard-part and hierarchical selection (IV C2b and V C3). The principle of burden will be very clearly confirmed in treating interdependence. Again I shall start from the simplest case.

1. *Parallel burden*. This can be expected whenever the action of a number of genes is *controlled* simultaneously (in parallel) by another gene, rather than being triggered by it. If only two genes are so controlled, and one of them needs to be further adapted independently, then a mutation of the regulator gene can disrupt the interdependence without involving negative selection pressure.

Things are different when, say, the action of 5 to 10 structural genes is controlled in a coordinated manner. Suppose it was adaptively advantageous that the phenes of gene 5 or gene 10 should evolve further independently. Then a disruption of the regulation would send 4 to 9 other phene groups out of control or out of coordination. The advantage on the one side would be paid for by bigger disadvantages on the other. Such a change could only be selected if the old regulation of all the other previously interdependent phenes had been replaced by new regulators produced by mutation and tried and tested. As already shown, the prospect of such a reform sinks exponentially with the number of single preconditions, and soon becomes inconceivable. In any case, the modification of one of the larger groups of interdependences will require long trial and error — longer than some of the original functional connections will be required.

The organism will thus lose a broad potential prospect of adaptation and will possess one unnecessary structural correlation more. It will need to try other adaptive directions and will lose some potential niches, until in the end it can find no more at all. If this is true, then every organism must be loaded with a great number of genetically established coordinations that seem no longer to have any point, being independent of any present-day function.

In fact pleiotropy shows many such apparently pointless coordinations (cf. Fig. 54a-c). Thus in the Kr/Kr mutant of *Drosophila* there is no visible functional connection between the dilation of the mid-gut and the partial loss of the tracheae, or between the swinging outwards of the mandibles and the whereabouts of the abdominal ganglion etc.

2. *Direct Burden*. This can be assumed when the action of a number of genes is *released* or triggered off by the action of another gene, rather than controlled by it. If two gene effects have become synchronized then the disruption of a predecision will cut out both subsequent decisions and both groups of phenes. Here, therefore, the situation is even graver. Suppose that selection, in response to the external environment, demands that a gene effect be made independent of this sequential switching. This can only be achieved successfully if the regulation *and* release of all the other gene effects has been taken over by new on-off switches *and* new regulators, all produced by mutational trial and error.

This difference between parallel and direct burden, itself has parallels with mosaic and relational pleiotropy (cf. Section VI B2*a*), i.e. in autochthonous and allochthonous gene effects. The burdens of autophenes are parallel to each other. But if autophenes have allophenes dependent on them, then they bear the burdens of these allophenes directly. I shall have to describe this mechanism in detail later as being the mechanism of traditive order (Section VII C).

c. *Superselection and canalization*

For a third time the mechanism of a superselection confronts us. For we have already met such mechanisms with standard-part and hierarchical selection (Section IV C2*c* and V C3). The principle is the same. In the last analysis the external environment decides, but only according to the possibilities which the internal environment permits. The result is also the same in principle. Yesterday's advantage of rapid adaptability is paid for today by a restriction in adaptability. The consequence of narrowing the play of accident is an increase in regulations and necessities, a canalization of available evolutionary paths.

Only the patterns of regulation are different. In all cases evolution acquires a sense of direction. But the pattern by which this directionality shows itself reflects the advantage got by self-design using the appropriate rules of play.

1. *Canalization by superdeterminacy.* Superselection according to the rules of interdependence should produce a form of determination by which equal-rank features are linked to each other. The constancy of the resulting linkage will greatly exceed that produced by the other two mechanisms (mutation rate and the protection by selection of functional advantages). The determination will be a superdeterminacy like those described already (Sections IV C2 and V C3).

The constancy of interdependences exceeds by many orders of magnitude what we should expect from the probability of mutation (multiplied by the number of individuals, the number of species and the number of generations). This excess of constancy is a measure for superdeterminacy. The interdependent features of mammalian hair, for example, are remarkably constant. Otherwise, the bat, flying fox, and flying squirrel would probably have adapted hair to form feathers in aid of a complete conquest of the air. The 9 + 2 pattern of the cilium is even more constant although the sperms of some turbellarians show that the 9 + 1 pattern is just as capable of functioning.[22]

The fact that most features are superdetermined far beyond the functional requirements is essential to the way we recognize and describe organisms. For adaptations only affect features relatively and within strict limits set by fixated interdependence. No anatomical concept could otherwise be created. An absolutely adapted organism would be a mass of pure adaptively convergent features and not open to systematic thought. Or rather, if a system of types of 'Lebensformen' were set up, nobody could mistake it for a system of inherited relationships.

2. *Canalization of thought.* The agreement with the structure of anatomical concepts is probably no accident. But how can we understand this agreement when our concepts in general are of the same structure? For concepts are defined by the interdependence of these few subfeatures which are most constantly correlated or where some form of integration or a statistic of correlation would best apply. In contrast think·of such surface analogies as star 'fishes' and jelly 'fishes', family 'trees' and 'boot' trees.

This brings us back to the universality of the interdependence principle. Like standard-part and hierarchy it is a principle of organic order and a reality of biological structure. It is most unlikely to be a projection of the fact that our thought functions with interdependences. This mode of thought will therefore be itself a product of evolution.

3. Interdependence in civilization

We therefore need to examine yet another problem of conjunction of features. Human civilization, and man's cooperative and competitive life in general, are enmeshed in a dense network of interdependences. The causes and results of these so closely agree with those of somatic evolution that a comparison will perhaps be useful.[23] I suspect that the forms of order in somatic and civilized evolution depend on identical determinative laws. However, as in previous chapters (Section IV C3 and V C4), I do not wish to force this suspicion upon anybody. As a scientist I am too well aware how much is still an open question. Moreover, it is uncertain whether people are yet ready to profit from it to make society more humane.

In any case, interdependent forms of order are the oldest in the social structure of man (Fig. 60). And they are the precondition for the origin of more complicated patterns like hierarchy or, as discussed later, traditive inheritance.[24] Interdependent forms of order must have arisen in the hunting packs of our prehuman ancestors and have been foreshadowed even earlier than that.[25] They are, besides, the first forms of order which science began systematically to consider in the subjects of ethology,[26] archaeology and

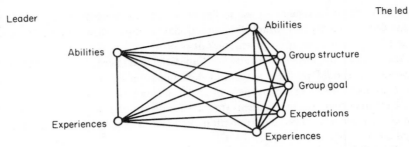

Fig. 60. Interdependence in civilization exemplified by the reciprocal dependencies of 'leadership qualities'. (From Hofstätter, 1959).

ethnology, social psychology,[27] sociology (Fig. 60), political economy and political science. The material is extremely extensive, which is the more reason to be brief.

a. Success and long-term success

As already explained, the highly significant adaptive advantages of interdependence depend on relative success today, paid for by disadvantages tomorrow. The relativity depends on the fact that 'success' can only be measured with reference to competing neighbouring systems. Even this success deserves attention since the interdependence pattern of the largest communities (of states and political blocks) has recently become worldwide. For success is measured by the neighbours' failure.

If the comparison can be trusted, the mechanisms during the successful phase involves achieving greater adaptability with the same information content, or the same adaptability with less information content, or in most cases probably both together. In terms of this second kind of evolution (i.e. within civilization), less information content means less knowledge, ability, farsightedness or wisdom; adaptive success, on the other hand, is the satisfaction of greed and of the demands for standardization, security, and power. Objectively speaking we ought to distrust this mechanism. But the experience of success, stemming from the successful phase, must be the reason why we assign approving labels to it such as 'specialization', 'technicalization', 'rationalization', and 'industrialization'. Without doubt the evolution of interdependence was a natural necessity for the protection both of the individual and of the group, even in the earliest communities. Nevertheless, the second civilizing type of evolution, freed from the genetic brake, has galloped away from this basic necessity as organization, communication, and belief in progress have all expanded.

b. Dependence and tolerance

A feature of the mechanism is that both producers and product become dependent and reduced in individuality. This goes with an appreciation that reduced individuality is scarcely compatible with human dignity. These drawbacks have arisen at a time which, biologically speaking, still belongs to the phase of success.

Evolution will not present its full audit until later. This will happen when the evolved interdependences have become matted together. Adaptive possibilities will gradually have become more restricted until conventionality seems like a necessity and canalization is seen as the goal of civilization.

To complete the comparison we should remember that the renaissance of any biological system demands the complete replacement of some dependences, in this case of implicit 'self-evident' social assumptions. Canalization makes such replacement the more difficult.

The obvious necessity of this mechanism may seem for subjective reasons unacceptable. But the mechanism of traditive inheritance, dealt with in the next chapter, will in that case seem easier to tolerate.

NOTES

1 The specialist will find this in Riedl and Forstner (1968).
2 Wainwright (1969).
3 Discussed exhaustively by Clark (1964).
4 This was surveyed as a principle acting even in the molecular realm by Wainwright *et al.* (1974).
5 These discoveries are well worth looking up, either in the works of the poet Christian Morgenstern or in the exploratory voyage of Stümpke (1964).
6 Excellent examples are given by Portman (1948), Cuénot (1951), Remane (1971), Heberer (1959a) on the lock-and-key combination, and Osche (1966).
7 Good examples in Simpson (1951).
8 In H.-J. Stammer (1959).
9 Thompson (1942).
10 Compare Hadorn (1945a) or Kühn (1965) and the survey in Hadorn (1955).
11 Hadorn (1955, p.191).
12 Kühn (1965, p.543).
13 Details and further information in Shatoury (1956), Waddington (1956, 1957), Lewis (1964). More recent literature in Kiger (1973).
14 These and many other relevant mutants have long been known and were summarized by Bridges and Brehme (1944).
15 Balkaschina (1929); see also Roberts (1964) or Gehring (1966).
16 Compare e.g. Lerner (1954), Waddington (1957), Stern (1968) and the abundant literature cited therein. See also the symposium volumes edited by Locke, e.g. (1966) and (1968).
17 Surveys in Hadorn (1955), Lerner (1953), Waddington (1957), Stern (1968), Hadorn (1966b), Waddington (1966) and other references cited in these works.
18 Even in 1927, Korschelt was able to make an extensive compilation.
19 The need for caution was already emphasized by Morgan (1907) and Herbst (1916) as cited in Korschelt (1927). References to still earlier works can be found in the first two authors.
20 Nicolai Hartmann (1950). See also Section VIII B4*b*
21 This refers to the well known Danfurth short-tailed mutant of the mouse (cf. Fig. 64a-d). This was studied comparatively by Grüneberg (1952) while its importance was particularly emphasized by Waddington (1957). See Chapter VII.
22 This question was surveyed by Hendelberg (1969).
23 Compare 'interdependence' in political economy as discussed by Kuenne (1969) and Georgescu-Roegen (1971) and in political science for example in Lehmbruch (1967).
24 For the origin of this interdependence and its dependence on civilization see Berger and Luckmann (1966). For its canalizing effect see Lorenz (1973).
25 Literature in Ardrey (1969) and Darlington (1969).
26 Lorenz (1963, 1973), Eibl-Eibesfeldt (1967, 1970), Wickler (1969).
27 Hofstätter (1959).

CHAPTER VII
THE TRADITIVE PATTERN OF ORDER

A. Introduction and Definition

The fourth pattern of universal organic order I shall call traditive inheritance (or tradition). It could also be called the 'order-on-order' principle in Schrödinger's sense. The principle in question is that of *tradere*, implying handing over or transmission. This traditive principle, so to speak, adds the time axis to the three principles of order already discussed. Correspondingly the evidence for it comes from the three developmental sciences of phylogeny, embryology, and development physiology. *The traditive pattern of order depends on the fact that events (whether features or concepts) are only understandable, recognizable, or meaningful because they can be traced back to, and depend upon, identical predecessors.* A whole series of sciences has concerned itself with this principle. It has been the particular concern of ethnology and ethology at one end of the series and molecular genetics and thermodynamics at the other.

In the first group of sciences[1] 'tradition' refers to the non-genetical inheritance of corresponding features (types of behaviour, usage, or fashion). These can become divorced from their original function, change their form, and even become ritualized to mere symbolism. In the second group of sciences it was Erwin Schrödinger[2] who proved that order must always depend on order and can only arise from order. This was a quantitative insight to which only the qualitative aspects now need to be added.

Traditive inheritance, is the sense used here, is universal. It ranges from thought to molecular-biological events and thus takes in structural biological events, from the ground plan to the reciprocal effects of two genes. The universality of traditive inheritance is easy to appreciate by trying to imagine a world where it is absent.

a. A fantasy world without traditive inheritance

The folk-saying has it that 'Nature does not make jumps'. This means that, if we are to have confidence in an ordered world, we expect that every thing in its place must have predecessors. And that, if we do not understand a thing at once, then by considering it more closely we shall recognize it as 'That is nothing other than a . . .'

Departures from expected traditive inheritance (Fig. 61) are known to us from the magic of sagas and fairy tales when a frog becomes a prince or a maiden turns into a flower. In a coarser form we know them from conjuring tricks, when a pair of guinea pigs turn into a chicken. Whether young or old, we distrust these departures from tradition to the extent that we trust in experience. Epistemologically this is the source of our distrust.

At the beginning of the book (Section I B4) I showed that order was the arithmetical product of law content times number of instances. Indeed, I defined it as such. To recognize a regularity, therefore, we must observe repeated instances of it. Our degree of conviction depends on confirmed prediction so that, for example, in a message with the

Fig. 61. An example of the breakdown of traditive order; the metamorphosis of Daphne by change in structure beyond any connection recognized by experience. Original based on the Greek myth.

events $2\,4\,8\,16 - 2\,4\,8\,16 - 2\,4\,8$ ... a further event will be expected and can be predicted as '16'. This predictability converts $bits_I$ (uncertainly) into the equivalent amount of $bits_D$ (certainty, or insight into the determinative occurrence).

The sequence of repetitions that is necessary to any insight can be thought of, also, as a sequence of the mechanisms behind these repetitions. This gives a general criterion for traditive inheritance — it is the handing on of identical determinative mechanisms. In this connection we would accept a gradual transformation in the sequence of events so long as a gradual change in the still identical regularity behind them was the most probable explanation.

This gives the surprising result that we can imagine a world without traditive inheritance but could not understand it. We can conceive the events of such a world, as a series of arbitrary perceptions, but we could not find our way among them.

b. Preconditions and forms

The handing-down of identical determinative decisions is therefore the precondition for the existence and recognition of traditive inheritance.

In measuring the probability that two similar events are instances of an identical law we can apply our stochastic solution of the homology theorem (cf. Section II B2*d*). This involves specifying the improbability that such a repetition would be expected by accident.

The way in which the expected regularity is established and repeated does not matter in the first instance. The question of causes should only be considered later, so as not to confuse the paths that lead to insight. This causal question involves considerable variety even as applied to systems with a mechanism of self-repetition, i.e. to living systems and their products.

1. *The decisions.* Sometimes the law content is established and transmitted by the genetic code. Sometimes, however, this happens with the aid of psychological functions involving behaviour, usage, and instruction. In these cases identicality of the replicas is monitored, not directly by the laws of chemical combination, but indirectly by the laws of the group, involving partners or teachers. But here also a mechanism has arisen by which decisions are defined and established. This second evolution, divorced from the genetic code, has developed a code of its own in the form of speech and writing. In this code decisions are materially recorded, established, and handed down, in the form of commands, prohibitions, and laws, just as in the first kind of evolution. Traditive inheritance is a universal principle.

It should be remembered that even the 'Yes' and 'No' of human thought only have a meaning because 'Yes' did not mean 'No' and never did. In principle this is the same as the back-and-forth movement of the balance wheel of a watch. We should also remember that we are confident of the identicality of the letters of the alphabet although their shape and indeed their sound has changed. We now see 'father' as different from 'pater' although formerly they were one and the same. Also a naval gunner knows what he means by 'torpedo' even if he does not know that it originally meant a fish.

2. *The events.* Nevertheless, the differences in degree of alteration are important. To take one limiting case, suppose that nothing changes. The decisions remain the same, whether in pyrimidine bases or the writing of Morse symbols. And the events also remain the same, as from one generation of mice to the next or one printing of a dictionary to the next. In such case we could, with a sense of relief, write repetition instead of traditive inheritance. Suppose, on the other hand, that the changes are perceptible. We then appreciate the truly strange quality of traditive inheritance which is that new events do not involve the introduction of new decisions, but the manipulation of old ones. At the extreme, the change in structure and function is so extensive that only scientific methods can show that it has happened. Consider, for example, the human appendix, the gill anlagen of an embryo, the modes of greeting, or the fact that people show their teeth when they smile. Special cases can be categorized as atavism, vestigialization or recapitulation of past stages (with Ernst Haeckel). An extreme change of function of a traditive action can be called ritualization. An extensive simplification of traditively inherited structures can be called symbolization.

The inclusive concept of traditive inheritance, therefore, justifies the remark that the handing-down of identical events, whatever their particular form, may involve extensive alteration in purpose. The causal connection may become so stretched or convoluted that it can no longer be discerned.

I shall now leave these dry definitions and return to the living phenomena. I shall deal first with the evidence and then with the mechanism, the necessity, and the explanation.

B. The Morphology of Traditive Patterns

Schliemann took Homer literally and discovered the Troy of King Priam. Following his example I shall take Haeckel literally. As is well known, the crucial biological insight of Haeckel's law resulted from the labour of several nineteenth century anatomists.[3] It can be expressed by the statement that: 'Ontogeny is a short recapitulation of phylogeny.'[4] But why ever should this be so? Why is it necessary that the life history of an organism should always repeat the deviations of the organism's evolution? This is the problem.

We are therefore trying to explain Haeckel's law.[5] But a critical reader might ask whether such an explanation is necessary. Has the law not been explained already? In actual fact, however, the two types of genesis have merely been ingeniously correlated. The 'why' is unsolved. It is not enough to say that 'Nature makes no jumps'. Folk wisdom is not sufficient when we are looking for a causal mechanism.

The more general question can be put as follows: Will ontogeny repeat not merely the pattern of events but probably the pattern of decisions also? But how are these complexes of determinative decisions organized and along what channels? (The existence of such complexes was established in Chapter VI.) I shall begin with the simplest case.

1. The conservation of ancient patterns

This is the general theme of what follows. The whole of evolution is a stratified structure made of such conserved patterns. The limits and contents of such patterns can best be illustrated by a particular phenomenon which is the sudden change of phene systems (cf. Sections V C1a and VI B2b).

a. Spontaneous atavism

Some individuals have spontaneously deviated from the usual outward expression of their species and produced structures simulating ones passed through in phylogeny. This remarkable fact has long been of interest. It is known as spontaneous atavism or the production of throwbacks. It is not always easy to distinguish such cases from mere mistakes but there are plenty of indubitable examples.

Two such undoubted examples are the two- and three-toed sports of the domestic horse[6] (as already mentioned in Fig. 62a-g). The side toes 2 and 4 have long been reduced in the normal horse to splints but in these mutants they develop with great completeness; they resemble the corresponding toes of *Merychippus* (Miocene, Fig. 62f) or even *Miohippus* (Oligocene, Fig. 62c). This exemplary case is examined more closely in the

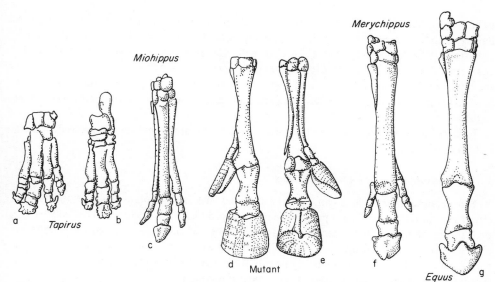

Fig. 62 a-g. Spontaneous atavism illustrated by the two-toed mutant of the horse (d and e) and forms that can be compared with it. (a-b) Fore- and hindfoot of a recent tapir. (c) and (f) Forefeet of two fossils. (g) Forefoot of normal Recent horse. (a) and (b) from Gregory (1951), (d) and (e) from Schindewolf (1950) and also Romer (1966).

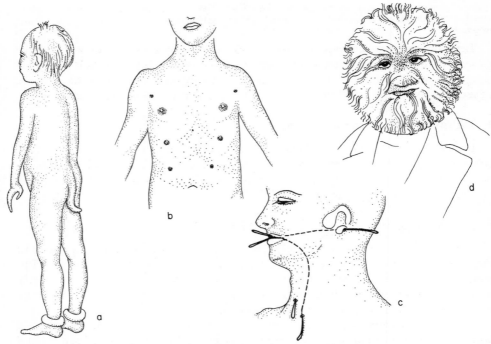

Fig. 63 a-d. Cases of atavism in man. (a) A tailed child; (b) A boy with supernumerary nipples (and alveoli); (c) A grown man with cervical fistulae — these are canals comparable with gill slits and are indicated by probes; (d) The face and body covered with hair — the so-called dog man. (c) After Corning, the others from Wiedersheim (1893).

next section (*b*). There are many other undoubted spontaneous atavisms — extra toes in llamas, hind legs in whales, wings in flightless insects, the limbs of insects and crabs, plaice coloured on both sides etc. There are even behavioural atavisms as with nest building in the house sparrow.[7] Atavisms have also been observed in man (Fig. 63a-d). For example, the tail can be conserved, or fur; four or more nipples can occur; the uterus can sometimes be divided; a cloaca may be present; even gill slits may be preserved as cervical fistulae.[8]

In hybridization, features of the common ancestor are sometimes expressed. This is hybrid atavism.

b. The cryptotype — relict homoeostasis

Stochastic considerations can again be invoked. We can calculate the probability that structures simulating ancient patterns could be produced by accident. This will be the probability of mutation to the power of the number of single features which need to be altered in a coordinated manner. This calculation shows that the effect of accident can be totally excluded. The only remaining possibility is a switching-over to a conserved archaic pattern of determinative decisions. This is most remarkable.

I can suggest a comparison. Suppose a fully automatic car factory, following prearranged plans, occasionally produced in a series of modern cars an example with carbide lamps, or with a mediaeval wagon wheel (complete with hand-wrought nails), or even with a perfect stone-age wheel on one of the axles? The factory would have made an extraordinary mistake. Accident could explain the place and time of the mistaken decision and perhaps also the choice of mistaken part. Accident could never explain, however, why the replacement should mimic a well-known part from the history of vehicles. Somewhere in the memory of the works there would need to be ancient plans which could be brought into use as the result of a single mistake. The factory's instructions would contain the history of its product.

1. *Functions.* Presumably we have here a cryptotype[9] or relict homoeostasis. There must be a pigeon-hole in the genotype[10] in which the old instructions are kept and there must be a mechanism that can switch them on again. But how do such building instructions, now functionless and superseded, come to exist in the modern genome? It is no surprise, of course, to find the functional pattern of phene systems copied in the genotype, for this fits with what we know of the selection of interdependence (Section VI C). Consider the periods of time over which they must have been conserved, however. Why *should* they have been conserved?

Twenty to forty million years have gone by since the Miocene and Lower Oligocene. A few to several million generations of horses have passed by since then. Nevertheless a complicated pattern of commands has been preserved which is coordinated within itself, still able to be integrated with the recent genotype and sometimes interchangeable with other patterns of commands. This pattern is, in itself, amazingly improbable. Enough is known of conservation prospects to show that, without constant restoration and massive protection by selection, such a thing would be totally impossible. Protection and restoration is never given unless needed, unless there are functions involved.

We are therefore forced to postulate such functions – relict or interfunctions. In the epigenetic system there must be true jobs for these archaic instructions. In the last analysis these jobs will be tested by selection which will act by way of the viability of the end product. We shall soon discover what these functions are.

2. *Distribution.* We are obviously dealing here with a widespread phenomenon, for the rarity of spontaneous atavism is, of course, deceptive. It must be remembered that many of these mutants are not expressed because they do not stay the course, disintegrating as young embryos. It is most striking that all the named spontaneous atavisms affect features at the distal ends of functional chains such as the end of the vertebral column or of the limbs. If the undivided heart, the pronephros, or the gills were retained they would obviously be catastrophic.

Naturally enough, little is known about atavisms of the internal organization. They would have to be very delicately balanced to be differentiated at all. Even here, however, there is one example. Danfurth's short-tailed mutant of the mouse[11] shows an alteration of the first two cervical vertebrae (Fig. 64a-d). The axis in this mutant has no odontoid process (Fig. 64d). The atlas, on the other hand, has an expanded articular surface in the centrum region, presumably in compensation (Fig. 64b). In terms of history, the odontoid process is the centrum of the atlas which has become separated from the latter and fused with the axis. This may, therefore, be a very delicately balanced throwback to an amazingly ancient stage.

2. The general characteristics of ancient patterns

The general characteristics of cryptotypes, such as are specially important here, have been studied for decades by developmental physiologists. They have been verified repeatedly so that we can be totally confident of them. These characteristics concern, first, the positions in the embryo in which the patterns of determinative decisions appear (see Section VII B2a). I shall also consider the characteristics of their functioning (*b*), of their temporal sequence (*c*), and of their phyletic relationships with respect to function and arrangement (*d*).

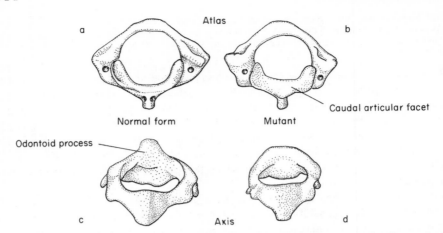

Fig. 64 a-d. A possible atavism of an internal feature. The first and second cervical vertebrae of Danfurth's short-tailed mutant of the mouse. Note the disappearance of the odontoid process of the second vertebra and the compensatory expansion of the posterior articular surface of the first vertebra. From Waddington (1957).

I must point out that this basic field of biology extends far beyond the modest questions dealt with here. It considers, for example, the chemical structure of the transmission of commands and their triggering and specific effects.[1][2] Lack of space forces me to be brief and the questions that concern us demand concentration on a few reliable results.

To summarize, we are again faced with the epigenesis problem as in Sections VI B2 and 3. This problem concerns how the flow of determinative decisions is organized from the genome to the completed organism. Is it methodologically sound to investigate this process by stochastic and morphological means, when in the last analysis it is molecular-biological in nature? This is a well-known controversy (Section II C1). The success of the synthetic flowing-together of chains of determinants is beyond doubt. The analysis of the complex entities that these chains produce has been equally successful.

a. The topography of the determination complexes

I have already shown that the decisions required for determination in differentiation processes are combined together to form organized complexes (Section VI B2). Where are these complexes situated in the anatomy of an embryo? Some could be active in the whole embryo, somewhat like the Bible with its commandments in all the houses of a town. (Hormones also act in this very general way.) In fact, however, the region of action of most determination complexes usually is very limited in time and space being like the plans for a new cathedral or the statutes of a society in the town. In developmental physiology such a region of limited action is called an embryonic formative tissue or blasteme (cf. Fig. 65a-c).

1. *Organizers or inductors.* During the actual period when a blasteme is the actual site of such a complex of commands it is called an organizer or inductor. It is called a primary inductor or organization centre if it is the first organizer of a whole chain. In some species, for example in amphibia which have been specially well studied, two dozen or more such organizers are known. It is assumed that some such arrangement of organizers must represent a general principle.

This principle, of course, is derived from organogenesis and is recognized in vertebrates and hemimetabolous insects. In primordial development, in extreme forms of larva, and in holometabolous insects there are deviations which I shall discuss later (Section VII B3c).

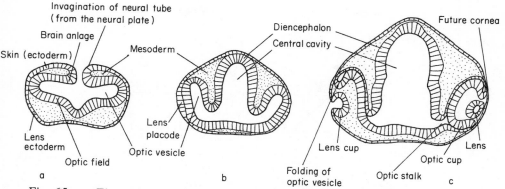

Fig. 65 a-c. The embryonic development of the vertebrate eye, illustrated by half diagrammatic sections through the heads of embryos of increasing age. Ectodermal blastemes are hatched. Mesodermal ones are stippled. From Korschelt and Heider (1936).

Characteristically these organizers exert their determinative, differentiating, and law-giving effect by induction on strictly localized blastemes, most of which are topographically adjacent. They may do this directly. Alternatively, having themselves become differentiated, they may act by chains and networks of organizers and inductors to attain their true goal, which is the differentiation of organs.

2. *The organizer and the interphene.* The fact that organizers can thus be topographically defined is important. Equally important is their clear individuality from the viewpoint of comparative anatomy. Because of this the organizers are named after the embryonic tissues and organs where they are located, e.g. the medullary plate, the notochordal and trunk mesoderm, the neural crest, the optic vesicle, and the otic vesicle etc.

The organizers are therefore sited not in random embryonic tissues, but in structures that were important in the ancestors of the species and are now obstinately recapitulated in the embryonic development. These structures, as I shall show later, are palingenetic (recapitulatory).

Thus the medullary plate must correspond to the scarcely invaginated nervous system of our Precambrian ancestors, which would have been somewhat like that of modern hemichordates. The notochordal anlage corresponds to our ancestors in the amphioxus stage. The optic vesicle would correspond to the first stages of the modern lensed eye etc.

Organizers thus correspond to phenes which, in our early ancestors, were the terminal goal of embryonic development. Such definitive phenes can be called metaphenes. Not until later were these built over by further differentiations and became preliminary stages in development. These preliminary stages can be called intermediate or interphenes for they are certainly identical to the metaphenes of the ancestors.

The organizers, or bearers of organic determination complexes, are therefore not random blastemes. On the contrary they are interphenes which are the functional systems of earlier ancestors recapitulated in embryonic development. This is a surprising correlation but a very important one. What are the present-day functions of these interphenes?

h. The functions of interphenes

This brings us to one of the most marvellous parts of experimental biology[13] and to a relationship which has become certain but still remains astonishing.

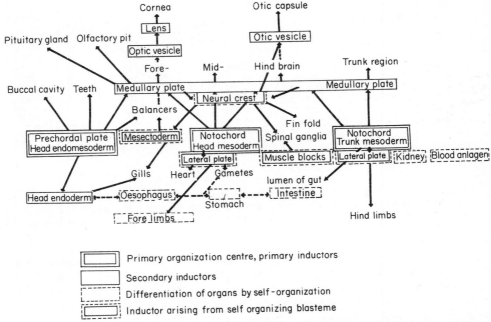

Fig. 66. The induction pattern of a vertebrate embryo. The inductors and the direction of the inducing effects of the blastemes are shown in their natural positional relationships. From Seidel (1953).

1. This relationship is that interphene organizers mostly contain the determinant *commands for one particular adjacent region* — the so-called reaction tissue. This is likewise an interphene, if it is not a metaphene. The classic case is the optic vesicle of vertebrates (Fig. 65a-c). At a particular time in development this will induce lens formation in the overlying skin of the head. The situation is best known in the amphibians. If the vesicle is removed, no lens arises. If the head skin over the vesicle is replaced by belly skin then the latter will be induced to form a lens.

It is true that in the edible frog *Rana esculenta* a lens will almost always arise even when the optic vesicle has been removed while this will sometimes happen in the toad *Bombinator* also. However, transplantations of skin from other species show that the optic vesicle still has an inductive effect in these two forms. The neighbouring tissue has merely become self-organizing, probably by way of some type of coordination.

A great number of such inductive effects have been studied together with other details such as gradients, polarization, and reciprocal dependence.[14] The distinction between induction and self-organization is important to us and is confirmed everywhere. Some blastemes differentiate into organs by self-differentiation. Why do not all of them do so? Would we not expect that the building instructions would always be found inside every building under construction?

2. *Commands to successors.* Why are the building instructions located in a neighbouring building, as if in the builder's hut next to the cathedral being built? Looking for a functional correlation between the organizer blasteme and the reacting blasteme, we always come to the same answer. *The organizer blastemes are the preconditions for the reacting blasteme and are also its phylogenetic predecessors.*

The chain of induction in eye development illustrates this clearly (Fig. 66). The development of the forebrain (being the first secondary inductor of the anterior medullary plate) is the precondition for forming the optic fields which grow out to reach the sides of the head (being the second secondary inductor of the optic vesicle — compare Fig. 65b). The optic vesicles fold back in a basin shape under the skin to form the optic cups and only after this happens would lens formation be expected, both functionally and phylogenetically (Fig. 67). Before that the lens would be a functional absurdity and a physiological impossibility. Only the lens (the third secondary inductor) makes it relevant to create a fully transparent cornea (cf. also Figs. 65c and 66).

Similarly in civilization a little country parsonage might be used as a builder's hut for the rectory and parish church built over it. And generations later the rectory might contain the plans for the cathedral that arose on the site. Planning and adaptation always act from old to new. The functions expand, without ever being interrupted.

There seem to be no exceptions to this correlation wherever induction and the direction of induction are known. The hindbrain is the precondition for the auditory vesicle, while only this permits the formation of the auditory capsule. The arrangement of the spinal ganglia is only meaningful given the neural crest and the muscle-block segmentation. The pharynx is the precondition for the gills, and so forth. In this respect the connection seems self-evident and indeed almost a triviality. But the relationship is again so constant and so complicated that it cannot possibly have arisen by accident.

c. The recapitulation of the determination patterns

The paths of induction therefore form a pattern extraordinarily like the steps of functional differentiation in the phylogeny of the animal in question. The agreement

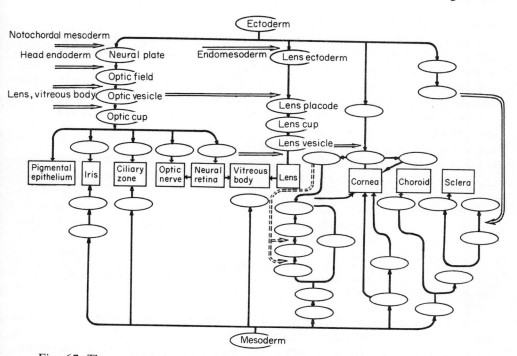

Fig. 67. The pattern of induction and of differentiation in the vertebrate eye. The embryonic blastemes (or interphenes) are indicated by ellipses and the definitive tissues (metaphenes) by rectangles. The paths of differentiation are represented by black arrows. Only blastemes mentioned in the text are named; simplified after Coulombre (1965).

between the two patterns cannot be explained by accident, so a causal connection can be expected. In such a case the older event must be the cause of the younger one.

At the beginning of the chapter I proposed to take Haeckel literally. Accordingly I now postulate that the biogenetic law holds not only for the pattern of events, but also for the decisions behind them. *The epigenetic system is a brief recapitulation of its own origin.*

I began with a postulate or working hypothesis. We have now reached a probability, though of very high degree. We have still not found the causal necessity which would prove the connection conclusively. We can foresee that this causal necessity lies in the connection between burden and constancy.

The fact that the paths of induction simulate the steps of functional differentiation in phylogeny is itself enough to solve three hitherto open questions. First, the cause for phylogenetic features being repeated, so that they follow Haeckel's law in the old sense, is that the decisions that trigger these features need to be repeated in the first place. Second, the remarkable fact that blastemes have no access to their own building instructions, but only to those of their neighbours, can be explained by the history of the origin of the blastemes. Third, the imitative epigenotype, which we deduced to exist from interdependence conditions (Section VI C1c), now reveals its mode of action. Primarily, it must correspond to the induction pattern in the system of interphenes (Figs. 66 and 67). This clears the path for a study of why the pattern of decisions is itself repeated (Section VII C).

The next step, however, is to test the hypothesis in all possible ways.

d. Decisions and phyletic relationships

If the hypothesis is correct we should expect, first, that the patterns of the epigenetic systems would become more similar with increasing phyletic relationship. Second, we should expect that related species would be able to read each other's inherited induction commands and that the legibility would decrease with increasing distance from the latest common ancestor. Both these expectations are convincingly fulfilled.

1. *Homologous patterns.* Induction patterns formed by organizers and reaction tissues are remarkably similar from species to species – so similar as to be virtually identical for large systematic groups, such as amphibians or even for vertebrates.[15] Figures 66 and 67 in effect present a space-time picture showing the universal significance of the morphotype. Our stochastic criterion of homology (cf. Section II B2d) leaves no doubt that we are dealing with homologous structures.

2. *Homodynamic effects.* The question whether the determination commands can be read, understood, and followed beyond the limits of the species has been clearly answered by xenoplastic transplantation experiments.

These involve, for example, the removal from one species of a particular reaction tissue and its replacement in the same position by the corresponding tissue of another species. Morphogenesis will then show how the new tissue reacts to commands from a foreign species. It will also show how far 'initiative' (or self-organization) can be tolerated on the one hand, or obedience to foreign commands on the other.

Experiment thus shows that the skin can 'read' the lens command from a foreign optic vesicle. This is true from species to species within a genus, from genus to genus within a family, and even from family to family. Thus the belly skin of the toad (*Bufo vulgaris*, family Bufonidae) can read the lens command from the optic vesicle of the edible frog (*Rana esculenta*, family Ranidae, both from the suborder Phaneroglossa of the tailless amphibians). The organizer induces the 'what' (or the 'text') while the reaction tissue decides the 'how' (or the 'pronunciation') Kühn states that: 'Such combinations indicate

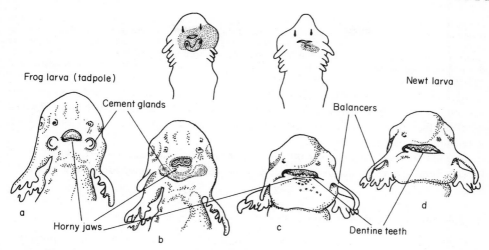

Fig. 68 a-d. Order chimaeras and the normal forms as shown by newt larvae with frog implants. (b) Big implant round the mouth region; (c) small implant behind the mouth region. (a) (d) Normal forms. The parts derived from the frog are stippled in the sketch-diagrams. (b and c from Seidel (1953)).

phylogenetic and developmental-physiological connections. They show to what extent general and specialized parts are present in the origin of an organ.[16] We are dealing here with homologous patterns of decisions, i.e. with homodynamic effects.[17]

3. *The limits of homology.* We encounter these in developmental-physiological decisions as we do in morphology. This is shown by the famous 'Order-chimaeras' (Fig. 68a-d) which are reactions to transplants successfully carried out between different orders of amphibians (Urodela and Anura). Here again commands are still 'read' as being identical, for example, in the optic and labyrinth regions, although both the 'texts' and the 'readers' have been replicated totally separate from each other since the Palaeozoic, i.e. for at least 200 million years.[18]

The legibility only alters when new larval features are superimposed on more primitive ones — for example, when the frog or toad features of horny jaws and attachment discs are imposed on the teeth and attachment 'balancers' of the newt *Triturus*. 'How far can the altered part of an anlage system make use of the unaltered remainder?'[19] Balancer epidermis of *Triturus*, when transplanted to overlie the mesenchyme of *Bombinator*, will produce a chimaeric balancer. The belly skin from *Rana esculenta* will produce a frog-tadpole's mouth when transferred to the mouth region of *Triturus* (Fig. 68b). *Bombinator* moieties xenoplastically transplanted to a *Triturus* host will produce chimaeric newt teeth even in the toad moiety. Kühn continues by saying: 'With a new formation the essentially new part of a normal reaction may be limited to a change in the potency of the epidermis. The induction system and reaction potential can be transferred.'[20]

It therefore seems that even 2×500 million generations of separate evolution do not detract from the legibility of the determining text. Only a new formation will add a new law text to the old one, but even so the old ones can be understood and followed without difficulty. The identical handing-down of patterns of commands can therefore be treated as another new fact.

A biologist will wonder how the peculiarities of primordial development and of holometabolous insect development fit into this replication thesis. In these cases it is true that self-organization predominates. The recapitulatory properties of the blastomeres in primordial development, and of the imaginal discs of holometabolous insects, are indeed dubious. For this reason I shall treat them only after considering the caenogenesis problem. Biologists know how much unnecessary controversy has resulted from 'the phylogeny of primordial development' and how impossible it would be to deduce the law of recapitulation from the Holometabola alone. I must leave these more complicated and advanced situations to the end (cf. Section VII B3*c*).

It is now certain, therefore, that determinative decisions are organized in complexes and that these can be preserved identical for periods of astounding length. The last question therefore remains: Why need this be so? Why is it necessary to preserve old determination patterns?

In what follows, evidence for the correlations in question (Sections VII B3 and 4) will be separated from the theoretical necessity for the causal connection (Sections VII C1 and 2).

3. Freedom and fixation of interphenes

Once again the evidence comes from a particular field of knowledge. Comparative embryology, with its applications in phylogeny and high-ranking systematics, has to be taken together with developmental physiology. It is an even larger field. It is more generally known, however, since it became part of the factual content of comparative anatomy in the nineteenth century, when anatomists sought to confirm Darwin's thesis. I can therefore be brief.

a. *The change from metaphene functions to interphene functions*

I shall take this first. In so far as we understand living structures we are forced to suppose that each one of them is necessary. This necessity may in the first place be the whim of an accidentally mutated decision, tolerated by selection, the phene of this decision finding some function in a distal position (as in the examples in Figs. 39-42, Section V B2*a*). But if the phene becomes constant within a phyletic group and takes on the value of a homology, then its functional necessity must have become more deeply anchored. When such a function belongs to the present day then it is easy to say what it is, although its constancy is surprising. An example is the sheath of horn on a bird's beak. However, if the function does not relate to the present day, as with the number of aortic arches in a human embryo (Fig. 69c), then we are at first content to compare it with the ancestors (Fig. 69a). We then say, rightly, that these four pairs of aortic arches must once have had a function in any case. They were the branchial vessels of our ancestors, resembling those of fishes (Fig. 69b).

1. *The theoretical necessity for a function.* It has long been known that an organ system is physically an extremely improbable condition and as such would never survive if selection did not protect it. Certainly it would not survive about 1000 million generations, as with the branchial vessels in the absence of gills. Necessary functions must always have been protected by selection. It follows that aortic arches must always have possessed a function. This would be true not merely when they were metaphenes, functioning as branchial vessels up to the latest gill-bearing ancestor of the mammals. They must always have had some function even as interphenes, up till the present day. The function may have changed, but there must always have been one.

I have already considered these functions (Section VII B2). They must be functions within the induction laws of organogenesis, being irreplaceable in the epigenetic system and therefore firmly anchored.

It should be borne in mind, to stick to the same example, how closely the gill slits and gill vessels must once have been coordinated and over what long periods. We should also remember that anlagen

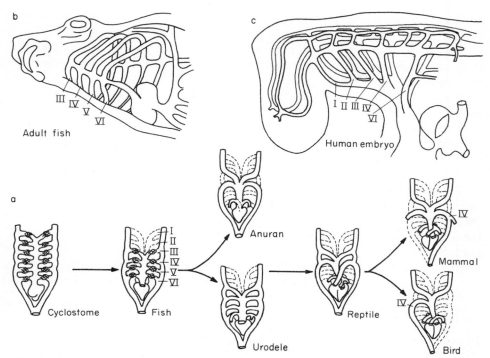

Fig. 69 a-c. A diagrammatic representation of aortic arches of vertebrates. (a) The differentiation of the anlage (dashed lines) to give the adult form (full line). (b) Situation in a bony fish. (c) Situation of a human embryo. Homologous arches are given the same numbers. (a) From Claus, Grobben, and Kühn (1932); (b) from Korschelt and Heider (1936); (c) from Corning (1925).

of these gill slits are still formed in birds and mammals and provide the starting point for important endocrinal glands (thyroid gland, parathyroid glands).

2. *Change in function.* In the foregoing example this has not been studied[21] and must therefore be postulated. In other cases, however, it is well known. These examples will show that the function of the original metaphene can be retained, or altered, or lost, without the interphene function disappearing.

The optic vesicle must have been a metaphene of Precambrian ancestors of the vertebrates, before the lens was formed, or the vesicle was folded back to form a cup. Even since the abundant fish-like animals of the Devonian, which probably already possessed lens and cup,[22] the vesicle has become an interphene of all vertebrates. Neither its optic nor its inductive function have changed.

The notochord is at least as old as the optic vesicle. It persists throughout life in appendicularians, in amphioxus, in some fishes and agnathans (cyclostomes, holocephalans, sturgeons, lung fishes) and in a few amphibians. In all other vertebrates it is broken up into discs while its supportive function has been taken over extensively by the vertebrae until only a tiny remnant is left as the nuclei pulposi in the intervertebral discs. The notochord in these forms has become an interphene, formed as a very early anlage in the embryo. Together with the adjacent mesoderm it has become the primary organization centre for all vertebrate development.

The kidney series is the third example. The pronephros has become an embryonic organ in all vertebrates except myxinoid agnathans and disappears during development.

The mesonephros, which develops later, remains a metaphene in fishes and amphibians. But in reptiles, birds, and mammals it too has become an interphene. The excretory function is entirely taken over by the superimposed metanephros. Nevertheless the whole series of interphenes persists, inductively interwoven in a complicated fashion with the function of the metanephros.[23]

3. *The two kinds of function*. Thus two kinds of function can be assumed for all phenes.

In the first place, there are functions with respect to the external environment. Their necessity can easily be understood 'by introspection', in view of the struggle for day-by-day advantage and of testing by selection. Often these are the first functions that a phene acquires. In metaphenes they may perhaps predominate.

A second kind of function is soon added relating to the internal environment. It is a function in the epigenetic system as soon as further features are superimposed on the phene. This function, as I shall later show, will increase with the burden carried by the phene in the embryological processes of development. It will not disappear even if the external function of the phene totally vanishes. In interphenes these systemic functions will predominate and will be among the last functions that a phene can possess.

In a similar way every functioning system of a house has a first function (such as support, heating or drainage) and a second function which is to help orientation when the house is being built. The foundations are orientated by means of the site plan, measuring the distances from the edge of the building plot. The walls are orientated according to the foundations, or ground plan. The fitted cupboards are adjusted to the walls, making allowance for departures from the ground plan. And the locksmith decides the positions of the locks and hinges by reference to the cupboards. Theoretically there would be no objection to fixing the positions of the locks by measuring from the edges of the building plot, but in practice that would be ridiculous. The same system arises as with living organisms and for the same reasons, it being the result of mistakes, tolerances, and the costs of information in $bits_D$.

4. *The transition of functions*. The change from an external to an internal function will be continuous and harmonious. It will be continuous because of the broad overlap of external and internal functions. It will be harmonious because the two sorts of function will be virtually identical at the point when the organ changes from one to the other.

Metaphenes will become interphenes because their external functions are taken over by substitutes. During the long period of transition the various subfunctions will form a nicely coordinated whole with an identical total function. They will therefore only be meaningful when integrated together. This can be seen, for example, with the change from an 'epithelial vertebral column' (notochord) as cartilaginization begins and the notochord passes into notochordal discs, cartilaginous anlagen etc. Consequently the optimal coordination of the way the organ forms in ontogeny (i.e. the genetically determined functions) will correspond throughout the organ's history, and indeed will depend throughout on the same pattern of commands. We have already met this as the phenomenon of the imitative pattern (Section VI C1).

b. Burden and fixation

In the previous three chapters I discussed the burden and fixation of the metaphenes or definitive features of organisms. I shall now compare the interphenes in this respect. The distinction is between phenes with and without visible external functions. This is somewhat like the customary embryological distinction between palingenetic and caenogenetic features.

1. *Palingenetic and caenogenetic features*. Palingenetic features include all true recapitulations of phylogenetic stages, such as the gill anlagen and transitory notochord of mammals. Caenogenetic features comprise all deviations from true recapitulation. They

are phase shifts, simplifications, and, especially, the additional phenes that appear during the developmental period by confrontation with the developmental environment. Examples of such deviations are the suspension devices of marine larvae, the protective devices of parasitic stages, the pupation of insects, and the umbilical cord and placenta of human embryos, all of which are abandoned before the definitive stage is reached. Other caenogenetic features are of totally different kinds and form a third group of features to which I shall return (Section VII B3c).

Palingenetic phenes are mostly interphenes. Some such recapitulatory features, however, have true external functions as with the tail and branchial vessels of a frog tadpole. If so they are subject to the same conditions of burden as are metaphenes.

Caenogenetic phenes are mostly metaphenes, for they have true external connection to the larval or embryonic environment. They can lose this connection, however, if on a change in the developmental environment, they themselves come to be handed down traditively.[24]

Vestigial organs and atavisms have a comparable functional relationship. At one time they certainly had a task to perform in the external world, but this has now become dubious and has often disappeared completely (cf. Section VII B4a). Here, therefore, I wish to consider the burden and fixation of epigenetic functions only.

2. *The degree of burden*. The burden with which an interphene is loaded in the inner environment of the organism can easily be estimated from its function in the epigenetic system. Using the method of quantification already applied. we only need to establish how many homologues are omitted or caused to disintegrate if the interphene in question does not fulfil its data-transmitting function. This can be shown, for example, by removing the interphene. It can also be shown by applying some chemical which blocks the inductional effects.[25]

Thus when notochord cells are treated at an early stage with LiCl they are prevented from dividing, although they remain fully viable within the notochordal mesoderm. The result is an unmistakable catastrophe. The organization of the trunk musculature and of its axial division into somites breaks down. As a result the order in the longitudinal differentiation of the dorsal nerve cord and of the spinal ganglia of the neural crest disintegrates. This disorganizes a region where many hundreds (or even a few thousand) of homologues would have been expected in the metaphenes. Division of notochordal material therefore has a burden of decision of 10^3 single homologues (within an order of magnitude).

This is a degree of order at which an accidental substitution, even of the most important functions only, is simply not conceivable, however many mutations there may be and however long the time available.

The network of inductions (Figs. 66 and 67) shows that there must be many more such examples. Moreover Fig. 66 only shows part of the pattern that has been known for decades, and even today only a fraction of the epigenetic system has been worked out. Much evidence is therefore to be expected in the future from this source.

3. *The degree of fixation*. The fixation must also be extraordinarily large. To show that interphene functions of this extent are immovable it is not necessary to treat them numerically.

Thus the transitory notochord is anchored absolutely in the ground plan of all chordates from the larvae of tunicates up to man. This is true although the notochord probably lost its primary function as a longitudinal support for the body as soon as the vertebral column had arisen. The fixation of this pure epigenetic function is enormous, being almost as long as the whole fossil history of macroscopic organisms.

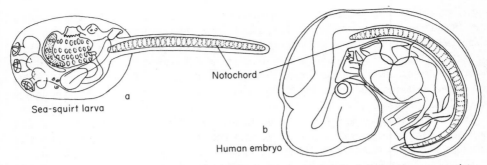

Fig. 70 a-b. Extreme representations of the notochord. (a) In the sea-squirt *Distaplia* it is purely a larval organ, disappearing completely in the adult; (b) In mammals it is solely an embryonic organ which breaks up later into the nuclei pulposi. (a) From Riedl (1970) and (b) from Corning (1925).

The notochords of a sea-squirt larva and of a human embryo are remarkably similar. Their identicality can be taken as proven purely on anatomical grounds (Fig. 70a-b). After all, homologues can differ greatly but their identicality still remains certain (Fig. 7a-e and Section V B1*f*).

Comparative embryology has compiled a huge list of fixated interphenes in the last hundred years. This is only what the network of dependences and burdens and the resulting fixations would lead us to expect. *The epigenetic system necessarily starts from a recapitulation of its own pattern of origin. For the most important decisions carry so large a burden that their accidental replacement is grossly improbable.*

Recapitulation seems to be a general principle, at least as a starting point. Simplifications, or shortenings of ontogenetic detours, can and must occur, however.

c. The freedom to simplify ontogeny

There are two fundamental differences between metaphenes and interphenes. In the first place, if a metaphene homologue has taken a detour in phylogeny, then that has happened irrevocably. This was true, for instance, when a jawbone became an auditory ossicle, or a fish-fin became the paw of a beast of prey, and then the flipper of a dolphin. But when an interphene homologue recapitulates such a detour then, at each reproductive step, it is subject to mutational experiment and selection. This is true when branchial arches become aortic arches or the notochord becomes the nuclei pulposi. In the second place, the functional demands made on a series of metaphene homologues will vary from niche to niche. So far as our methodology is concerned this will happen accidentally, constrained only by the physical possibilities of the system. In this way, a jawbone will finally be used for hearing, a swim bladder for breathing, or a fish's fin for writing a manuscript. But the functional demands on a series of interphene homologues are defined by the 'niches' in the epigenetic system — this is a determination system whose development and change is to a large degree shielded from accident. It is a feedback mechanism in which the individual function under test is itself part of the system.

Therefore we should expect that the interphene sequence, also, would be continually under experiment. Proportionate to the number of replications, functional improvements will be introduced, though only towards the goals set by an intrinsically conservative epigenetic system. It is usual to count these improvements as caenogenetic. However, they are not additions produced by external stimuli, but simplifying novelties caused by internal stimuli. They can be summed up as follows.

1. *The disappearance of burdenless features.* The deletion of phenes of small or moderate burden is probably the most striking characteristic of all embryos, even of older ones. All features which are recently evolved, variable, species specific, and functionally terminal (i.e. distal in position or late in development) are apt to vanish from the epigenetic system. Indeed in the case of accessory features, they need never be taken into it (cf. Section V B2*a*). Such features only become important in free-living larvae. Because of this deletion of transitory features, embryos are to some extent generalized, tending towards the morphotype, as shown in Section VIII B2*b*.

2. *Adaptive shifts of phase.* These are well known in embryology. Sometimes a developmental process is brought forward compared with the time of origin in phylogeny. This happens when the production of a particular organ needs more time than that of other organs. An example is the grossly accelerated development of the brain in the fish stage of a human embryo. Sometimes, on the other hand, a process is postponed, if this gives some advantage, as with the retarded opening of the eyelids.

3. *New short cuts.* Sometimes an ontogenetic detour is so extreme that it is hard to dismantle all its special features when metamorphosis takes place to the definitive condition. In such cases a short cut may arise, rather as a meandering river will cut off a loop.

These ontogenetic detours may greatly exceed anything that has happened in phylogeny, representing extreme larval or caenogenetic specializations. Such forms are specially likely to show abbreviations when metamorphosing to the imago. Examples are found in the metamorphosis of the pilidium larva of marine nemertines and with the ophiopluteus, echinopluteus, brachiolaria, and bipinnaria larvae of echinoderms. Among terrestrial groups it happens with the caterpillar or maggot of holometabolous insects and the subsequent pupa (Fig. 71a-g). As a rule, development away from the extreme larval form happens in these cases by means of invaginations of the skin. These form amnion-like internal environments within which embryonal discs arise and from these discs the definitive form originates. Despite these entirely new short cuts some remnants of palingenetic (recapitulatory) features may persist, though these are of course much more difficult to interpret than if the development were not modified.

The same difficulties of interpretation do not happen with a second group of cases (4 and 5, which are extreme examples of 1). In these the recapitulatory detours are traditively and gradually modified in function.

4. *Schematization.* This is, so to speak, a sort of diagrammatic simplification which is often observed in the recapitulation of very ancient features and therefore in the anlagen of the most fundamental parts. In my opinion such schematization is at work in the frequent occurrence in ontogeny of epithelial tissues, foldings, and cavities, although the organisms in question later show such features only weakly or not at all. Contrary to usual opinions I see such schematization in the way in which many primary and secondary body cavities are formed, and also many archentera and unfolding organ anlagen. These are the phenomena of blastula formation, gastrulation, the subdivision of the mesoderm, and the differentiation of the so-called germ layers (Fig. 72e-h).

The germ-layer doctrine shows excellently how the origin and flow of building materials and of their determinative commands can be ensured in a simplified manner. The germ layers are a diagrammatic scheme, not a representation of the final functions of the phylogenetic ancestors.

If these 'diagrams of organisms' represented functional ancestors they would prove the paradox of a teleological evolution. For their parts always strive towards functions, without being able to possess them during their formation. Like orderly piles of bricks or building timber they do not yet have a

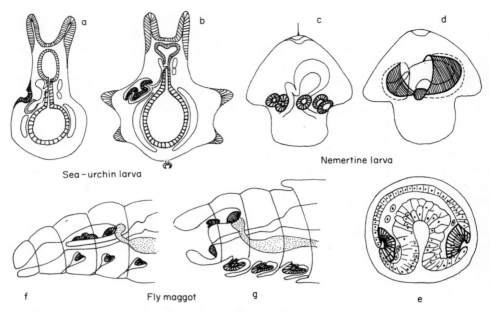

Fig. 71 a-g. Short cuts at metamorphosis by the production of embryonal discs (heavy lines). (a) Pluteus larva at the beginning of disc formation. (b) Pluteus with a completed amnion. (c) Pilidium larva in lateral aspect at the beginning of disc formation. (d) At the end of disc formation. (e) Desor's larva in transverse section inside the egg membrane. (f-g) Stages in forming imaginal discs in a holometabolous insect (lateral aspects). From Korschelt and Heider (1936).

function. In the same way scaffolding may indicate the shape of the future building, though it would fail any test of thermal or noise insulation, not to mention habitability.

The interphenes are command posts. They are the archives for organized determination complexes and carry the sequence of building instructions. They logically repeat the differentiation and development of the determination complexes. And they schematically repeat the ancient design process.

5. *Symbolization*. This is the extreme end-point of simplification. It would be expected to act on the oldest features and is found in primordial development, among types of cleavage and groups of blastomeres (Fig. 72a-d).

Position-structure is correlated so well with representation in such cases that accident can be excluded. The reign of identical regularity is certain. Examples are the spiral cleavage including the destination of the 4d blastomere or the formation of the molluscan or annelidan cross. However it would be naïve to imagine that the first molluscs crept around with a cross on their backs, or that the Algonkian seas were populated at one time with four-cell stages, and later with eight-cell stages and that these finally decided to let cell 4d sink in, so as to form the future musculature from it (cross-formation in Fig. 72d). (This would be like using the architect's drawings as a tent or biting into the sentence : 'This is a juicy apple.')

The blastomeres are sorting trays for the material, and for the determination complexes. They have a greater or lesser power of regulation to which end each pigeon hole also carries the total plan. However, as with schedules, more and more is crossed out as inapplicable, beginning with the master builder's job, passing to the window joiner, the glazier, the boy who puts the putty in, and finally the insurance man who comes later, when a football from the garden happens to go through the glass.

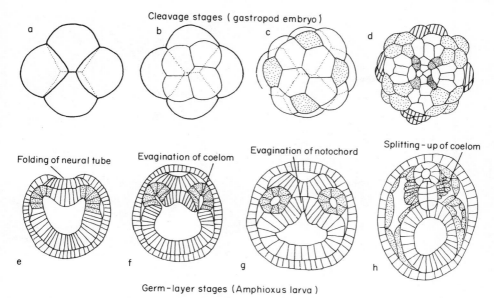

Fig. 72 a-h. Simplification in development. (a-d) Symbolic simplification as shown by the arrangement of blastomeres in spiral cleavage of *Trochus* at the stages with 4, 8, 16, and approximately 64 cells. Cells with the same determination have the same shading. (e-h) Schematic simplification shown by epithelial folding in sections of amphioxus larvae. The lines indicate contacts between cells. The coelom walls are stippled. From Korschelt and Heider (1936).

The pattern of blastomeres and blastomeres themselves are identicalities in their content. They are also flow patterns of traditive determination and are reduced to symbols so as to give the greatest certainty in the differentiation of commands.

4. Traditively inherited metaphenes

This brings us again to definitive features. The argument will depend on the two different types of function that these have, related respectively to the external environment and to the epigenetic system (Section VII B3a). The initial function of a metaphene is probably always a small external function, but the metaphene will be loaded more and more with internal functions as time and responsibility increase.

Suppose, however, that the external functions disappear but the phene is conserved in the definitive organism. We then have a metaphene with no visible function at present, and therefore not visibly protected by selection, but which can persist an extraordinarily long time. This is the next problem.

a. Atavism and the slowness of vestigialization

In this connection it is not the disappearance of organs which is difficult to explain, but their persistence. Neodarwinism has produced sufficiently good reasons why functionless organs should disappear. But how can we explain the many remnants of organs which no longer signify in the life of the organism?

Examples in man are the ear muscles, the fold in the inner corner of the eye, the appendix, and the caudal vertebrae of the coccyx which still carry traces of vestigial tail musculature (the ventral sacrococcygial muscle, Fig. 73a). Our ancestors have entirely

221

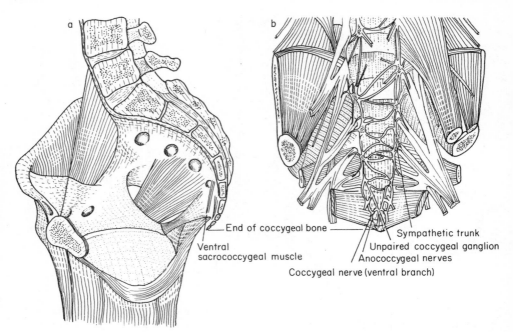

Fig. 73 a-b. Vestiges in the internal organization of the human adult tail musculature. (a) Position of the caudal muscle of the right half of the body (sagittal section through the pelvic region). (b) Position of the system of immediately associated nerves in the dorsal part of the pelvis, opened from the central surface. After Hochstetter (1940-46).

lacked a tail for a few million years, but this little muscle has managed to slip by for more than one hundred thousand generations. In the old sense it must certainly be functionless. Why should it be so persistent?

It has been supposed that such traces, doing no damage but costing nothing, have no selective disadvantage and would therefore be tolerated. Against this is the improbability that improbable organization will survive if selection does not protect it. Moreover such harmless vestiges are not the only ones. Others, such as the human vermiform appendix, are under considerable negative selection pressure. After all, many people die from a faulty appendix, if it is not cut out quickly enough.

1. *The causes*. It has been suggested that such features, in the recent words of Osche,[26] are: 'not functionless at all, but have important jobs to carry out in the complex events of development e.g. as organizers'. Or, as he continues, it could be that the features of organ systems: 'are so interwoven with each other by polygeny, that the anlagen of "superfluous" structures, because of their many subordinate effects, are not simple to remove by mutation'. This indeed must be the answer, as already established in Section VI B2 and VII B3. It is the internal or systemic functions of phenes which forbid their removal by mutation.

Even in the perplexing case of the sacrococcygial muscle (Fig. 73a) the coordination of vertebrae, muscles, and nerves must have been very finely balanced, given the functional importance of the tail from primitive fishes up to and including the primates. As a result it would have been imitated by the induction system of the notochord, notochordal mesoderm, and somites, so as to ensure the correct subdivision of the spinal and sympathetic nerves, and thus handed down traditively. Even the surviving orderliness of the musculature may be a precondition for the organization of the nerves which are functionally very important in supplying the skin of the anal region (ventral branch of the coccygial nerves, the anococcygial nerves, and the unpaired coccygial ganglion, Fig. 73b).

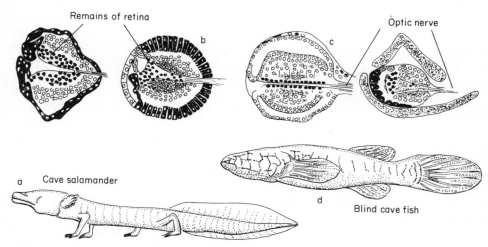

Fig. 74 a-d. The course of vestigialization as shown by the reduction of eyes in cave vertebrates. (a) *Typhlomolge* — above, eye in longitudinal section; below, adult specimen. (b-d) blind fishes or Amblyopsidae. (b) *Amblyopsis spelaeus;* (c) *Typhlichthys;* (d) *Amblyopsis rosae* — above, eye in longitudinal section; below, complete specimen. Note the reduction of the vitreous body (b), the retina (c), and the optic vesicle (d) till the optic nerve is almost the only surviving part. From Vandel (1964).

The slowness with which burdened phenes become vestigialized is a necessity, so far as the present interpretation is concerned. If it had not long been known, we should have had to postulate it.

2. *The course of vestigialization.* This gives a further confirmation. Because of the pattern of traditively inherited internal functions, vestigialization runs backwards up the chain of induction (or sequence of ontogenetic formation). The best evidence comes from cave animals.[27] In vertebrates (Fig. 74a-d) it is the cornea which is dismantled first, the lens second, and then the nervous part of the optic vesicle. The oculo-motor muscles degenerate correspondingly, but the optic nerve remains preserved. The inductive sequence[28] is convincingly shown, but backwards (Figs. 66 and 67).[29]

The pelvic-girdle system of whales is likewise dismantled from out inwards. The python shows the converse, however, for in it the terminal claw retains a function.

b. Adaptation and toleration

If we consider the concept of vestiges further, examining the necessity for subordinate features, then every organism is full of vestiges.

For example, why are human arms innervated from the neck, although they lie in the shoulder region? Why are the muscles that open the jaws so complicated and those that close them so simple? Why are the great descending arteries asymmetrical, while the smaller vessels that they supply are symmetrical? Why are the receptor cells in the eye inverted? Why do the respiratory and alimentary tracts cross over each other? The list could easily be made one hundred questions longer.

Anatomists always give the same answer. The facts can always be explained by the long history of the to-and-fro of past functional conditions. All of them are survivals. Are they also vestiges? They are not, in the sense that the whole still functions. They are in the sense that many subfunctions have been given up and many subfeatures have become functionless, counter functional or even preposterous, but have still been continued. Every organism is a historical being and can only be understood by remembering everything that once it was. Wiedersheim showed in 1893 that only fifteen organ systems in man are evolving progressively, including the hand, the laryngeal muscles, and the brain. Ninety, on the other hand, are in retrogression.

Evolution is a continual compromise. Its results are an omnium-gatherum of obsolescent structures and functional half measures. Most biologists will find it difficult to agree with this, since we have all been brought up, justifiably, to admire the miracle of adaptation. It must be conceded, however, that if evolution was planned, it was planned with great contrariness. A fish fin is used for playing the piano; an archaic olfactory brain has been adapted to produce logic; and an original torpedo design was manipulated in tetrapods into a bridge design and finally came to be balanced on two legs, these being the hind pillars of the bridge. Indeed evolution is not planned, and not teleological. What exists in it is what selection still tolerates out of the reservoir of historically given facts.

Evolution is as full of pointless history as the frontiers of Europe. It has the unplanned contrafunctionality of one of our venerable old cities (even merely as regards the traffic), as distinct from the unatmospheric functionalism of those that were planned last week and finished yesterday. In the first place, evolution is *only* traditive inheritance, or 'order-on-order' as Schrödinger said (1951). Only when this tradition has been organized, does the adaptation of what has not been adapted come into question.

This is important, because our heads were certainly made in the same workshop. Our thinking apparatus must be just as full of non-functional, and indeed harmful, survivals which will bring considerable difficulties in the future. This justifies our worry about the 'well adapted progressive individual', who, while admiring adaptation, is also admiring his own ostensible completeness. How far he has brought us is shown by the mess that humanity is in. It is a dangerous mistake to think that *we* deal with evolution. On the contrary, evolution deals with *us*.[30]

However, let us return to the orderliness of structure. I have given the evidence for the traditive patterns. I now need to present the mechanism that gives these as necessary results. Most of it can already be foreseen.

C. Traditive Selection

The mechanism which produces traditive inheritance in the first place, and causes it to prevail over obstacles, assumes self-reproduction as a precondition. It depends on selection and pays with tomorrow's disadvantages for the crucial advantages of today. It is a causal mechanism and totally non-teleological and it results in a further directive component of evolution. Only this component allows the other three selective mechanisms to achieve their full effect.

1. The advantages of traditively inherited data

As Schrödinger said, order must depend on order, which is to say on determinative regularity that has already been applied. It could not depend on anything else. It could not depend on the effect of accident and, apart from necessity and accident, no third category is known. Of course, every heritable alteration must be ascribed to accident. It is known, however, that the prospects of success for such an alteration depend on the restriction of its possibilities (Section IV C1a VI C1a). And these prospects increase exponentially with this restriction.

a. The necessity of traditive inheritance

From the stochastic viewpoint, life is an extraordinarily improbable condition of its elementary component parts – a condition that can only be improved adaptively by means of accident. This is in no way a contradiction. For the accidental improbability is extraordinary – meaning the extent of selection, again determinative, from random possibilities within the realm that remains accidental.

How should the thousandth part of a half-ordered jigsaw puzzle be put in position? Not by shaking the whole puzzle in its box. How should a wrong letter in this book be corrected by accident? Certainly not by mixing all the letters together in a drum. The only way would be to conserve existing order and try to fit a single piece or letter, chosen at random with respect to evolution, into its correct position by an extraordinary amount of trial and i.e. by an extraordinary amount of selection.

One of the most basic laws of life is expressed here. It is called self-replication, identicality, or molecular memory. The selective advantage (equation 26) is the reciprocal of the mutation rate, to the power of the number of new decisions spared, times the prospect of success, to the power of the number of events, i.e. $(1/P_m)^R \cdot (1/P_e)^E$. Even in the simplest biological systems this advantage is astronomical. Order-on-order is a basic precondition, and indeed self-evident.

b. Traditive inheritance of past metaphene laws

But what if this order needs to be altered? What if phylogeny required that a lens be added to an optic vesicle, a mesonephros to a protonephros, or that a fin should become a hand? This will require the preservation of old determinative regularity and the addition of new regularity to it. Accidental prospects and functional necessities go hand in hand.

The optic vesicle, or its equivalent Precambrian metaphene, must have been able to see all the time, so that the invention of the lens should have positive selective value. The pronephros must have functioned without interruption, so that the organism could remain alive to experiment with the evolution of the mesonephros. And so forth.

The determinative laws leading to the optic vesicle or pronephros must have been firmly anchored in the genetic code. The half-finished puzzle must be held steady and the composition of the book remain undisturbed. Accident could make alterations only here and there. Any weakening or mixing up of the laws relating to the optic vesicle or pronephros would be selected out. So would any attempt to produce a lens that was not connected with the optic vesicle, or to produce a mesonephros that was not connected to the pronephros (ridiculous as that would be). Only if the attempted lens had a definite position in space, time, and 'memory', centred in front of the optic vesicle, could it withstand selection; and only an attempted mesonephros with even more extensive functional connection to the pronephros metaphene (which it would change to an interphene).

But the command 'precisely in front of the optic vesicle' is not a trigonometrically exact position. What point of reference is there, except in the laws of the optic vesicle itself? The same goes for the mesonephros with respect to the pronephros, or the case of the definitive ureter with respect to the pronephric duct. Where would the lens form if the optic vesicle and its laws were removed? Where would the cornea arise, if the lens were taken away?

The old laws of determination are anchor points for the new ones. This is crucial. The external functions of old phenes, which once required them to exist, might disappear, but the internal functions of the old commands would remain untouched. The phenes caused by the new rules may be successful and necessary for life, but the internal functions of the old phenes are also indispensable. *The laws of such former metaphenes need to be traditively inherited.*

225

They can be replaced only by better, more reliably workable data. But this replacement is risky, for the prospects of successful substitution must sink exponentially with the complexity. We know this well enough already. Remember the web of interwoven gene effects and the degree of polygeny, with more than 40 genes showing in the eye of *Drosophila*. These show how improbable successful replacement is.

c. Copies of old genotype patterns

The whole army of necessarily conserved decisions is under a bombardment of mutational changes. However, only those alterations can be tolerated which lead finally to the same goal. And we already know that the reject rate for experimental alterations is extremely high (cf. Section V C3b).

The only alterations that can be successfully carried through will be minimal, well fitting, and balanced changes. They will not alter anything of principal importance. We should expect such patterns of commands to recur over long periods of time and would recognize them as identical, i.e. as homologues. Just this is demonstrated by the miracle of homodynamy, acting through hundreds of millions of generations (cf. Section VII B2d). *The conservation over long periods of time of old patterns of determinative decisions is a necessity.* Atavism proves it sufficiently. The bivalent function of these decisions — morphological and epigenetic — demonstrated the same thing. Everything confirms the stochastic necessity of recapitulating the sequence of old commands, as I postulated at the outset (cf. Section III C3c,d and Fig. 20, Section III C3c). The necessary repetition of epigenetic functions carried the connected morphogenetical functions with it, in several grades ranging from simplification to symbolization.

d. The archigenotype

The most recently superseded form of an organism is therefore necessarily repeated, purely because of chance considerations. This is a repetition of the last metaphenes which have become interphenes. But this process of repetition would itself be repeated so that the new metaphenes retreat to the position of secondary and then tertiary interphenes. This leads us to a second group of predictions about the structure of epigenetic systems (cf. the corresponding Section VI C1c).

(1) The first prediction is that there will be a stratification of old patterns which will overlie each other in ontogeny according to the sequence of their phylogenetic origin. These will materialize one after another in the data-processing of the epigenetic system, during the so-called sensitive phases, and will then vanish again. (2) The localization of the pattern of commands in the interphenes will be recapitulated. (3) The induction paths of the flow patterns of additional commands and data preparation will be recapitulated. (4) These patterns will correspond to the palingenetic (recapitulatory) connections. (5) Degree of simplification will depend on burden and age. (6) The degree of similarity and of identicality will correspond to the pattern of phyletic relationships. *The essence, or morphotype, is traditively inherited in the sequence of decisions.*[31]

The predictions thus made possible allow the thesis to be tested in almost every respect.

The mechanism of traditive selection, which forces the epigenetic structure to be built up, simply depends on the preference in every case for the most probable solution, i.e. the first solution to persist (Fig. 20, Section III C3c). Selection will exclude the bearers of mutated genes if the latter hinder the production of the required definitive pattern. Archetype selection, as conceived by Waddington in 1957, is probably identical to this traditive selection.

2. Canalization by traditively inherited organization

The advantages of traditive inheritance are great and the mechanism that builds up order on pre-existing order is indispensable. Consequently, the same laws of probability require that the adaptive advantages which they shaped yesterday, must be paid for today by a restriction of adaptability. The ratio between prospects and the range of accident cannot change.

Suppose that in a dice game only a six would win. A player who doctored the die so that odd numbers could not be thrown would double the probability of all even numbers, including the six. However, the possibilities of the die would be halved. If the rules of the game changed (i.e. the ecological niches) so that only a 'one' would win, then the internal functions which the player had open to him would scarcely meet the external requirements.

For the fourth time we have a mechanism which decreases the adaptive prospects as regards the external environment, if these turn in a direction which had previously been excluded. Traditive selection, which had formerly been a shield against the limitlessness of accident, becomes selection against breadth of adaptive possibilities. It becomes canalization.

a. Functions, burden, and fixation

I should certainly exhaust the reader if I again explained the forms and effects of burden as already recognized in their standard-part, hierarchical and interdependent forms (Sections IV C2b, V C3, VI C2b). I shall therefore concentrate on the effects which traditive selection introduces because of the time axis. The previously mentioned forms of burden still act, for standard-part, hierarchical, and interdependent patterns of order are all traditively inherited.

1. *Bivalence of function.* This is the essential new fact. It depends, as mentioned, on a possible opposition between the epigenetic and the external functions of gene effects (Section VII B3). The functions of a phene, as they are tested by selection, may begin as external functions, which at first adapt themselves to the pre-existent internal functions. A new form of burden thus appears in the argument — that of internal versus external functions.

2. *The burdens of epigenetic functions.* As already mentioned, interphenes of central bodily position have large epigenetic burdens. For example, so many vital phenes depend on the establishment of the embryonic notochord that successful replacement of its internal function is grossly improbable. It does not matter whether such a burden is intensified by external functions, as in the case of the lens (Section VII B3a), or whether the external functions diminish or disappear, as with the notochord and mesonephros of mammals. The epigenetic burden on its own is great enough to ensure huge periods of fixation, irrespective of external function.

3. *The fixation periods due to epigenetic burden.* These are the longest known to us. Interphenes of central function occur without exception and identically in all vertebrates. Examples are the medullary plate, the neural crest, the notochord anlage, the lateral plate and the subdivision of the muscles and coelom (cf. Fig. 66). They must have been fixated since at least the Ordovician (4×10^8 years) and perhaps for a half a billion years (5×10^8).

It is important that internal burdens produce internal fixations, for the latter are separate from external function and indeed may act against it.

b. The dimensions and direction of superselection

The epigenetic system thus becomes independent. It acquires its functions, so to speak, under the guidance of the environment. But it develops these functions into a sort of self-regulation, independent of the demands of the environment in which the completed organism must survive.

This acquisition of independence is like that of business organization or business selection as against market selection. Thus in some car factories the running board has survived from the times of the stage coach. It is no longer necessary for climbing into the vehicle and is ignored by the market or taken for granted, although it attracts dirt and rust. It has taken on its own life. Perhaps it has acquired a function in construction. Or perhaps it has merely become fixated in the tradition of the coachwork department.

1. *The extent of traditive superselection.* This can be estimated by specifying the number of possibilities of change for a feature since it was relieved of external function. In other words, how often could it have disappeared if it really were free of all selection?

The pronephros is an example. It is an embryonic organ of all vertebrates. At least since the origin of the gnathostomes, at say 2.5×10^8 years and 10^8 generations, it has lost its function and in the adult is dismantled and replaced. Assuming that a mutation causing it to disappear had only one chance of being successful per generation, in every species it ought to have disappeared 10 000 times already ($10^8 \times 10^{-4} = 10^4$). With more than 40 000 species it should have disappeared 40 million times ($10^4 \times 4 \times 10^4 = 4 \times 10^8$).

Selection must have protected its epigenetic functions with at least this intensity.

2. *The main points of action.* Superselection predominantly affects phylogenetically ancient, functionally central, and fundamental organ systems. In vertebrates, for example, there is absolute stability and strong protection of the basic patterns of coelom formation, of the dorsal position and the invagination of the dorsal nerve cord, of the subdivision of the spinal nerves, of the aortic (or branchial) vessels, of the pronephros in principle, and of the notochord. This is clearly independent of whether there is an external function or not. Indeed the functional necessity of the invagination of the nervous system or of the metamerically subdivided coelom is not totally evident.

These are just the features particularly suited for characterizing the morphotype of a group of animals. Because of their complete constancy they are used in systematic differential diagnoses.

3. *The results of superselection.* Traditive superselection produces organs which have withstood all change of function and biotope and all the transformations and substitutions that the vertebrates have passed through in 300 million years, from the sawfish to the humming bird, and from the bat to the humpbacked whale. They have survived almost as if they had no function, or rather their schematically simplified variation shows no connection with external function. Organs with these characteristics are valued in morphology and high-ranking systematics as particularly reliable features of the ground plan and especially important. They have escaped the changes of function and the functional transformations which comparative thought needs to exclude when distinguishing analogies and accident from homologues and essentials.

It is no surprise that anatomists a hundred years ago, because of this correct insight, looked for the organ without functions as the most reliable guide. Nor is it surprising when they thought they had found it in the coelom or secondary body cavity. This approach was physiologically mistaken, morphologically correct, and methodologically well proven. Their conclusion was remarkably close to the truth.

These are features not deceptively transformed by external functions. They are based on the stabilization of important, highly burdened, unalterable internal functions.

c. *Counterselection and canalization*

A further fundamental directional component is forced on evolution by this rigidification of building instructions — or rather of the building instructions for building instructions, being the interphene laws in which metaphene laws are anchored. Every stage in fixation thus becomes still more firmly cemented, so that a vertebrate can only become a vertebrate and a mammal only a mammal.

Such irreplaceable fixations explain why organisms evolve along paths; why individual systems stiffen up; why return beyond fixed points cannot happen; and why we can recognize stratified phyletic relationship at all.

Together with what has been said before, this would be enough about the directional effect. But functional bivalence, by which traditive selection is protected, implies an even more deleterious mechanism — that of counterselection.

1. *Counterselection*. This must act as soon as the two functions of a feature begin to work in opposite directions. Without doubt this must happen often, but so little is known about the selective values of most organ systems that the situation usually remains obscure.

However, the human appendix can be taken as an example. Without surgery, the death rate from appendicitis would be so high that the external function must be under negative selective pressure. The appendix is known from the reptiles upwards, however, and persists in man. The epigenetic function which causes the fixation of the appendix must therefore exert a positive pressure. This positive selective pressure has prevailed completely, perhaps for a million generations already.

In this case therefore a feature is preserved because internal selection insists on keeping it, while external selection is just able to tolerate it. The evidence permits this conclusion (Section VII B4a) already. Even quantitative values for these forces (i.e. for the tolerances of canalization) could perhaps be worked out.

2. *Canalizations of structure*. These are by no means exceptional. Indeed they are universal. All the well known half-measures of human structure, for which a planner or teleological designer would be accused of plain botchery, are stuck more or less deeply in the culs-de-sac between the internal and the external selectional fronts.

Consider the contrariness of the structure of the retina; the fact that the fertilized egg passes through the body cavity (!); the fact that birth takes place through the only ring of bone in the body; the crossing-over of the respiratory and alimentary tracts; the interconnection of the urinary and reproductive tracts and so forth. Consider further the bridge construction, based on a design for a torpedo, which finally comes to be balanced upright. And remember all the so-called constitutional complaints which must be the result of such lack of planning — vertigo, slipped discs, inguinal hernias, haemorrhoids, varicose veins, flat feet, splay feet etc. Such things are not confined to man. It is merely that more is known about them in him. An experienced veterinary surgeon recognizes them in all his charges — the brittleness of a horse's leg, the fragile skulls of many birds, the delicacy of the flight membrane of bats. All these are specializations which have sealed their own fate.

Canalization is universal. The only exceptions are functionally new strata. These add additional burdens to the older structure, but retain evolutionary freedom till they themselves sink downwards as bearers of additional responsibilities. This universal canalization is why we are sure that 'Nature makes no jumps' and why we can say 'That is nothing other than ...' although almost every aspect of a thing *can* be otherwise. It is why we are confident that identicalities can be identical by 'inner necessities' far beyond the limits of their forms and functions. Indeed, to be convinced of the existence of these nesessities, we do not even need to know what they are. Where does this prescientific conviction come from? Thus yet another circle is completed.

3. *The agreement with the patterns of thought*. This is based on the human tendency to look for identicalities everywhere and to assume the existence of hidden properties in common, even where these cannot be proven. We think in terms of the transformation of

presumed identicalities up to the limit where the causal connection is inextricable, the forms not comparable, and comparison totally loses its function. This is a wide field, containing within itself the motives for research and for rational thought in general. Caution and scepticism about the search for identicalities is called nominalism — an overcompensation which supposes that general features have no real validity at all.

Consider how, when comparing a sea horse and an ant-eater, we are ready without hesitation to disregard a vast number of differences so as to extract the hidden identicality of the 'vertebrate'. Nominalists protest against this. In the field of organisms, however, the method is justified or self-confirmed by its own success. Not only does every organism show identicality, but it shows it at many different levels, so as to share some identicalities with the group of all living organisms. And identicalities are arranged in a system so self-consistent that it has long been confidently referred to as the 'natural system' of classification. The Simon-Pure nominalist doubts even this reality; for where is its necessity?

We have now established this necessity or cause. The identicality of homologues and the system of phyletic relationships built upon this identicality are real natural objects. One of our crucial presentiments is confirmed. Morphology is right, together with Darwin and all his successors[32], and nominalism is wrong. The agreement between natural patterns and thought patterns can once again be explained only by the lessons that the traditive order of selective evolution has imparted to our brain.

4. *The canalization of thought.* This can be seen by considering how we react to apparent disorder, where there is no present hope of proving postulated common features nor of showing the traditive inheritance of identical regularity. Consider, for example, the arbitrary extrapolation of known conventions into the unknown. Formerly there was the chariot of the Sun, Atlas supporting the Earth, and the vault of heaven bearing the stars. Now there is the all-pervading All.

The hypothesis or thought-necessity of a traditive universe is, in my view, a limitation. Perhaps it is often a source of error and nominalist caution would apply. It is probably an extrapolation from the world best known to us, which is the world of living organisms. Once again it is demanded by the need to economize the required information.

3. Tradition in civilization

The reign of traditive order is particularly obvious in the field of civilized productions. Darwin himself pointed out that functionless letters in words were vestiges. He said: 'In all fields which have had a history at all, written or unwritten, we shall be able to demonstrate remains of elements which formerly were viable and useful, but are now more or less obscured, worn out or obsolete.'[33]

Concerning the origin of this agreement between traditive order in thought and in organisms, it was already clear fifty years after the *Origin of Species* that: 'The various forms and types conform to the same laws as govern the organic world.'[34] At the present day this great subject is being methodically developed.[35] I shall therefore merely point out those laws which I expect, from my viewpoint, to be identical in both fields.

a. The necessity of adopting old practice

I consider that previous order is taken over in civilization so as to save coordinating new decisions. This agrees with what is known of organic evolution. In other words, transmission of old order avoids the risks associated with inserting new decisions. Continuity ensures the best success at the least cost.

For instance, the first cars were extremely like the horse cabs of their period. This was necessary because, apart from the actual motor, all the inventions which have turned the cabs of that period into the cars of today were still to develop and could only be fitted into the whole by trial and error. Likewise, at the time when a letter becomes silent in a word, or an expression or usage loses its function, its future complete redundancy is still uncertain.

With change in function, there arise marginal uses in the process of transmission which only have meaning within the traditive system, rather like the epigenetic function. Silent

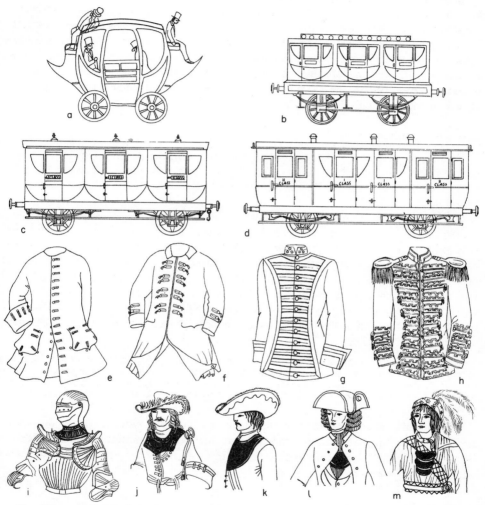

Fig. 75 a-m. Examples of traditive inheritance in civilization. (a-d) A stage-coach pattern by traditive inheritance becomes a symbol of quality: (a) The first railway carriage (England, 1825); (b) Somewhat later with three compartments (England); (c) All the window curves preserved (Sweden); (d) Curves preserved only on the 1st class compartment. (e-h) The sewing round button holes becomes ornament for a uniform: (e) A soldier's tunic of 1690; (f) Tunic with 'blind' button holes of 1756; (g) A court service tunic of the Royal Imperial Household Guard; (h) That of a Captain in the same (Austria). (i-m) The change of the gorget from functional armour to a symbol of rank: (i) Armour of about 1500, (j) Officer with a large gorget of about 1690; (k) Officer with a small gorget of 1688; (l) A symbolic gorget of 1710 (Brandenburg); (m) A Seminole chieftain with a triple decorative gorget for impressiveness. (a-d) From Leche (1922); (e-m) from Koenig (1970).

letters indicate pronunciation. The embroidery of hole-less buttonholes becomes ornament (Fig. 75e-h). The neck-piece or gorget becomes a symbol of rank (Fig. 75i-m). The oblique windows of a stage coach vestigialize to make a comfort symbol for a first-class railway carriage (Fig. 75a-d). A bare-toothed snarl, originally directed at a supposed threat, becomes a smile. A bugle call to attack becomes the fanfare of an academic festival. A war-cry turns into the cheering of football supporters.

As to its function in individual life, a traditively inherited feature therefore finishes as symbolism in fashion, custom, convention, or life style. As to its constancy, however, it comes to seem self-evident. It belongs to an area where questions about cause are greeted with surprise, while questions about meaning provoke indignation. Tradition is often a substitute for explanation and is the static predetermined quality in civilization. It is an explanation for the inexplicable and a limitation of the play of accident. As such it would, if developed to the extreme, neither have allowed understanding to arise, nor civilization, nor any culture at all.[36]

b. Tradition and tolerance

As already predicted, it is an important result that tradition, or the comfortableness of adopted order, is paid for by various degrees of new loss of freedom. No complaining will help, since the mortgage has already been used up. Reasonableness is the best that we can hope for.

We pay for tradition by being immersed in taboo prescriptions and self-evident 'truths'. These have lost the rational objective functions by which they may once have tended towards tolerance and humanity. We thus become enclosed in a *perpetuum mobile* of value systems which are both artificial and expansive. I do not object to the reign of order or to the diversity of its expression. But tradition causes a confusion by which what was formerly practical is supposed to be inevitable and canalization of thought is mistaken for the goals of civilization. Against this confusion I do object. I decidedly object also to the notion that one of the competing value systems should strip the others of their rights.

Let me add that I am not speaking of the pluralistic morality which hinders us, with mischief and corruption, from paying our mortgage back. What I am speaking of is the false-floored morality which Thomas Huxley and Ernst Haeckel already reckoned as 'hoax and humbug'.[37] So far as I know only Nathan the Wise has solved this problem and he likewise did it, in Lessing's version, by discovering relationship.

I have tried to show that traditive inheritance is a universal principle. It is as universal a principle of organic order as the standard part, hierarchy and interdependence (Section IV C3, V C4, VI C3). If I am right it is useful to understand both the necessity of its origin and its grievous results.

———————   ———————

Four chapters ago I finished my chapter summary of the order pattern of molecular biological decisions with a note of reserve (end of Chapter III). This is appropriate for a biologist who knows that 'in an evolutionary mechanism comprised of accidental decisions and necessary events, one cannot be the sole cause of the other.' In discussing the order pattern of morphological events I must finish in the same sense.

The agreement between the order patterns of events and those of the decisions behind the events almost certainly represents a causal relationship. But even supposing that I could convince the reader of this well documented connection, it would remain a one-sided approach to the facts. For no unidirectional approach to cause can embrace the total complex interconnectedness of evolution.

However, the order patterns of decisions are in fact visible. Moreover, the extent to which they agree with those of events would indicate that the phenomena behind both are identical. If the morphological patterns were not visible they would need to be postulated. The circle of theory completes itself. The systemic conditions allow morphological effects to be traced back to molecular causes, just as molecular effects can be traced back to morphological causes. The circle of systemic, two-directional causality is closed.

NOTES

1 For example Eibl-Eibesfeldt (1967), Wickler (1970), Koenig (1970), Lorenz (1961, 1973).
2 Compare Schrödinger (1944) with the second German edition (1951).
3 Meckel, (1821), von Baer (1828), Müller (1864), Haeckel (1866).
4 This is known as Haeckel's law, the biogenetic law or the law of recapitulation. It is quoted here from Haeckel (1866).
5 This law can be expressed more exactly, but less clearly, by saying that: 'dependences exist between phylogeny and ontogeny'. Looking more closely we could say that 'The hen came before the egg' which led Garstang to suppose that the egg came before the hen (1922).
6 Surveyed by Plate (1925), Korschelt (1927), Rensch (1954), Osche (1966).
7 Wickler (1961).
8 Leche (1922), Plate (1925), and Remane (1971).
9 This was introduced into evolutionary theory by Osche (1966).
10 This was considered from the genetic viewpoint by Lerner (1954).
11 Surveyed by Grüneberg (1952). See also Waddington's comments (1957, p.155).
12 For background see Weiss (1939), Seidel (1953, 1972), Kühn (1965), and the symposium volumes edited by Locke (1966, 1968).
13 Spemann was the principal pioneer in this field (1936).
14 Summarized in Seidel (1953).
15 Coulombre (1965) summarized the details of the developmental pattern of the eye for the whole vertebrate group (as shown in simplified fashion in Fig. 67).
16 Kühn (1965, p.541).
17 As is well known, the concept goes back to Baltzer (see Section II B2*a*).
18 The main original literature is Baltzer (1952) and Chen and Baltzer (1954). There are even class chimaeras since the medullary reaction tissue of amphibians can still read induction commands from notochords of birds, fishes, and cyclostomes (examples in Hatt (1933), Oppenheimer (1936), Bytinsky-Salz (1937)). In these cases the power of understanding is as old as the whole vertebrate ground plan, i.e. 400 million years, or 500 million separated generations.
19 Kühn (1965, p.541).
20 Kühn (1965, p.544).
21 A survey of the literature up to 1965 is given by Shepard.
22 The size of the orbits in Stensiö's reconstructions (e.g. 1958) strongly suggests this.
23 The present situation and controversies over the undecided details were summarized by Torrey (1965).
24 I would interpret the Desor larva of nemertines in this way (cf. Fig. 71e). In principle it has direct development but it shows imaginal discs and amnion formation within larval ectoderm. These seem to make functional sense only as a shortening of metamorphoses of the related pilidium larva. Surveyed in Korschelt and Heider (1936).
25 A particularly convincing survey of this evidence is given in Kühn (1965).
26 Osche (1966, p.846) in his survey of the problem of evolution.
27 The most recent survey is in Vandel (1964, p.504).
28 This was already suspected by Cahn (1958).
29 In consequence the course of eye reduction in arthropods is different, just as the inductive pattern differs from that of vertebrates.
30 Goethe's Mephistopheles expressed this thought clearly but our 'successes' have obviously made us forget it.
31 I shall compare the details of our respective viewpoints in Chapter VIII but must point out here that Waddington (1957) has almost completely anticipated my own thoughts about traditive selection.

32 These successors extend from Romanes (1892) to Lorenz (1965c), to name but two. It should also be remembered that these results, deduced four times here, in Chaps IV–VII, from the facts of comparative anatomy, were reached independently by Lorenz, starting from the facts of comparative behaviour (1973). See also notes 25 (Chapter I), 7 (Chapter IV), 49 (Chapter V), and 24 (Chapter VI).

33 Quoted from Darwin by Leche (1922, p.229).

34 Leche (1922, p.229).

35 Compare for example Koenig (1970) and Lorenz (1971, 1967).

36 There is an important agreement here with Lorenz (1973), as I can add while in the press. My Fig. 75a-m is even independently repeated in his Figs. 3-4.

37 Quoted from a letter from Huxley to Haeckel on this subject (sources in Hemleben (1964, p.84)).

CHAPTER VIII
THE THEORY AND ITS CONSEQUENCES

A theory without new consequences would have no meaning. Even consequences, if they did not offer some more inclusive explanation than before, would be valueless. Logic and epistemology have long ago demonstrated that the usefulness of a theory depends on the additional explanatory value which we gain by introducing it. For every phenomenon of interest has long had some sort of explanation affixed to it, whether this be sound and accepted or neither. In surveying the problems before us (Sections II B3 and IIC) I showed that the so-called 'open questions' did not lack attempts at explanation. Rather they were marked by unfinished controversies as to whether the attempts were important or sound.

I shall therefore end by considering what the present theory explains better than the synthetic Neodarwinist theory on which it was built. 'Better' here means more precisely or more extensively. I shall consequently summarize the causal mechanism which the theory predicts to be a more universal explanation than anything proposed in previous theories (Section VIII A) and confront the mechanism with its consequences (Section VIII B).

This final survey is the more needed because I have not merely been trying to explain an already defined phenomenon. Indeed I had to begin by defining the problem itself as comprising the high-ranking aspects held in common by a whole series of individual problems. These latter varied greatly in how they had been formulated and included all the open questions of transpecific evolution and morphology. The phenomenon to be explained comprises all the things in macroevolution that are covered by the term necessity, in its meaning of determinacy, predictability or order.

A. A THEORY OF SYSTEMIC CONDITIONS

What does the theory say, therefore, and how does it fit in? The theory asserts that evolution is excluded from accident to a much greater degree than hitherto supposed and that this shielding is the necessary result of a selection dictated not only by environmental conditions, but also, and chiefly, by the functional systemic conditions in the organization of organisms themselves. In terms of feedback it is a *selectional theory*. It assumes the correctness of the Neodarwinian synthesis as a precondition, but supplements this by causal explanation of the transpecific phenomena of evolution.

The theory thus demands only the known mechanisms of mutation and selection. It proves, however, by using a *probability theory*, that these mechanisms will be linked together in a system of reciprocal dependences. The concept of evolution has up till now implied a linear, unidirectional causality. But the new theory recognizes a reticulate or feedback causality, such as is required in general as knowledge increases. Structurally speaking such a biological concept is a *systemic theory*.

The theory asserts, moreover, that the continual sequence of unequal prospects of accident and necessity leads to determinacy. Restriction of accident will increase the prospect that new law will be successfully invented during evolution. By establishing conformity to law, however, restriction of accident leads to a canalization of possibilities and to fixation. This will be true from molecular decisions, through morphological events, into the realm of thought and civilization. The theory is therefore a *determinacy theory*, in the sense of predictability or determinability.

1. The survival prospects of molecular decisions

One of the peculiarities of the rationally understandable world is that we can divide it easily into accident and necessity. And one of the remarkable peculiarities of the evolution of organisms is that both of these drive it forward. For any improvement of the required decisions can only be attempted by accident. But the success of the events released by the decisions can only be judged by their improved prospects of adaptation, which are matters of necessity. The survival prospects of molecular decisions are decided by the necessary results of accident. (The equations can be found by looking up the key words in the index.)

a. *Accident and necessity of decisions*

The whole information content of every limited system is made up of the determinacy content and indeterminacy content together ($I_I = I_D + D$). In other words, the sum of accident plus necessity is constant. Gain in determinacy corresponds to a loss of accidental possibility. This holds both for subjective changes in determinacy caused by deeper study of a system and also for real changes when order increases.

If the prospects of occurrence of one particular accidental event increase, then the possibilities which accident can choose from will diminish. Accident is the lack of fixation and the converse is also true ($I_D + D = constant$). The range of accident is therefore decided by the number of possible accidental decisions. If a decision passes from a free to a determined state, then the range of possible events decreases to the extent that the realization prospect of the remainder increases, i.e. $D = \log_2 (P_D/P_I)$.

The decisions that enter into the determinacy content of a nonteleological system are not of equal value, however. Some are necessarily required in defining the content of the message. Others can be left out, given a suitable arrangement of switches, without decreasing the content of the message. A determinative system therefore consists of law decisions and redundancy decisions ($D = L + R$). Law content here means the decisions which cannot be left out of the message. Redundancy content means prolixity and repetition. It specifies the number of instances when a law is repeated. Order is law content times the number of instances where the law applies ($D = L \cdot a$).

In this connection the number of instances, i.e. the repeated or visible redundancy is at the same time the means by which we become certain of the reign of law, i.e. $P_L = P_D/(P_D + P_I)$. This depends on the fact that, as the number of confirmed predictions rises, the probability that we are dealing with the reign of accident gradually disappears entirely.

b. *Economy, redundancy, and burden*

All material systems of determinacy include a principle of economy because the creation (or rather the accidental discovery) of decisions and their conservation and decoding increase both the failure rate and the expenditure of energy. This is the more important because, even in simple systems, the redundancy content can exceed the law content by several orders of magnitude.

In logic, in the design of apparatus and in the genetic code the redundancy content comes to be dismantled. And always in the same fashion, by ranking some decisions above others, which is called the position effect or switching. In principle there seem to be four ways of bringing about such ranking, all of which are realized in all three fields. Repeat switching is accomplished by self-replication of the nucleic acid system; the superimposed on/off switching is realized in the operon system; synchronous switching in the regulator-repressor system; and sequential switching in the order-on-order system of traditive inheritance.

The decisions required to establish the events of a system decrease sharply in number as a result of systemizing the determinacy content. This produces a crucial increase in adaptability but, on the other hand, the individual decisions become enmeshed in a network and burdened with responsibility for more than a single event. The decisions, such as those in the genetic code, may have equal prospects of being changed by accident. But the determinative events changed by these decisions will begin to differ from each other in their prospects of success.

Dismantling of redundancy will make a system easier to modify in any sense compatible with the switching pattern. But this increase in modifiability in particular directions will be paid for by losing directions of modification that are not compatible with the switching pattern. For if no more decisions that had previously been left to accident in the system can be turned into determinative decisions, then the total ratio between determinacy and the prospect of change must remain the same. Dismantling of redundancy by systemization only alters the prospects of particular sorts of change. Increase in regularity must be paid for by decreasing the range of accident. The sum of accident plus necessity remains constant.

2. The prospects of morphological events

Characteristic of the progressive evolution of organisms is an increase in the number of events (E) and in their reciprocal dependences. This is an increase in differentiation and organization, in complexity, and coordination. We have come to take it as self-evident, but it has profound results on the prospects for adaptive change of the individual events. That is to say, it deeply affects the basic morphological transformations which are endlessly required of all organization by the ceaseless change in external conditions and external possibilities.

a. Accident and necessity of events

If determinative events are triggered off by determinative decisions equal in value to each other, as I shall again assume at the outset, then these events have equal prospects of being changed. Heritable change in the events depends in organisms on the gene-decisions that specify them, i.e. on mutations, which are ruled in principle by equal accidental prospects of happening. It is a totally different question whether the changes so produced in an event will be accepted by selection or not.

Here it is functional necessities which determine the strictness of selection. The organism is like a lock-and-key mechanism for which it is required that the key be changed by accident so that it can also open the lock to the most profitable subsequent ecological niche.

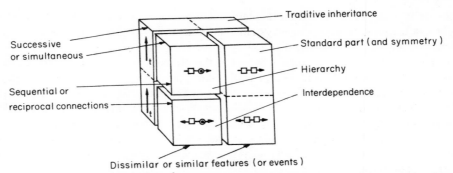

Fig. 76. The interrelationships between the different qualities of order. The three coordinates represent the time component of dependency, direction of dependency, and similarity of feature. Accordingly there are in principle eight qualitatively different patterns of order. These are combined to give the four concepts of order used here.

b. Burden and superdeterminacy — the double burden

Organization of events (i.e. of features) implies their mutual dependency in particular patterns. Features can be at the end or in the centre of some functional connection. Marginal features will carry the responsibility only of their own function. Central features will carry the *burden* of all those features which depend on them like links in a chain.

The burden of an event is specified by the number of those dependent single events whose functioning will be affected by a change in the event. Selection will therefore test all these dependent events collectively. If the prospect of successful accidental change for a particular event is P_e, then this prospect will decrease for the collective as the power of the number of dependent events (E) which would be affected at the same time (P_e^E). Even with just a few dependents, the prospect of successfully changing the dependent events that rest on independent decisions is virtually nil. These systemic conditions therefore lead to superselection and fixation. They exceed by many orders of magnitude the determinacy of a single feature (the reciprocal of the prospect of alteration, i.e. the reciprocal of the mutation rate) and also the prospects of altering a single feature successfully.

Unities thus arise where a functional pattern results in a pattern of burden, of superselection, and of superdeterminacy.

We should therefore expect to find that functional patterns of a basic sort would occur. There would be simultaneous or successive dependences of identical or non-identical parts with sequential or reciprocal relationships (i.e. dependences either in series or mutual, Fig. 76). In actual fact the entire diversity of functional connections in organisms conform to these basic patterns. Four out of the eight (Fig. 76) deserve their own name. Among simultaneous patterns the dependences of identical parts constitute the structural phenomenon of the standard part (and the positional phenomenon of symmetry). Non-identical parts with unidirectional connections constitute the phenomenon of hierarchy. Non-identical parts with reciprocal connections constitute interdependence. And all dependences that follow on each other in time constitute tradition, or traditive inheritance.

3. The cycle of adaptive prospects

The prospects of successfully changing decisions by accident are not independent of the necessities of events (or features). On the contrary, together they form a system that

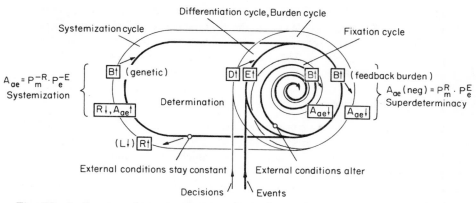

Fig. 77. A diagram of the feedback causality of the systemization mechanism on the basis of the three cycles of organization to gene decisions and phene events. It also shows the increase (↑) and decrease (↓) of burden (**B**), determinacy content (*D*), number of events (*E*), of law content (*L*), and redundancy content (*R*), adaptive advantage A_{ae} and canalization ($A_{ae(neg)}$) of the determinacy.

conditions itself mutually. The increase of particular necessities entails a decrease in accidental possibilities and this decreased range of decisions produces a canalization of the possible events (Fig. 77).

As a result the concept of unidirectional causality, which has dominated the analysis of evolutionary mechanisms must be replaced by 'feedback or multidirectional causality' which considers also the feedback of an effect on its own causes. This widening in the interpretation of causality began and was established in physics. It was surveyed by Eder (1963).

a. Decisions copy the patterns of effects

Progressive evolution reacts to new conditions in the environment by means of progressive additions, differentiations, and coordinations. This increase in organization results in an increase in features (events *E*) and the required gene decisions ($bits_D$ of the determinacy content *D*). With the increase in feedback systems the range of accident increases likewise, while the prospect (A_{ae}) of successful (P_e) accomplishment (P_m) of a change sinks exponentially, i.e. $A_{ae(neg)} = P_m^R \cdot P_e^{E'}$. The *functional burden* of many features increases and so does rejection by selection. At the same time, if the adaptive goals remain the same, many decisions become redundant. For these mutually repetitive decisions ($bits_R$) are those which define the events that are functionally dependant on each other.

As a result, with a selective advantage of more than 10^5 per decision and event, it is necessary to reduce redundancy (*R*) by ranking decisions one above another. The genome becomes systemized to give the epigenetic system. The adaptive advantage (A_{ae}) increases exponentially with the number of decisions that can be avoided, i.e. $A_{ae} = P_m^{-R} \cdot P_e^{-E'}$ The increase in the prospects of accomplishing a particular condition will correspond to a decrease in the accidental possibilities (compare the systemization sector in the cycle of decisions in Fig. 77).

The second result of this systemization is a *copying* of the four patterns of functional dependence of events by the switching pattern of those decisions responsible for establishing the events. This happens because the coupling of dependent events is demanded as $P_m^{-R} \cdot P_e^{-E'}$. While the uncoupling of events which need to adapt separately is demanded as $P_m^R \cdot P_e^{E'}$. This leads to the building-up of an epigenetic system that copies the pattern of events according to their importance and time of application, using four systemization mechanisms. This leads to an imitative epigenotype, traditive along the time axis.

b. Events follow the patterns of decisions

The adaptive possibilities of morphological events correspond in the first place to the burdens that these events have acquired. They also correspond, in the second place, to the switching patterns that have been built up, onto which the difficulties due to functional burden have been shifted.

The switching pattern of decisions has a *feedback* connection with the prospect of modifying events. This connection corresponds to the fact that, as possible accidental changes in decisions become restricted, there is also a restriction in the range of possible events. Suppose first that the functional pattern of events does not need to change, or needs to change only compatibly with the systemization of the genome. In that case its adaptability or prospects of realizing the desired events will be higher in so far as the prospect of realizing undesirable events is excluded. For the prospect of success will correspond to the reciprocal of the remaining range.

But suppose, on the other hand, that the patterns of change demanded by the environment take on a new direction not corresponding to the built-in switching patterns. At that moment there will be a drastic change in the prospect of successful realization, i.e. in the prospect of establishing successful new decisions by accident. The signs of the exponential relationships will change. $P_e^{-R} \cdot P_e^{-E}$ becomes $P_m^R \cdot P_e^E$ and the adaptive advantages previously attained turn into their reciprocals. The previously required rejection of preposterous events becomes, in proportion, the preposterous rejection of what is required. Yesterday's freedom becomes today's canalization. The rules of chance cannot be avoided. Nothing is got for nothing.

This *canalization* is monitored by well known selection. The loose fits that have to be eliminated, however, always depend less on the accidental change of the external environment than on the determinative conditions of the internal environment. That is to say they depend on conditions in the switching pattern of the epigenetic system. This leads to a constancy in selection, to a sort of superselection or superdeterminacy which exceeds by many orders of magnitude the degree of determinacy and constancy that was originally ascribed to the evolutionary mechanism $(P_m \cdot P_e)$. This superdeterminacy manifests itself in the four patterns of order to which it owes its origins. However, it becomes increasingly rigid and less and less touched by change in external conditions. Moreover, with increasing systemization of the internal environment, inalterable directions begin to prevail for every evolutionary path. This explains the directional and orderly phenomena of transpecific evolution as well as the causal background of morphological and systematic laws. It explains the phenomenon of predictability which we experience as order.

The *result* is a self-ordering of living organisms, a regularity that can be called self-design. Where we are able to comprehend it we respect it and name it harmony. However, it is not a prestabilized but a poststabilized harmony. It is not entelechy but causal autonomy. The difference is that the laws behind such evolution are not in principle unknowable.

The play of accident in evolution thus seems even further restricted. If we, as biologists, can still believe that God plays dice, He can do so only in two fields. One is the unpredictable instability of the matter in which the laws of life are codified. The other is in the encounter between the predictable creature and the unpredictable meanders of the ever-changing environment.

As already said, this is both too little and too much. The credibility of a theory can only be measured by its agreement with the facts. The usefulness of a theory, on the other hand, can be measured by its explanatory value and the correctness of the new predictions which the theory permits.

B. THE CONSEQUENCES

The explanatory value of a theory, which needs to be treated more fully now that we are approaching the end, contains one more important feature. That is the possibility of verification. This must depend on confirming predictions which can only be made with the help of the theory. This point is particularly important to the experimenter and deserves a few more words. They need only be brief since predictions that have been confirmed are usually reckoned as part of the theory's explanatory value, while those still unconfirmed will only be encountered in the future.

The theory of systemic conditions should be verifiable in almost all the fields that it touches. This can be foreseen, together with the sorts of possible predictions, both descriptive and experimental.

The distinction between experiment and description is customary but epistemologically naive. It results from the arbitrary separation of disciplines. And the value judgement behind it, of 'pure experiment' versus 'mere description' has done all sorts of damage. A natural experiment, that we describe, is no less significant than an experiment which we ourselves set up, and whose results we describe in precisely the same fashion in terms of coinciding events. However this is not our present theme.

Descriptive verifications are to be expected in the morphological disciplines (which are syntheses of events). For every still-undiscovered species and for every fundamental morphological transformation and mutant, predictions can be made about how homologues, ground plans, and trends will behave. With each new form it will be possible to make deeper organizational and structural connections beforehand, and to test these connections afterwards. Though still undescribed, such are the steady methodological tools of every experienced palaeontologist, systematist, and comparative anatomist.

Experimental verifications, on the other hand, are to be expected in the physiological disciplines that deal with the epigenetic system, i.e. with the systems of decisions. For every species, the present theory permits a prediction about the imitative pattern of the epigenetic system and the traditively inherited results, as well as about the phyletic relationships of the palingenetic features. These will count as predictions particularly because the 'logic' of how gene effects are systemized is still unclear from the viewpoint of developmental physiology and because the existence of the postulated switching pattern in the molecular genetic systems of higher organisms still waits to be discovered.

I do not doubt that this verification will come. Just as I also do not doubt that the theory is still immature and awaits testing.

However, I return to the situation as it is now. In setting up the problem (Sections II B3 and II C) I surveyed the relevant open questions. Now we can turn to solving them. In doing so I follow a grouping of scientific disciplines which corresponds well to the four

Table G. *A survey of the problems and controversies that the present theory could solve*

Division of biology	Problems and controversies	Subject — patterns of order in the solution
Morphology Standard part	*Anatomical Plurals* Homology (as duplication) Homonomy Identicality of homonoms *Anatomical Singulars* Homology (as fixation) Identicality of homologues The limits of homologues The causes of homologues (Idealism) Morphotypes Ground plans	*Phylogeny* Traditive inheritance *Anatomical Repetition* Atavism Vestigialization Heteromorphoses *Ontogeny* Traditive inheritance Haeckel's law The induction pattern The organization pattern Homodynamy Symbolization
Morphology Hierarchy	The reality of systematic groups and of the systematic classification. Weighting *a posteriori*	*Genetics* Traditive inheritance The epigenotype, the number of morphotypes Cryptotype Genetic or spontaneous atavism Relational pleiotropy Homoeotic mutants Phenocopies The archigenotype
Systematics Hierarchy	*Anatomical Directedness* Synorganization Coadaptation Parallelism Trends Orthogenesis Cartesian transformation Typostrophy Stasigenesis Additive typogenesis Typostasy	*General Biology* All the patterns of order *Anatomical Coordination* 'Internal' mechanisms Homoeosis, regulation Irreversibility The organic nexus Post-stabilized harmony The number of realizations Self-ordering Fixated drawbacks
Phylogeny Hierarchy and standard part		*Epistemology* All patterns *Anatomical agreements* with the patterns of thought and with the patterns of civilization.

The problems and controversies are given as headings and arranged according to subject. On the left are given the relevant chief divisions of biology. The patterns of order most concerned in the solution are inserted underneath these subjects (details in Section VIII B).

patterns of order (see Table G). However, I must limit myself to the basic controversies because of space.

1. Principles of evolution

This group of questions involves the nature and position of the mechanisms which drive evolution. In consequence it deals with the problem of whether these mechanisms are accessible to scientific method or whether we must assume a transcausal, vitalistic residue of the natural harmony, this residue not being open to explanation.

a. The units of measure — law and the number of instances

I have given the formulation : order is law times the number of instances where it applies. This formulation explains in the first place those paradoxes and contradictions which have emerged in biophysical analysis of living matter.

The *paradoxes of information* were illustrated by the riddles of the Einstein theorem and of the toothed wheel[1] (Section I B4). These were solved by distinguishing between indeterminacy and determinacy content involving the question : information about what? The crucial insight is, that the objective sum of accident and necessity of a system is constant, as also is the subjective sum of uncertainty and predictability. This insight is the starting point of the present discussion and also its synthesis. This will be dealt with again below (Section VIII B7).

The *paradoxes involving number of instances* are that of breakdown of order by increase in information (as with an organism *plus* a virus)[2] or that of increase in order with constant information content (as with the reproduction of a protist). These paradoxes can be solved by formulating order content as the product of law content times content of relative redundancy.

The *paradoxes of order content* can be solved by applying the same formulation[3]. They can be illustrated by comparing a human germ cell and an adult person. The law content here is obviously the same, while the difference in information content amounts to 20 places of decimals (Section II A2). This can be explained by a correspondingly large repetition of the identical law content of the genome in the definitive organism. The formulation gives a self-consistent method of describing law and order. The next question is how to apply the method.

b. Molecular or morphological synthesis

The controversy between reductionism and holism includes, besides matters of method and principle, the second most important question in explaining the mechanisms of evolution. One side holds the view that all the laws of life have a molecular foundation. The other side believes that the phenomena of life cannot be explained in terms of molecules alone.[4] (Section II C1).

The solution advocated here is somewhat related to holism and involves feedback causality, as already summarized in Section VIII A3. This allows us '. . .to examine simultaneously all the actions and reactions and to comprehend their regularity'. For ' . . . in every natural science the reciprocal effects of various components are centrally important but not explicable in terms of unidirectional causality.' (Eder 1963, p.208). See also Thorpe (1970).

The solution advocated here is as follows: the system achieved in morphology is not conceivable without the corresponding molecular system. But neither is the system of molecular sequences conceivable without that of morphology. The starting point is the distinction between decisions and events. The explanation, however, involved the systems of reciprocal effects which decisions and events form with each other. For decisions are produced by accident but events correspond to necessities. If the probability of attaining a necessity by accident is too small, the range of accident will decrease correspondingly. Accident will hit upon the decision, but necessity will select it.

Monod has a similar concept (1971, p.122). He says: Accident is captured by the mechanism of invariance and transformed into order, law, and necessity.' But likewise, necessary order is copied by the possibilities of accident. In this way accidental changes in the possible messages of a source are prevented from affecting the decision mechanism, since selection only accepts those messages from which such accidental decisions have been excluded.

There is no total solution in terms of morphology nor in terms of molecules. The basic transformations to which morphological types are subject are limited by the epigenotype and its structure, i.e. the limits depend on present and previous morphotypes. In the last analysis, even decision and event are one and the same, as discussed again later under epistemology in Section VIII B7*b*. They are distinct only in terms of how the question is put. Objectively they differ in complexity. Subjectively they differ in the possibilities of human perception.

c. Are there inner mechanisms?

This controversy, as we have already seen, is as old as Darwinism and is of central importance, probably because of its obvious implications. One side maintains that we must postulate an ordering principle within the organism itself, since blind mutation and myopic testing by the external environment could not of themselves produce the orderliness of evolution. The other side maintains that a third principle is not observable, and indeed is unlikely; further, the two known principles of mutation and selection are able to explain all the phenomena.

I am happy to say, in this as in the controversies dealt with later, that the opposing viewpoints adopted by three generations of the most far-sighted biologists can both be confirmed. The solution postulated here is that mutation and selection are the only factors that act, but selection does not work from the outside only.

1. *Burden.* There is an 'environment' that penetrates deep into the internal structural conditions of an organism. This is shown by considering: (i) the burden associated with features; (ii) the organizational gap, measured in single homologues, between a mutated part as observed and what would be needed for the mutated part to be fully operative; and (iii) the changing course of selection as evolution proceeds. In this internal environment selectional conditions arise which continually become less related to the world outside and more related to the functional systemic conditions within the organism. These represent the rules of selection under conditions inherent to the organism — selection of the sort that Stern and Schaeffer,[5] Waddington,[6] Haldane,[7] and Whyte,[8] clearly anticipated. Leading representatives of pure Neodarwinism (i.e. the synthetic theory) likewise naturally saw the force of the 'developmental requirement' (Entwicklungszwang)[9] and the reduced prospect of successful change for 'deep-seated' features[10] which appeared early in development.[11] But now it is possible, by knowing the functional position within a system, to predict the degree of fixation or freedom which fate has allotted to a feature.

2. *Organization of reciprocal gene effects.* I postulate that gene effects will be reciprocally organized so that the epigenetic system, because the prospects of adaptation decrease with burden, will be forced to imitate the phenetic functional systems. Such a systemization will be brought about at first only by the pressing selective conditions of the external environment. But it will come to form a system obeying laws of the organism's own organization. At least the location of this internal, canalizing, restricting

mechanism has been anticipated by geneticists.[12] But now it is possible to make verifiable predictions about its special structures and governing laws. Indeed we can see the reason for the four types of molecular mechanism themselves, as sketched in Section III C.

The relationship of this 'internal' selection with 'external' selection is like that between selection within a business and market selection. In the last analysis, selection within a business arises by the demands of the market, these acting by way of the functional conditions of the product and the organization of the business. Selection within a business, however, achieves its own laws of testing and tolerance — its own autonomy.

3. *Absence of a Lamarckian principle.* The evolutionary mechanism involved has little to do with any Lamarckian principle[13], since events cannot act directly on decisions. The functional pattern of events or features is imitated only indirectly by the systemization of decisions, up to an organizational gap that always remains, and the systemized decisions then act backwards to adjust the events. There is a fundamental difference from Lamarck's environmental theory. It lies in the fact that the determinative principle, in the last analysis, acts by a mechanism running completely counter to what the external environment demands. Indeed a theory of the environment has to be developed which is diametrically opposed to the old concept.

I fully accept Weismann's doctrine and the 'genetic dogma' of today, in the sense that no direct action of the phenes on the genes can be expected. An indirect feedback, however, can be postulated from the laws of probability. As with the laws of entropy, the causality defined by the genetic dogma is not broken by organisms, but evaded. Neolamarckism postulates that there is a direct feedback. Neodarwinism postulates that there is no feedback. Both are mistaken. Truth lies in the middle. There is a feedback but it is not direct. The consequences of this insight will be dealt with again below (Section VIII B7 *f* and *g*). For the moment only the ruling principle is of interest.

d. Pre- or poststabilized harmony

Finally I have to deal with the vitalist controversy involving an 'inner principle'. One side asserts that random change and opportunistic selection cannot explain how evolution can be increasingly directional, as if striving towards a goal; there must be a component that supplies such a goal — an entelechy or vital force. The other side maintains that the force proposed by vitalism is methodologically inaccessible and therefore scientifically irrelevant.

Forces directing towards invisible goals are certainly not scientific facts. Nevertheless, the directional, harmonious transformations of evolution cannot objectively be gainsaid. In this situation it is a matter of basic philosophical viewpoint whether to deny the existence of the dilemma or to believe in a plan of creation or prestabilized harmony. But, on the other hand, perhaps it is possible to conceive some causal mechanism which would give a significance to evolution (though only a directional significance), harmony (though only a constancy of equilibrium) and also a goal (though only provable by the identical order of the actually realized paths).[14]

In point of fact, the theory here advocated fills all these demands. Directionality, harmony, and identicality of order are the results of the encroachment of determinacy into the total range of accident. This encroachment involves a crucial precondition, which is realized in fact. This is that order, whatever accidental constellation it may arise from, works backwards in an order-creating manner on its own fate. The harmony in the creation of living organisms follows a law of nature. But the consequences of this harmony are not predecided, as one would at first believe. On the contrary, they arose with the harmony. The orderliness of evolution is a consequence not of prestabilized, but of poststabilized, harmony. Once again, previous workers were very near to the nub of the matter.

2. The basis of a causal morphology

To an outsider this second group of controversies often seems to be a subordinate methodological discussion, which moreover is outdated. We know, on the contrary, that they concern one of the pivots of the whole evolutionary problem. Moreover, there have been, say, 200 years of scientific biology, and for 150 of those years morphology was its backbone. But after scarcely 50 years of experimental study we are on the point of losing this backbone entirely. This would mean losing the method which gave scientific proof of relationship, descent, and phylogeny in general.

This remarkable and discouraging turn of events must have to do with the introduction of the causal principle into biology. For, except for functional anatomy, a causal principle has played no part in the basic questions of morphological research. Thus morphology, comparative anatomy, and systematics have seemed to be second-class sciences. Many regarded them as outside the bounds of strict scientific method.

Besides this, the unsurveyable accumulation of facts isolated these fields of study and allowed them to fall into hundreds of specialisms. For there are 2×10^6 species with at least 5×10^3 features each. Moreover, as shown in Section II A3, the law content is two orders of magnitude greater than this, while the event redundancy is several orders of magnitude greater still.

The breach between 'old' and 'new' biology resulted from the morphotype problem and on grounds of consistency ought to extend into the homology problem also. From the viewpoint of the history of science, it was more serious than any criticism brought against reductionism or Darwinism. The morphotype is: 'a consequence, a law, according to which Nature will be expected to act'.[15] Since Goethe's first formulation it has remained an abstraction, which could scarcely be measured, was difficult to show in a figure, and which was not easy even to think about. Its cause was inaccessible and consequently it was sagely limited to a thought principle rather like a Platonic idea. Science soon restricted itself to measurement and the immediate causal nexus. It soon came to be asked, therefore, if science had anything to do with almost inconceivable, unimaginable ideas of shape whose causes and epistemological bases were uncertain. The answer came that these had nothing to do with science, and this was difficult to refute. Science downgraded morphology as being idealistic, and tried to exclude it from the guild as being a sort of literary art.

However, understanding the causes of living order is the same as understanding those of living form. The cause of order must, at the same time, contain the epistemological basis for a causal morphology. Consequently the causal principles of morphology constitute the other side of the present theory. I shall survey their consequences here, in sequence of increasing complexity.

a. *The law of homology, from molecules to behaviour*

The homology theorem has been increasingly criticized in the last decade and indeed called completely into question. Its recent form, as expounded by Hennig, Remane, and Simpson, is the basis for the view of homology presented here. But even this has been accused of vagueness and subjectivity — in particular by Sokal and Sneath (1963) and their followers in the American school of numerical taxonomists. This criticism starts from the problem of weighting features in systematics. It asserts that the classical method has confused question and answer, or phenetics and phylogenetics, so that judgement is prejudice and *a posteriori* is *a priori*. It also asserts that no mechanism exists by which features can be given unequal rank. If this were true, it would be a death sentence on morphology. However, it is false.

I shall come back later to the question of phenetics and *a priori* judgements (Section VIII B3c). Here, however, I must concentrate on homology. For we should be in chaos if homology were not entirely reliable within the whole field where it applies.

Numerical taxonomy, however, has done an undeniable service by focusing disquiet concerning natural order without a cause. But we now know just what this cause is. Nothing therefore prevents an objective analysis of its consequences.

On the other hand, it did no service by establishing 'operational homology' as a substitute. By definition this starts by confusing analogy with homology, as for example, 'head, legs, and leaves'.[16] It finishes by homologizing non-homologues such as leaf length.[17] Let us return to the main subject, however.

1. *Identicality*. As already shown, if a message is sent out only once, it is impossible to say how far it depends on accident or necessity. (Section I B1e). The situation changes, however, with repetition of what is received and with the range of the source. With increase of the number of identical repetitions (the collective of species that show a particular homologue) and with increasing complexity of these repetitions (positional-structural events of the minimal and cadre homologues) the probability with which law can be expected increases exponentially. Even with middle-sized systems (Fig. 12, Section II B2d), the improbability of accident greatly exceeds the number of possibilities in the universe. It makes no sense to doubt the presence in such cases of identical conformity to law, nor to doubt that these conformities are anchored in the genetic system. The general limit to the validity of such deductions is that of low accidental improbability. The special limits are next to be considered.

The expectation that an instance conforms to law depends on a minimal number of features being present. This is a consequence of the interpretation of homology given here. When Goethe defined the limits of morphology he recognized one of our methodological limits. He said: 'Anything that destroys the form of the part, dividing a muscle into its fibres, or turning bones into jelly, will not be applied here.'[18] This corresponds to the homonomy limit, separating single forms from mass homologues, and separating identicalities of the anatomical singular from those of the anatomical plural. Beneath this limit, however, homologues continue in histology, cytology, and biochemistry. In another direction they continue as modes of behaviour.

2. *The limit of homology in the realm of micro-phenomena*. This lies at the molecular level of complexity. At this level it follows the probability theorem here proposed, as Florkin has shown convincingly. For isology of molecules does not necessarily imply homology to the biochemist,[19] any more than similarity implies it to the anatomist. We need only assume that molecules are homologous when the cause of complex similarity 'is not consistent with the effects of accident'.[20] This viewpoint is beginning to be adopted for cytological features also.[21]

These methodological considerations hold for 'indirect homologues' or episemantids. These are distinguished in biochemistry from 'direct homologues' or semantids,[22] such as nucleic acids etc. With the latter we have direct insight into the presence of identical commands. This bring us to the problem of homologous genes, with two relevant questions.

First, in considering two homologous cistrons, is the first triplet homologous, or indeed the first base, or the first hydrogen bond of this base? Is there a meaning in which we can speak of homologous atoms, for atoms are indefinitely interchangeable. The answer given here is that, in homologous parts, the building blocks take up identical positions.

Second, corresponding cistrons of siblings can certainly be called homologous. I question, however, whether this is necessarily true for the commands giving rise to homologous events. Are the commands involved in constructing part of the jaw of a shark necessarily homologous with those involved in producing an ear ossicle in man (cf. Fig. 7, Section II B2a)? It may seem strange, but the homology of these decisions is neither certain nor necessary. Thus it is neither necessary nor probable that all the commands involved in producing the series of letters *father* and *père* are identical. Nevertheless these events are homologous in view of the primitive sequence *pater*. Decisions do not have primacy over events, for these themselves are the sum total of the consequences of the decisions. The presence of identical conformity to law is decided by considering the majority of all the characteristics of a system.

The agreement in interpretation is therefore complete, whether the features are molecular or morphological. The only difference is that the particular component parts alter with the hierarchical level of complexity.[23]

The ratio of accident to necessity can always be calculated. In any case as Florkin says: 'The opposition that is sometimes argued as between the organismic and molecular ways of looking at evolution is completely meaningless.'[24] I am happy to be able to agree with Florkin.

3. *The limit of homology in the realm of function.* This likewise is fundamental, easily recognizable, and distinct. In any case the distinction between functioning structures, and functions dependent on structures, is irrelevant as regards recognizing the probability of law. Schneirla and his followers[25] have criticized Lorenz's attempt[26] to homologize functional temporal entities, but this criticism is not well founded.

There are people, including myself, who usually recognize the regular identicality of a piece of music more easily from the funtion than from the struture, i.e. more readily from the sound waves than from the score. There can also be alternation between function and structure. Thus consider the telegram 'start to build house'. This reaches us as a structure. But it produces functions which result in new structure. And these have the purpose of taking over functions which produce new structures etc.

The homology limit once again lies at the minimal number of features at which explanation by accidental possibilities can be invoked. This is true for a body movement as it is for a product such as a melody, a nest or a spider's web. The contraction of a biceps naturally stands beyond the limit of homology.[27] But, contrary to Atz's criticism,[28] the lack of the positional criterion of homology in itself signifies nothing. On the contrary, a behavioural homologue can be repeated indefinitely. In the ground plans of organisms this would place it as one of the mass homologues. And we know that such standard parts can achieve considerable higher constancy than single parts.

All my results further support the concept of homology in Wickler's (1961) sense. They likewise confirm the decisive and indeed dramatic significance of this knowledge for man himself as maintained by Lorenz (1963) and Eibl-Eibesfeldt (1970).

4. *The cause.* It remains to consider the cause of the extraordinary constancy of these laws. The constancy is several orders of magnitude greater than the duration of the molecules that encode them and many times greater than would be expected from environmental selection. Earlier in this book I showed that the cause of this constancy was the functional burden of the feature (or event) and the systemization of the epigenetic system (or decisions) — a systemization by which the relations between decisions tend to imitate the functional relations between features. Thousands of millions of such regularities form systems according to four basic patterns, each occurring from giant molecules[29] up to individuals (Figs. 8b and 14, Sections II B2*a* and II B3*a* respectively). These regularities can be described according to their constancy and their basic morphological transformations ('metamorphoses') and they are open to experimental analysis. Thus the phenocopy gives insight into the course of decisions and homodynamy helps to show how the decisions are phyletically related.

Homologues are the mode by which complex determinative laws show themselves. Within the realm of living organisms their precision and constancy is the highest known. This precision and constancy is maintained by the structural conditions of the homologues themselves.

Homology, like all recognition of law, is in all its forms a probability theorem. It is based on the ratio between successful or unsuccessful predictions concerning expected, identical determinative laws. Any experience within a group of similarities (i.e. of hypothetical relationship) affects every other experience.[30] This is true both of position-structure and of conjunctions of features. Consequently every gap in possible experience signifies a gap in our competence to judge. And whenever we renounce such experience[31] we are to that extent renouncing this competence.[32]

I use our old example from I B3c to reiterate the importance of such experience. If I ask the printer to set down the sequence of letters 'Ohan' then nobody would be able to perceive what I am talking about — whether 'Shan', 'than', 'Oban' or whatever else makes half-sense to the thought-decisions of the reader. Suppose, however, that we add on the 4 million $bits_D$ of the *Canterbury Tales* so that we find this sequence of letters in the first word of the first line, i.e. 'Ohan that Aprill with his shoures soote'. Then everyone will know what I am talking about and also what to think about Chaucer, the printer, and the proof-reader.

The provability of identical regularity in the whole realm of homology cannot be doubted. All human experience, from the phases of the moon to the theory of evolution, rests on the same method.

b. The necessity of morphotype and ground plan

Anyone who doubts the methodological purity of homology will deny the reality of the morphotype above all. This is grave, because the morphotype represents the synthesis arising from investigations of homology and is the basis of high-level systematics.

1. *Criticisms of the morphotype concept.* The first accusation made against the morphotype is that it is teleological. It defines the content of structural regularities arising from unknown 'internal' causes. The known mechanisms, by contrast, only effect a continual alteration, or indeed dissolution, of these regularities, by the action of external causes. In this book I have explained what the internal causes are. The second accusation is that the morphotype concept is pre-Darwinian, and therefore precausalistic. Indeed the evolutionary cause *was* discovered afterwards, but this criticism involves a confusion between recognizing similarity and explaining it, as discussed again in VIII B3c. The third accusation is that the morphotype is difficult to conceive. This is true, but it is baseless to conclude that therefore it is a fiction.

2. *What the morphotype is.* The morphotype is the necessary totality (or basic pattern) of homologues within a group. The group in this connection is a framework within which similarities can be compared. If this framework has escaped from the play of accident it is a phyletic group and to this the concept of the morphotype applies more particularly. The degrees of freedom and fixation for all the homologues involved can be found and defined empirically. Consequently by objective methods it is also possible to work out and describe how far the morphotype is free or fixed. The representatives of the morphotype are related to it as are instances to a law, as Goethe stated in 1795. It would be ridiculous to regard the instances as more basic or more real than the law on which they depend.

Thus the morphotype, with its special pattern of freedom and restriction, is the consequence of its functional burden and of the network of its own epigenetic dependences. It is no surprise that the morphotype is difficult to show in an illustration. It differs from the latest common ancestor, or from a primitive member of the group, in the same way as the archives of the building regulations of an ancient town differ from one of its old houses. Ever since Goethe's 'proto-plant' most morphologists have not thought it possible to illustrate the morphotype. A picture of it can only be approximated. These attempts to picture the morphotype have received various names[33] and compete unnecessarily with each other for a notional alternative correctness.

1. The *diagrammatic morphotype* is one in which only the minimal constituents are defined, as in a structural formula (Fig. 78a). It is the most cautious expression of the morphotype, leaving out a large

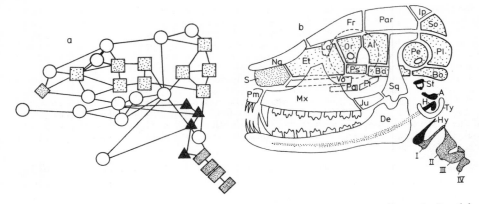

Fig. 78 a-b. The morphotype illustrated by the mammalian skull. (a) Diagrammatically simplified, showing only the positional relationships and five groups of features. (b) A simplified explanatory diagram giving additional indications about shapes. Membrane bones are white; replacement bones and remains of the chondrocranium are stippled (light stipple for the neurocranium, intermediate stipple for the remains of the chondrocranium and dense stipple for the branchial arches). The hyoid arch and auditory ossicles are black. (b) From Kühn (1955); (a) original.

part of relevant experience. The explanatory morphotype, which Remane calls the disguised diagrammatic morphotype, tries to include the most general structural principles (Fig. 78b). Both diagrammatic and explanatory morphotypes correspond only superficially to the diagnoses of systematic categories, for they do not mention any sort of deviation.

2. The *generalized* and the *central morphotypes* begin from these deviations only to subtract them again. For everything that has to do with functional specialization is left out, or else the middle value is taken for all the conditions of form of each constituent part. In diagnoses the characteristic features of these morphotypes are described with the words 'as a rule' or 'usually with'. The kind of deviation is not mentioned.

3. The *systematic morphotype* considers the primitive characteristics of the homologues by taking into account what is known of the group of next higher rank. The diagnoses say 'primitively' or 'originally with'. The directions of deviation are not mentioned.

Unfortunately the first described species of a genus is referred to by taxonomists as the type. This has, of course, nothing whatever to do with the systematic morphotype.

If in the morphotype we wish to include all the empirical results attached to a group, it will take account of homologues and trends, the position-structural relations of parts, the basic morphological transformations ('metamorphoses'), and the various conjunctions of parts. It will include how members of the group deviate, and the directions of deviation. The construction of such a type is a multidimensional process, for these conditions can only be thought of one after another and cannot be reflected by a single picture. Admittedly this is complicated. However, this complexity says nothing against the reality of the morphotype but only reflects the limits of efficiency of our thinking apparatus. All morphotypes have already been conceived approximately. They are laid down in the maximal diagnoses of animal groups. The morphotype defines what is possible in a group, taking burden and the epigenetic system into account. The extraordinary thing is that the possibilities of the epigenotypes have thus already been described in their major aspects, without knowing what the structure of the epigenotype was. Recognizing the natural classification is one of the most baffling of mankind's achievements.

3. *The ground plan.* In a group of animals, the ground plan can be thought of as constituted by the series of morphotypic characteristics of the hierarchically ranked

phyletic subgroups. It comprises the epigenetic total prescriptions which have up to now been followed by the group in its evolution.

No morphotype stands alone. Instead it receives its meaning from the morphotypes of higher rank and receives its content from those of lower rank, as is known for the hierarchical pattern.

c. Science or art?

Finally I wish to deal with a question which goes beyond what the superempiricists seem to understand by 'science'. (I shall return to them in Section VIII B3c.) The true basis of their own activity is the study of homology and its application in comparative systematics and anatomy. But they have brought it to pass that this same basis is widely denounced as an art form. In this connection, of course, they have confused the recognition of law with quantifiability (which is some misunderstanding!)

Some great systematists have spoken of the study of homology as an art within a science.[34] In referring to such evaluations, however, the superempiricists have committed a still worse misunderstanding. They obviously do not know that art within science means the craftsman's feel for his craft or the meaning behind the act. They also do not realize that without thought (or without the heuristic idea) there is no theory. And without theory there is no scientific framing of questions. And without framing questions in this manner nothing can be achieved (except the monthly pay cheque from a public purse that understands this even less).

It is satisfying to speak out against this phantom. The study of homology is the key to recognizing regularity of form and this regularity is indubitably causal, being the reign of necessity over accident and the precondition for human knowledge in general. Art may be confined to great men, but laws will be accepted by small ones. In any case the classical morphologists saw completely correctly, although denigrated as artists or idealists. Everything essential in their predictions was true. The huge libraries of systematics and anatomy contain obvious laws of Nature. Systematics describes the most complex and relevant events on this planet, including the rules that governed the rise to humanity — both our vicious inheritance and the canalization of the hope that remains to us of a higher human order. To say this gives me personal satisfaction.

Attack and counterattack are obviously penetrated by emotion, so I wish to finish on a note of hope. I hope to counterattack the decay which is the cause of our present troubles. For we use our huge knowledge to defend mass slaughter; we overspecialize so that the details shall be hidden in the Babel of scientific dialects; we have retreated into the entanglement of an atomized thought which no longer permits us to see the meaning of our activities. One of the most inclusive fields of human thought is in danger of being lost in this fashion. This must not be allowed to happen.

I hope that the study of form may, in future, make use of the cause of that form. I hope that the young people of today will retrieve the old books from the warehouses and the writings of the old teachers from the box rooms where they have been put and prevent them from being flung on the fire. For the search for tidiness usually leads to this sort of incendiarism.

3. The nature of the natural system of classification

This is a consequence of what has been said. It is no longer possible to doubt the reality and conformity to natural law of the hierarchical system of similarity of organisms. For corresponding homologues prove the presence of identical and real laws, which constitute the morphotypes reflected in the diagnoses of the groups. And the morphotypes necessarily show hierarchical connections. Such doubts have been extensively expressed however. The present theory can help to counter these doubts since the lack of a visible cause for order was what awoke them.

At first sight this seems to be a second-rank controversy. In actual fact, however, the fields of comparative anatomy and systematics have been severely hit by it. Only the 'new systematics' has undergone a renaissance, above all through Mayr's work.[35] This was because it brought intraspecific **phenomena** back into the realm of causal research. The transpecific fields of evolutionary study, on the other hand, threaten to be lost. They have often been devalued as an auxiliary science or the mere pursuit of tidiness, so that their basic requirements are beginning to fail. Scientific schools, or even competent teachers, begin to disappear. The extraordinary body of knowledge, which even now takes up half the space in biological libraries, is being got rid of in many places, although one of man's deepest insights depends on it — the recognition of evolutionary descent. The controversy must therefore be tackled again. There are three unequal groups of problems involved.

a. Nominalism

This is the most extreme of the controversies. Gilmour (1940) denied all reality in Nature apart from the individual, but this has not been widely accepted. For the laws of inheritance and speciation[36] have clearly proven the reality of the species. Crossing allows it to be tested experimentally.

With the genus and higher categories, on the other hand, the situation is different. No doubt these are also connected by inheritance. But are the groupings not artificial, for who drew the limits and how can they be tested? And what is the meaning of the 'correlation effect' which is as puzzling to us as it was to Darwin in 1872. How can we understand, for example, why the notochord is given up in all ascidians but retained in all vertebrates?

In this book I have given the reasons for this effect. Features gradually became fixated along the different phyletic branches, at first by burden, and later through epigenetic interconnection. Consequently they remain preserved in correlated groupings which are characteristic for the groups of the natural classification. If constancy is taken as the criterion of reality,[37] then higher groups are more real than species.

For example, suppose we made an educational film about 17 minutes long (10^3 seconds) of evolution since the Cambrian (5×10^8 years). Then the evolution of a species, with a turnover time of 10^6 years, would go by in 2 seconds on average. The species-specific features, lasting 10^5 years, would take 1/5 second of projection time and could no longer be recognized at all. In such an extraordinary whirl of events we should see nothing of the species, not to speak of the 'existent' individuals lasting one millionth of a second. But the higher systematic groups and their diagnostic features would emerge with increasing constancy from the turbulence and with everlasting immobility, persisting almost the whole length of the film, would form the unmoving core of all events. We overrate the reality of our own peculiarities. The master builder of evolution needs to see them otherwise.

Of course there is no sense in arguing about degrees of reality. But one thing is certain. This is that the regularities which solidify in the epigenetic system become successively more basic, more inclusive, and more immovable with their hierarchical position. In the clock of evolution the species are the balance wheel, the individuals are molecules, but the higher systematic groups are the big wheels in the works. The nominalists are far from understanding the laws that govern these groups.

All scientifically accepted systematic groups are therefore realities, from the genus to the kingdom. The names are conventional but the mutual order is objective. The homodynamy of related epigenetic systems allows this to be tested experimentally. Systematics and developmental physiology are working to recognize this homodynamy, with ever-increasing agreement.

b. Causalism

This is the source of a less extreme but withal more important and indeed central controversy. We should still be in a fog of tacit assumptions and prejudices in this matter if Hassenstein had not examined Goethe's morphology with this in mind.[38] What he found was discouraging enough, but can now be fully explained. According to Goethe, the morphologist says nothing about causal connections, but asserts that forms determine mode of life (Goethe's *Metamorphosis*) and mode of life acts backwards on the production of form. The subject of morphology therefore has the task of specifying the pure 'general phenomenon' of the morphotype (*Typus*) and this exists only as a Platonic idea. Morphology therefore became idealistic. An explanation was much required but its position was usurped by Goethe's 'esoteric [i.e. mysterious] property' of form (*Gestalt*). Where does this leave us? It brings us to the edge of something important, but no further. For the laws of form would be a projection or necessary result of the laws of human thought – this is a situation which demands clarification, for Goethe himself believed in an agreement between the principles of Nature and the forms of perception. And in that case a natural system of classification is a self-contradiction.

Concerning this criticism the answer given here is complete (summary in Section VIII A). It says that the causality of form lies in the reciprocal action of accident and necessity in the molecular determinative decisions that cause that form. The retroaction of form on its causes (Goethe's 'esoteric property') depends on increasing the prospects of necessities by decreasing the range of accident. Necessities which came to be excluded yesterday from accident (to an extent exceeding the constancy of the constituent matter) are protected by the selective conditions of today. The natural system of classification is indeed a system – a pattern of natural laws. If this is so, then the agreement between the patterns of human thought and this pattern of natural laws must itself be a product of selection acting on the evolution of our perceptive apparatus.

The result is the rehabilitation of two centuries of morphological and anatomico-systematic thought. The total correctness of the intuitively applied method is a matter for wonder. I shall return to this (cf. Section VIII B7*d*).

c. Superempiricism

As applied by the so-called pheneticists this attitude has the least validity of all, as will be shown. However, as usually happens, it has already produced a large controversy. The opposed viewpoints extend right through the fields of minor and bacterial systematics. Why is this? It is a question of putting speculation back into its cupboard.

Pheneticists[39] are a group of Anglican taxonomists who hold the defensible opinion that new theories should not be erected without foundation, that facts should be kept separate from interpretation, and that we should strive towards an objective, quantitative analysis of similarity.[40] They oppose themselves to the phyleticists, which would include all other systematists (i.e. those that have phylogeny in mind), and they accuse them of disregarding all these requirements. A knowledge of the cause of morphological order is important in countering all three points of attack. I shall therefore go into this question, but not at length.

It is not necessary to explain everything here since the many misunderstandings depend partly on ignorance of the theoretical structure of pure morphology, even of the important works of Remane (1971), Hennig (1950), Troll (1948) and going back to Tschulok (1922), Naef (1919), and Haeckel (1866) and even to the morphological writings of Goethe. This has already been pointed out by Kiriakoff (1959).

1. *Theory and explanation.* In the first place, theory and explanation always have different fates and functions. Theory presents what needs to be explained. It is indispensable, whereas explanation can change.

The result of the morphological *theory* of a connection between similarities is independent of the explanation affixed to the theory. This is shown by the fact that the groups recognized by the idealistic system of classification were not altered by Darwin; that Darwin created Darwinism as a Lamarckist; that Darwinism made sense even without genetics. It is likewise shown by the fact that high-level systematics, with the morphotype and homologues, applied without the present theory of systemic conditions.

On the other hand, even the simplest question cannot be put without theory, even if the latter is only of the most trivial sort, such as 'something will happen' or 'I shall meet that again'. There is therefore no purpose in trying to avoid theory, for at most it can only be unrecognized, or formulated in a useless form. Every observed feature includes an expectation; every correlation includes a supposition; every ordering in sequence implies a theory that awaits confirmation (instances of a law (= *a*) are the precondition for every insight; Section I B4*b*). A very inclusive theory of this sort is that of homology and it holds without an appended explanation or with one, and whether the explanation is esoteric and occult, or deterministic.

And finally, a theoretical expectation, if confirmed, will be the basis for formulating the next theoretical explanation. This is self-evident. Suppose, for example, that I expect that, within the tetrapods, the first cervical vertebra will be more constantly represented than the twelfth caudal vertebra and this expection is confirmed. I would then be a fool not to expect the same of the thirteenth caudal vertebra even before verification. By this sort of method it would be possible to make gradually more inclusive and significant predictions.

The *theory* that homology can be expected can therefore be replaced only by a theory that is even more to the purpose. Otherwise the possibility of insight will be lost. The *explanation* of the theory, however, can give place independently to a more plausible explanation.

2. *The processes of recognition and of explanation.* It has often been supposed that a systematics which has phylogeny in mind, using similarity as the basis for relationship and relationship as basis for similarity, is using circular argument, mixing up cause and effect. Now that we have a causal explanation of the transpecific phenomena of evolution, this criticism can be refuted. It emerges that two different things are meant by 'basis' : there is the basis for expectation and the basis for cause (cf. Fig. 79).

Supposing a bone is gradually uncovered during an excavation. We make predictions (as Cuvier is known to have done in the lecture hall). And as the uncovering continues we find agreement with expectation, or we are disappointed (i.e. surprised), we correct ourselves and predict again. Finally,

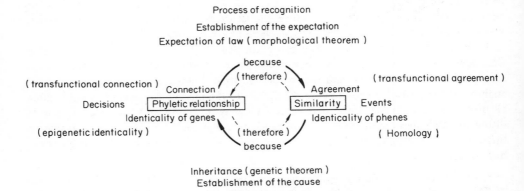

Fig. 79. The connection between the recognition and the explanation of identical or homologous similarities. The concepts that respectively replace each other are shown as mirror images of each other.

when the specimen has been uncovered completely, it has already been allotted to a position in the system of similarities. The cause of this similarity, however, will be explained by a mechanism which is not accessible on the basis of this fossil bone. And this mechanism will lead us to expect identical replications on the basis of totally different experiences.

This is no circular argument, therefore, but by a legitimate cycle of mutual verification. This cycle is as necessary as when we conclude the presence of identical determinative decisions on the basis of identical messages. The human sensory apparatus also requires this cycle, for it can perceive the events but only reconstruct the molecular decisions, since the apparatus itself is constituted by such.

The question remains whether the process of explanation has an influence on the process of recognition. In fact it has such an influence but only in a very restricted fashion — it increases certainty by satisfying expectations (of such remarkable things as monozygotic twins, for example, or atavisms). In the first place, it does nothing else. We should not change the perceived pattern of similarity (i.e. the natural system of classification) even if we were inclined to explain it according to Lamarck or according to Cuvier's catastrophe theory. However, we should only gradually wander into nonsense and contradiction. We would give up, but we would abandon the process of explanation, not that of recognition.

3. *Weighting of features.* This is an insignificant controversy but it is the test of the homology theorem in actual practice and again it was set off by the fact that there was no visible cause for homology. It has a quantitative aspect, and an aspect that depends on principle.

The accusation comes once more from the superempiricists and can be stated as follows. It cannot be understood why features in themselves should have unequal weight for judging phyletic relationships. It must therefore be suspected that those unequal weights have been assigned to them by systematists *a priori*, that is subjectively. Consequently morphology argues in a circle and objectivity can only be reached by using a random series of features which are simply assumed to have equal weighting.[41]

In actual fact, however, an experienced systematist weights a character *a posteriori*. He does it by its conjunction or counterconjunction with other characters in other systems, when possible homologous. The imputation has been clearly refuted by others.[42] If the comparative material is wide enough, then even the basic transformations of position-structure enter the equation, i.e. functional analogies come into the probability equation, so to speak as the denominators, while the functionally independent constant features of the morphotype are the numerator. The causes of the different weights are, as we know, primarily the burdens in the phenotype and secondarily the interweaving into the genotype. As Mayr has said weightings are: 'manifestations of original, highly integrated gene complexes'.[43]

The positions and terms of a weighting formula remain to be specified. So far, however, this is not possible. The construction of such an equation can only be approached through the homology theorem. Remane saw this already but added that it was still uncertain: 'How far, for example, the value of the third criterion will increase with an increasing number of intermediate stages, how it should be assessed for ontogenetic and morphological intermediate stages, and how the different criteria reciprocally reinforce each other.' Consequently, as Mayr advises, it is: 'still best left to the computer in the brain of an experienced systematist'.[44]

This computer has, up till now, created the whole miracle of insight into the natural relationships of organisms without knowing the cause of these relationships. Indeed it has done it without knowing the computer well enough to describe its own functions. This is the true miracle, as must be emphasized again later (Section VIII B7d). The explanation is probably a matter for psychologists but would be useful to the systematist.[45]

I do not doubt that a quantitative morphology will one day be written.[46] I hope that I have helped towards it by synthesizing the customary six dimensions of homology, by proposing the minimal homologue as a unit of measurement, and the improbability of accident as a scale of measurement.

Some further points of contention, or misunderstandings, are of subordinate importance being consequences of those already discussed. I shall limit myself to reviewing two of them from the present perspective (under 4 and 5 below).

4. *The construction of phylogenies without fossil evidence*. This has been condemned as methodological sin, on the grounds that, apart from fossils, no other authority could give information about the structure of an ancestor. However, we already know that the condemnation turns the matter upside down. The sole authority is in fact morphological theory. The only exception, never attained, would be the presence of an observer at some point where the transformation series changed, if only from one genus to the next. Only morphological theory can predict the structure of the nearest relative or of the nearest but one. This is true not only for the graded relationship of ancestors. It is true also, in identical fashion, for mutual contemporaries, whether they are contemporaries of the Triassic or of the 1960s. Up till now every fossil has been placed according to the theory already in existence (or hurriedly made up in face of the fossil). No fossil carries a name.

There is one exception to this – a fossil named 'Beringer'. It was buried in 1725 in a Muschelkalk pit in Würzburg by the over-playful students of the antiquarian Johannes Bartolomäus Beringer. The story is so instructively comical that, for the interested reader, it can help to illuminate even the confusion that I am now discussing.[47]

A piece of fossil evidence proves only a particular conjunction of features, like a piece of recent evidence, and in addition supplies a more or less precise date. This date is the only additional information which a recent organism cannot in principle supply. It allows us to correct the relative positions in time of branching points in the theoretical phylogenetic tree, in this one place. This is the only gain, important though it be, which exceeds what morphological theory on its own can do.

5. *The study of phylogeny viewed as research into fortuity*. This view of phylogenetic studies is held by those who regard all the features not connected in function as being associated only by accident. Why is the retention of the right aortic arch correlated exclusively with the possession of feathers and retention of the left arch with hair? However, we now know what to think about accident. At first it describes, in the same fashion, both lack of regularity and lack of knowledge. Beyond that it is a bad morphological adviser.

It has already been suggested, on the basis of great experience, that such features: 'may have originally started as a functional complex in which the genetic integration was retained even after the functional correlation had broken down'.[48] Without doubt this approaches the essence of the matter. As already explained, it corresponds exactly to what the present theory requires. In both the mentioned examples, the cause for the correlation between aortic arches and skin covering would have arisen in the Trias, and I do not doubt that it could now be found in the epigenetic systems of birds and mammals. In the case of the inverted retina and the notochord (usually broken up into nuclei pulposi) it is even possible to suggest a cause for the correlation. As shown in Fig. 66, both are connected in a degree with the building instructions for the central nervous system. Any break in the connection would lead to certain death in the very embryo.

Nothing in these correlations is accidental. There must always have been necessities of life and livelihood to force them together so constantly.

4. Ontogeny and developmental physiology

This leads to a group of questions dominated by the traditive pattern. In the last analysis they involve the epigenetic system and the structure of the internal causes of order in the morphogenesis of the individual. The first question is why the morphogenetic stages of an organism resemble the sequence of its ancestors. The second question is what mechanism can produce the astounding degree of goal-directed regulation and purposiveness − of reason, so to speak − in a system that needs to function even before the environment can begin to select its products.

Apart from the vitalists all the scientific schools faced with these questions have invoked the influence of selection on how the reciprocal effects of genes are organized. But how can this influence be understood? Hartmann suggests that: 'it is most unlikely that we are *not* confronted with the action of another form of determination, previously totally unknown − a special organic nexus.' The same thing has repeatedly been emphasized by Baltzer.[49] This unknown form of determination has been described as a mechanism of superdeterminacy. This increasing functional burden of events goes with a reduction in the prospects of adaptation so that the functional pattern is copied by the pattern of mutually systemizing decisions.

At a more detailed level, there are as many individual problems as there are individual phenomena, as already discussed in Section II B3*d*. I shall now explain the solution of the more important of such problems from the present point of view (see Table G. Section VIII B).

a. The cause of Haeckel's law

Haeckel's biogenetic law has given meaning to embryology for more than a century and added a new dimension to biology. To the practising research worker it will therefore seem strange that, as here asserted, the law still has no causal explanation. Must we therefore ask whether the law was wrong. For has it not confirmed itself countless times and does it not explain all recapitulatory (palingenetic) conditions? Indeed it does have explanatory power, it has been confirmed and is justifiably called a biological law. (Some consider it a 'rule' but the title of 'law' is much more appropriate.) The mechanism that causes it, however, is still unexplained (cf. Section II B3*d*).

The formulation of a connection describes how the parts are related to each other but does not explain the relationship. The connection can be explained only in a wider context, of which it is itself a part. Thus Kepler's planetary laws explained the planetary orbits. But they were themselves explained by Newton's law of gravitation. This wider framework explained the connection between mass and distance. But it was itself explained by Einstein's general theory of relativity.

Applicability must therefore not be confused with understanding. The practical man should remember, for example that electricity has been known for 370 years, has been applied for 170 years, and indeed whole towns are driven by it. Nevertheless its cause has never properly been understood. Application does not presuppose understanding. Electrical fishes prove this, as the reader will agree.

This does not mean that an explanation for the law of recapitulation has not been sought. Dobzhansky asked: '... what advantage would accrue to the organism from such radical alterations of its development?'[50] Kosswig[51] concluded that embryonic features would be conserved because the new additive features were built using them as a genetic basis. Finally Mayr showed that genes that determine the phenotype of the definitive organism will be the youngest ones, underlain by older ones that cannot change: 'Even if we assume that the structures which are produced by the embryonic gill arches could be produced directly, it is, so to speak, far simpler for the organism to retain these unnecessary aspects of the phenotype than to destroy the harmonious gene complex that controls development.'[52] All this is convincingly confirmed by the present theory.

What is added here, to this somewhat neuralgic part of biological theory building, is not perhaps much. It concerns origin, pattern, and extent. The origin of the systemization of gene effects can be understood from the functional patterns of burden whose structure is copied by the genes. The extent of the adaptive advantages achieved is finally overtaken by the corresponding extent of superdeterminacy (Sections VII C1 and VIII A3). Indeed the phenes which previously belonged to the adult and which, as such, necessarily received the final information in ontogeny, are built over and thus become embryonic phenes, functioning as switches in the transmission of data. Haeckel's law is a consequence of the chance relationship of decisions and events. If it were not known it would have to be formulated.

Most of this has been anticipated by others and to that extent is scarcely worth a new treatment. In other connections, however, it constitutes the centre of the problem, as with the morphotype (Section VIII B2*b*) or orthogenesis (Section VIII B5*b*).

b. The epigenotype, its structure and consequences

This brings me to the second group of questions associated with the traditive pattern. The problem is whether epigenetic systems follow a corresponding pattern of feedback, similar in cause, origin, and phyletic relationship to the morphotype. And if they do, then why do they? In the first place, this is evidently only the converse of the problem about the function of the patterns of decisions. However, we usually discover functions by way of structure, and explain the structure by means of functions. This changes the way of looking at the question.

The enormous amount of knowledge and theory of the epigenetic system can only be mentioned in passing here. It ranges from the molecular level, (Section III C) through the cellular level,[53] to the events at the organ level (Section VII B). Here we are dealing with the cause and structure of a presumably identical mechanism at all these levels, so it is desirable to know what the general principle may be.

The crucial insights here are supplied by developmental physiology. Even the concept of the 'epigenotype'[54] signifies not merely complexity but also a unifying principle. Baltzer[55] appreciated that the epigenotype must include the primitive features on which new features were dependent. Kühn[56] showed that related species must possess identical basic characteristics in their developmental physiology which, by modification, would be made use of by later characteristics. And, finally, Waddington produced another fundamental insight − the concept of the archetype. This predicts that: 'There are only a certain number of basic patterns which organic forms can assume.'[57] This was the last link in a remarkable series of anticipations. Only a small amount needs to be added.

This new contribution is the deduction that there is a feedback mechanism produced by the reciprocal effects of phenes and acting back on the organization of the corresponding gene effects.[58] Thus the gene system copies the patterns of the phene system. This amounts to saying that the organizational pattern of reciprocal gene effects contains in an abbreviated form the process by which it arose. This leads to the feedback definition of the law of recapitulation which is that *the epigenotype (ontogeny) corresponds to a shortened recapitulation of its own history (phylogeny).*

The following palingenetic (i.e. recapitulatory) characteristics can therefore be postulated; (1) There is a limited number of archigenotypes arranged in ranks that correspond to degree of phyletic relationship; (2) Organized complexes of determinative decisions are built up according to the functional patterns of the complexes of events; (3) Old patterns of decisions will be conserved; (4) The locations, lines of communication, and switching sequence of these patterns will be conserved;

(5) Harmonious simplification will take place, graded according to burden, time, and phyletic relationship. The evidence for these postulates is based on the phenomena listed below. The problems raised by the postulates can be resolved by means of the overriding principle.

1. *The number of archigenotypes and phyletic relationships.* The evidence for this postulate comes, for example, from the success of xenoplastic transplantations, i.e. the fact that developmental chimaeras can be produced between different orders or even different classes.[59] These demonstrate a connection between phylogeny and developmental physiology and allow the broad phylogenetic components in the patterns to be separated from the specific components.[60]

One of the consequences of the theory is homodynamy. This comprises homologous patterns of decisions, which are understood and followed from species to species, or even from class to class, to an extent graded according to phyletic relationship. Structural homologies constitute the structural morphotype of the morphologists. In like manner, the total number of homodynamic functions, constitutes the equivalent 'functional-type' in the epigenetic system — this is the archigenotype as proposed by Waddington.[61]

From the etymological viewpoint 'epigenotype' would correspond better to 'morphotype'. However, in genetics the root 'type', as in phenotype or genotype, unfortunately has a totally different sense to what it has in 'morphotype'. Archigenotype is therefore preferable as it has historical implications, referring to conserved ancient elements in common.

2. *Purposive interrelationships of genes, corresponding to the functional relationships of phenes.* The evidence for this comes from the phenomena of homoeotic or systemic mutations; from many phenocopies in the developmental process; and, in regeneration, from heteromorphoses, doublings or more-than-doublings, and replacements of one part by another.

The organization of patterns of gene effects to imitate the organization of phenetic patterns explains several strange facts. First, we can see how whole complexes of gene decisions will produce inherently 'purposive' structures in the wrong place, in erroneous duplications or as replacements. Second, it explains why complexes of all levels of complexity are so bundled together that they can be controlled by a single decision (as shown by mistakes due to a single such false decision). Third, we can understand how these complexes of gene decisions are able to produce regulatory alterations in the phenomenon known as homoeosis (homoeorhesis), homoeostasis[62] or regulation. The most extreme of such phenomena are the xenoplastic chimaeras.

3. *The conservation of archaic patterns of organization.* The evidence for this comes from the phenomena grouped together as spontaneous atavism. In this a change in a single decision (i.e. a point mutation) triggers the production of phenes which are often meaningful within themselves, well balanced, and very complex. They are lacking in the normal recent organism and appear to be meaningless for it, but were represented in its ancestors.

No function can be recognized for the conservation of these well organized archaic complexes of commands which are pointless for the definitive phene system of the organism. We have here gene atavism or cryptotypy. This conservation, together with the mechanism that produces it (relict homoeostasis), is totally enigmatic until we recognize that the complexes of commands need to be conserved, unconditionally, as the necessary bearers of the more modern pattern of decisions which is built on them.

4. *The conservation of the locations of decisions and of their lines of communication and sequences.* This postulates the conservation of what once were final decisions, which have sunk down to the different depths in the epigenetic system to become preliminary or early intermediate decisions. The evidence for this conservation is convincing. The

locations that house these preliminary decisions (and give them out) are known morphologically as blastemes. From the developmental point of view they are called organizers, being of primary, secondary, or tertiary order. The patterns of flow for decisions are the paths of induction. The sequence of decisions is the induction sequence. It is crucial to recognize that homodynamic decisions can be proven: 'not only in the primary positions of induction, but also in subsequent inductional positions.'[63]

The homology of the locations and paths of homodynamic decisions and their similarity graded according to phyletic relationship can be understood in terms of the feedback formulation of the biogenetic law.

5. *Shortenings, according to burden, time of development and phyletic relationship*. Under this heading come a group of phenomena illustrated by the germ-layer theory and the types of cleavage.

The schematic quality of the folding and separating embryonic epithelia is explained by this type of simplification and also the symbolism in cleavage patterns of early development.

c. *Systemization in the molecular realm*

Up to now we have accepted the systemization of decisions in the genome as an observed fact (Section III C). However, it could be asked why whole systems of preliminary decisions are built up when an equally complex volume of events can be determined by a code of very low rank — as an organism can be coded in DNA or the *Canterbury Tales* in Morse. We have established, however, that this systemization prevails because of crucial selectional advantages.[64] Even the most recent genetic models demand '... that selective factors can influence the integrative configuration in which an organism uses its genes'.[65]

Our model predicts that, as phenes become more differentiated, there will be an increase in the number and the ranking of preliminary decisions. Likewise, in the molecular model the increase in DNA (from $< 10^4$ in a virus to $> 10^9$ nucleotide pairs in a mammalian nucleus) depends 'mainly on an increase in complexity of regulation, rather than on an increase in the number of structural genes'.[66] The same holds when decisions that already exist are promoted to the rank of preliminary decisions and even when a certain degree of higher-order ranking comes to exist in which: 'the effect of the integrator gene is to induce transcription of many genes in response to a single molecular event.'[67]

Even in the molecular realm, therefore, our theory explains, in terms of selectional conditions, the necessity of systemization and the forms that it takes. It is confirmed in this by the most recent results and models. The real contribution is the prediction that imitative patterns will arise, what their selective advantage will be, and the four basic forms that they will take — as necessary results of the DNA, operon, regulator, and order-on-order systems.

5. Phylogeny

This brings us back to the starting point of the controversy with the question: Is there a dirigism in phylogeny or not? The group of problems which we found to be the nub of the matter has not previously disquieted people — this includes, for example, the lack of a reason for the law of recapitulation, for massive numbers of standard parts, and for hierarchically arranged systematic diagnoses. Even the existence of homologues was seen as a freak of Nature. On the other hand, ever since Darwin the scientific world has been worried by the many problems resulting from the observed directive components of phylogeny. However, the reader will have seen already that this directive quality is one of the most obvious consequences of the determination process. The cause of homology, as given here, or of the law of recapitulation, will explain all the forms of dirigism in phylogeny.

We are once again confronted with a group of questions whose existence has never seriously been doubted. But the individual problems were enlarged or diminished by different workers. This depended on their initial opinion, i.e. whether they thought the principle of environmental selection would probably explain the matter, or not explain it at all. Here we are dealing not with rigidity of opinion but with different starting points. This indicates the numerous invisible hurdles that we have already taken.

All the individual cases of directionality can therefore be used as evidence of the total phenomenon. And contrariwise they can be explained by its mechanism. They are listed in Table G (Section VIII B).

a. Synorganization or coadaptation

Some features are known to have arisen separately but show purposively coordinated evolution thereafter.[68] We have already seen this as one of the basic problems (Section VI B1c). However, Remane himself predicted the reign of an internal principle of dependency. Rensch[69] likewise supposed that there must be: 'a particular condition in the harmonious construction of animals' by which: 'every change is guided by special laws which affect the organism as a whole'. Osche supplemented this by saying that these rules: 'produce selective advantages at the moment when features combine'.[70] The cause of this internal principle can already be glimpsed.

Only the mechanism needs to be added. The decreased prospects of coordination of individual events (Section VI C1) require the setting-up of synchronizing decisions (in the regulator-repressor system Section III C3a). This mechanism confirms the reality of the problem and also of its solution. However, the effect of such systems in epigenesis which arise for imitative reasons is not limited to the coordinations of phenes that originally were independent. It will prevail generally. Thus the whole of evolution is synorganized from the coordination of the largest articular surfaces to that of the small vessels. Or, more correctly, the problem of synorganization is one of the more striking extremes of the universal phenomenon of homoeostasis or interdependence.

b. Trend, orthogenesis, and Cartesian transformation

These headings cover three variations of the controversy as to how far and why phylogenetic sequences are oriented, apart from directive external factors. There has been discussion about how straight a 'genesis' has to be in order to be orthogenesis (*orthos* = straight). This led to the less definite term 'trend'. On the other hand, Cartesian transformations have not caused much argument. They excellently describe these miracles of coordinated change and have been taken as mere forms of illustration.

However, the conservation of a directional meaning, whether the same directions or a single direction, can be observed everywhere in evolution. It extends over hundreds of millions of generations and over innumerable contrary possibilities offered by environment and selection. Justifiably this has never ceased to cause discussion. Some have assumed 'internal factors'. I mentioned the view of these workers in Section II C2 and would like to mention again the courageous survey that Whyte (1965) gave of the matter and his demonstration that the subject is now thoroughly alive again. These

Fig. 80 a-e. Unusual and aberrant antlers. (a-b) Reconstructions of early and very early forms of antler. (a) *Syndyoceras* with two pairs of bony processes on the skull. (b) *Cranioceras* — three-antlered with long bony processes and short terminal antlers. (c-e) Errors of growth in recent forms: (c) Three-antlered fallow deer following implantation of an antler bud; (d) Wig antlers following castration, similar to what occurs after injury in the hunt; (e) Many-pointed white-tailed deer resulting from several years sojourn under warm conditions — the annual loss of antlers by frost has been prevented. (a) After Müller (1970) and Portmann (1948); (b-e) from Goss (1969).

workers correctly believe that orthogenetic controversies are the essence of the matter. The pure Neodarwinists, led by Simpson, Rensch, and Mayr, take longer to find this essence. They suppose that: 'the evolutionary chances of the phenotype owing to natural selection are limited by the possible amplitude of response of the epigenotype.'[71] However, does this not say the same thing? Certainly it does, and as a biologist I find it satisfying that neither of these two great groups of workers has been on the wrong road. There are regulations inside the epigenetic system which give the cause to this meaning or direction. The experimentalists have anticipated this result with their 'canalization',[72] 'archetypal selection'[73] and 'network of connections with the remainder of the genotype'.[74] The mechanism rests on burden, the traditive inheritance of an imitative epigenotype, and on superselection which finally acts counter to adaptability. As soon as the universality of this mechanism is recognized the phenomenon will also be seen as universal. Indeed it is difficult to find varying features unaffected by dirigism.

Even examples of maximal evolutionary freedom (as shown in Figs. 39-42, Section V B2*a*) contain less freedom than determinacy. It is the freedom of a leash. The end is free but the base is totally fixed. Thus with horns and antlers, there are numerous features which have been fixated. They are always paired, project from the frontal bone, are free at the distal end, and tapered. The ancestors show relatively more freedom in these

Fig. 81 a-e. Unknown and impossible horns. (a) Divided horns are impossible because of mode of growth. (b-c) Sagittal horns and horns in the position of teeth are both totally improbable. (d-e) Horns thickened at the ends or fused together are improbable on developmental grounds. (Original).

respects, naturally being less constrained (Fig. 80a-b). Developmental errors are likewise more free than normal recent animals (Fig. 80c-e). A greater degree of freedom would be shown by the non-existent antithesis of the position-structural features quoted above (see Fig. 81a-e). For a still further increase in freedom see Hieronymus Bosch, as shown in Fig. 52, Section VI A*a*.

Antlers have never been found to carry eyes, nor to grow on toes, nor on fishes. They have never been found mobile with suckers nor broken into twigs and covered with leaves nor covered with cilia and swimming in drops of water. However, the last trace of interdependence is lost only when it is no longer possible to speak of antlers even in the most attenuated meaning.

So long as concepts can be formed, then basic fixations exist and, in the time axis, basic directions also. All evolution is directed and mostly to a totally amazing degree. It is not easy for anybody to fit this fact properly into our biological world-view since for a century we have been brought up on the 'miracle of adaptation'. Moreover, it seems to fit badly even with the freedom that we ourselves demand. However, let us return to biology and its laws.

c. Typogenesis and typostasy

This theme is related to what has just been discussed. In particular, however, it concerns the major evolutionary paths and the change in direction of evolution in the temporal axis, i.e. the cessation of all change or what Mayr called 'the hollow curve'. There is no doubt about existence of the phenomenon. Discussion concentrates on three questions, however. First, do all paths lead to a typostatic phase, as Rosa's law requires?[75] Second, is the change from a typogenetic to a typostatic phase a necessity, being a typostrophe as many have supposed since Schindewolf?[76] Third, what would the cause of the process be?

In this book I have shown that fixation results primarily from burden and secondarily from the imitative epigenotype. Freedom and a new increment of morphological distance depend on newly added features, of low burden. This agrees with the commonly accepted notion of a stabilizing stasigenesis[77] and an additive typogenesis.[78] It also agrees with the resulting mosaic evolution[79] according to which a single evolutionary path may combine typogenetic and typostatic characteristics. This was deduced in detail in Sections V B4 and V B5 (cf. Fig. 50). It was also shown that newly added features will contribute to the fixation of the features which form their substratum. At the same time the new features require these more basic features as a precondition. The mosaic that arises is meaningful.

Both phases are necessary in view of the selectional conditions in the system. Obviously, however, typostrophes may sometimes be incomplete.

Many extinct groups were not lucky enough to achieve the next typogenesis that they needed — either it was never begun or it ended unsuccessfully in a so-called typolysis. Many recent groups are so young that the last features added are still very little fixated. Also the adding of features takes place at unequal intervals and in unequal amounts, which shows itself both in the course of phylogeny and in the systematic categories defined on these features. In principle, however, both phases are necessarily to be expected.

d. Homoiology, parallelism and irreversibility

These phenomena also support the interpretation here advocated and provide additional reasons for the views now current. Parallel evolution is the process which leads to homoiology as the achieved result. The question that arises is why phene systems repeatedly possess evolutionary freedom and evolutionary possibilities in the same direction.[80] The usual answer is that: 'it is due to response of a common heritage, to similar demands of the environment.'[81] Similar anlagen contain similar phene functions and correspond to similar gene decisions.

To this we can now add that the pattern of burden for the phenes and the pattern of interconnectedness for the genes will also be very similar. This is confirmed by the phyletic limits of the phenomenon and by its obviousness and the necessity of its occurrence.

New storeys of similar function built on top of similar buildings of similar function will most probably lead to similar forms. The prospects of going back to old functions, on the other hand, are totally different. If a hunting château falls into neglect we should not expect that the form of the preceding hunting lodge would reappear, nor the gamekeeper's hut that it once arose from. This is an illustration of the phenomenon of the irreversibility of phylogeny. The enormous advantages gained by coupling decisions together (Section V C3c) are not merely lost when the environmental requirement changes, but transformed into their reciprocals. Only during the very youngest preliminary stages is it possible to go back. With all older ones the interconnections are too great.

Old patterns can here and there appear. They can have no success, however. On the other hand, late ontogenetic stages can persist as neoteny. Many structures survive better in an incomplete state.

e. Vestigialization and atavism

These are two aspects of the phenomenon of traditive inheritance. They are respectively the slowness of dismantling and long conservation of features which, as metaphenes, are either functionless or scarcely functional. They have long been an evolutionary riddle.[82] It is nowadays believed that extended conservation depends on the persistence of functions in the epigenetic system. Indeed, the interpretation presented in this book has been anticipated in detail.

This is shown by the following lines from Osche (1966, p.846): 'Some of those anlagen which seem at first sight pointless are by no means without function. They have important tasks to fulfil in the complex process of development. Some, for example, are organizers which induce particular developmental processes in adjacent regions of the embryo, as the notochord induces the neural tube. Others act as 'stencils'; for example the branchial arches serve to orient the blood system. Others represent routes for the transport of material.'

What I add here is relatively little. The cause for the location of particular organizer effects in particular blastemes is deduced. It is the selective advantage that results from copying functional connections. The difficulty of finding substitutes for these organizer blastemes is also explained and the cost in time needed to simplify them.

The transmission of organizational commands and their mutual coordination must be universal in all component parts of organisms. As soon as this is recognized it will also be obvious that traditive inheritance is universal in the evolutionary process. These apparently functionless structures therefore prove to be only a special case, though a particularly striking one, in the great number of structures which cannot be justified on grounds of function alone. Indeed such structures probably form the greater part of every organism. They are the historical part of the organic form.

For one hundred years there has been a one-sided, though justified, admiration of adaptation (which is indeed a miracle). It has therefore become difficult to convince people that adaptation is only a relative matter. When compared with consciously purposive planning, the historical component of organic form is oblique, confused, and full of detours, as Helmholtz already appreciated. This can only be understood by knowing the detours that every construction has passed through and by realizing how improbably difficult it is, or indeed miraculous, to alter the given dispositions. Man himself is just such a compromise. He is a rag bag of vestiges. He includes the greatest number of fixated lacks and evils which selection will tolerate.

6. Ecology

The evidence in this book has been taken almost solely from systems within individuals. It might therefore be asked how super-individual systems come into the matter. However, I have concentrated on morphology on tactical grounds, as being the methodologically most reliable field. The identicality of regularities of individualities is by no means coterminous with the concept of the individual — a concept whose limits are in fact very vague. The reign of order is also to be expected in systems made up of individuals. Very little is known of such systems, but their existence is important. The deficiencies of order in our own individual systems have long been a matter for pathologists, psychiatrists, and judges. But man has evolved no trustworthy authorities to judge the deficiencies of human social systems.

I shall once again restrict myself to the consequences of the theory and to a few basic problems of research into ecosystems.

a. Superindividual order

This consists of the determinative laws of the individuals united in a biocoenosis or organic community and the laws that connect these other laws together. The second group of laws are very much complicated by the reciprocal effects of events and decisions, since systemic events have their common cause in many different gene systems (or gene pools). They only indirectly feed back on the determinative decisions established in the gene systems.

Nevertheless it is possible to recognize standard parts (or units) and interdependences and also hierarchical dependence and tradition. The latter for example is seen in the phenomenon of 'Lebensorttypen'[83] which is the tendency for morphotypical features to be correlated with biotype features in a manner which does not depend on a direct functional connection but on long causal chains, some of which are probably due to developmental physiology. These are the historical constituents of a biocoenosis. They confirm the type concept, as does the morphotype, and are in turn confirmed by an insight into the causality of this concept.

Again there is an evolutionary tendency for the instances of order to spread into all positions still unoccupied. Also the primitive forms of order, recognizable by a particularly large proportion of repetitive events, tend to transform into higher forms. In ecology this is the phenomenon of diversification. Everything tends to show, once again, that increase in order will increase the prospects of realization, while the attainment of higher forms of order[84] will increase the stability and survival prospects of the systems, which in this case are communities. All this emphasizes the general validity of the present interpretation, on which these special conclusions are based.

b. Order, energy, and the biosphere

The above-mentioned correlations give hints about the future prospects of organic communities in general. However, their relevance is still more obvious when they are applied to the biosphere, which is now approaching the limitations of a space ship. The universal application of the principle of energy flow[85] has shown that all biosystems are selected for a continual increase in the throughput of energy. This holds for human communities and their products, as well as for organisms and biocoenoses. It has two immediate applications in the limitations of energy sources and in the morality of power, the latter involving crops, reserves, influence, capital, armaments, and all the various civilized variants of energy. It has also two immediate and converse connections with order which is the reason for emphasizing it here.

First there is the convertibility of order and energy as already discussed (Section II A1). We can translate this and say that all biosystems — whether organisms, or biocoenoses, or human societies and their products — are selected towards an increase in order content. This sounds more hopeful. Moreover it includes the mutual confirmation of both theories. The only remaining question, if it is one, is which of the two mechanisms contains the primary cause? In the mechanisms of evolution is increase in energy a consequence of increase in order, as might be hoped? Or is wisdom, as might be feared, a mere corollary of power?

On the other hand[86] order and energy flow in opposite directions, as we have already seen (Section III A*a*). The conservation of order demands not merely energy, but a particular quantity of energy, suited to the structure of the order (cf. Fig. 16b). Too great a throughput of energy destroys the order of the system by overheating. This is the cause of the present environmental problem. In the second or civilized evolution, success-oriented societies have become too powerful for their own biotopes. They are sawing at the bough that they sit on. For survival, the quantities of energy and of order must again be brought into equilibrium. Besides the brake on increase in population, there must be a brake on consumption of energy, but also a more efficient breakdown of energy, and a transformation, at the correct time, of lower into higher forms of order. This will be discussed again later (Section VIII B7*e*).

Thus even on the scale of the Earth, order can be recognized as a phenomenon that conditions existence. Here, however, I am again at the limits of my particular speciality so I shall turn now to the most general consequences of the theory.

———————   ———————

In its gross features (and at considerable length) I have therefore depicted the consequences of the theory within biology. I shall now step back from the individual facts, so as to summarize the connections. There seem to be four main consequences.

1. In the study of evolution no discipline has primacy, nor is there any solution in terms of unidirectional causality. The results of the postulated reciprocal process between decisions and events can be proved equally at all levels.

2. There is no real contradiction between the synthetic Neodarwinist theory, on the one hand, and those who demand an 'internal principle' as implied by the morphotype or orthogenesis. We can anticipate that even the most apparently irreconcilable standpoints will prove compatible.

3. There is, however, a general law of structure (to take the static aspect) of which the present theory explains the basic phenomena.

4. There is also a general law of transpecific evolution (to take the dynamic aspect) and the present theory explains the phenomena of its orderly cause.

At this point a cautious biologist would probably finish. I still have something to say however – as the reader can see. This is for two reasons. First, because biology as a subject extends from the evolution of molecules to that of man. Second, because it is one thing to save one's own skin but quite another to stand up for the consequences of one's own handiwork.

7. The perception of Nature

We are now verging on natural philosophy. An expert will only need to read a few more lines, however, to see that they are written not by a philosopher but by the anatomist that he already knows too well. For just this reason, I shall not stray far into philosophy itself. However, the philosophical questions are relevant because they can be defined as a consequence of biological experience, though not solved without physics.

a. Law, knowledge, and explanation

It seems we can know nothing for certain. For all conditions and events depend only on accident and necessity. This smooth separation in the world is smooth, in the first place, by definition. It ceases to be trivial, however, when we notice that our thinking apparatus is built completely on this pattern.

1. *Order and experience.* Order is the repeated occurrence or realization of identical law texts, or patterns of determinative decisions. The origin of order corresponds to the origin of experience and order is the precondition for experience. This agreement between order and experience can only be explained, as shown below (Section VIII B7*d*) as a product of selection.

In the first place, we are able to draw conclusions only by observing determinative decisions which are complex enough in their effects to be appreciated as what is commonly called an event. I return to this later, (also in Section VIII B7*d*). The reason for this second separation of the world into decisions and events is very simple. All decisions that can be encountered finally depend on atoms entering the next most stable condition, whether they do this individually, in masses, or in complex systems. Neither the finest of our sense organs, however, nor the elements of the human data-processing apparatus, can perceive atomic states.[87] For even an individual sensory cell includes many millions of such states. We are therefore constrained to perceive complexes of decisions and to reconstruct their constitution experimentally. Since we lack a special sense organ for atomic states, this detour is forced onto us. We can avoid it only by transferring the individual decision to the macroscopic world, as when we use a relay switch.

2. *Perception of the presence of reality*. This depends on comparing the probabilities with which the repetition of an event could be expected as a product of accidental or of determinative decisions. The precondition for this is the storage of an equivalent of the event (memory) and the ability to compare the equivalents and to consider the results of the comparison. Experience is confirmed by the recurrence of events within the framework of predictions which are already possible. In this connection 'event' stands for conditions as much as for processes.

It would be totally unjustified to expect a difference 'between structural and causal laws' as it is commonly expressed. The degree of certainty concerning the presence of regularity depends, rather, only on the extent of the features and the number of identically repeated observations. This fact, as shown above, plays an essential part in the homology theorem as it does in observing physical experiments, for example.

Likewise it is wrong to distinguish in principle between laws and rules. For there is no evidence for the existence of a third condition somehow intermediate between accidental and determinative decisions. Rules are imprecisely recognized laws. They include border regions where either accident reigns, or where further laws of unknown kind are at work. Of course the difficulty of defining such limits between accident and necessity increases with the complexity of the object. In biology the difficulty is therefore very large. In every case, however, we can gain insight only into the degree of improbability that identical repetitions of an event will contradict our prediction. This prediction can verge on certainty, but this is just as true of structural laws as it is of the laws of the lever.

3. *The explanation of an event*. This has nothing to do with predicting an event. Indeed, with increasing knowledge, one and the same regularity can receive one explanation after another, without being changed in itself. Examples of this are known from all sciences. The lack of an explanation, or the proof that the only known explanation cannot be right, does not alter the prediction, i.e. the regularity itself is not at all changed by it.

This at first sight is a strange situation. However, it is based on the fact that the description of a regularity (i.e. the definition or formulation of a correlation, dependence, or mathematical function) is not in itself the basis for a higher-ranking, more fundamental principle. But it can only be explained as an instance of such a principle. Explanation requires laws of higher rank. And these laws again form a hierarchical system rising up to the universal laws of space, time, mass, and consciousness. These highest laws in consequence escape explanation. Much of this, of course, has long been known[88] but, as concerns morphological laws, it has been overlooked.

b. The system of decisions

At the centre of the present theory there are, as will be recalled, two statements. First, as events come together to form a system, their prospects of successful change come to differ. Second, the prospects of successful change will be improved when the functional pattern of events is copied by the systemization or ranking of decisions. This results in an increase in the prospects of adaptation and of survival or, in general terms, of stability. Even outside biology, we should expect that systemization would advance the intercoordination of stable internal conditions (or laws) under defined external conditions.

1. *Increase in redundancy.* The tendency for decisions to become redundant is therefore a basic condition and depends on two preconditions. Starting from the assumption that a biosystem at first includes no redundant decisions, then such will only arise when decisions of equal rank come to be arranged in a system of ranking. The second precondition lies in the rejection of particular permutations of events — combinations which would be possible in terms of the decisions contained in the system but which, during a sufficiently long phase of evolution, have not been accepted by the external conditions. If all events could take on all forms of their range, then no decisions would be redundant.

Both these preconditions, however, seem to be fulfilled — perhaps so constantly that no decision remains long in position if it becomes redundant. Systemization would therefore largely keep pace with the differentiation of the system.

However systemization very drastically reduces the range of permutation, i.e. the number of permutations which would be expected merely from the number of parts of the system.

2. *The discrepancy between possible and realized systems.* This is extraordinarily large, as has already been shown from thermodynamic considerations. However, the reason for this discrepancy, according to Morowitz, is: 'at the moment quite beyond the realm of thermodynamics since we have no real explanation of why, of all the possible quantum states, the biosphere is restricted to such a small subset'.[89] The explanation for this fact seems to be given by the present theory. The cause is systemization, i.e. the dependence of decisions on each other.

Whatever rank they have, all decisions correspond in the last analysis to stable quantum states. A general description of the phenomenon accordingly becomes possible. Whenever the decisions that determine a system are arranged in ranks, the number of possible permutations will be drastically reduced. This reduction will be a function of R = the avoided absolute redundancy. Even in simple systems this will be very large.

Differentiation under systemic conditions seems always to tend toward canalization of possible states. It tends to a canalization pattern according to the possible pattern of systemization. This explains why natural phenomena, both organic and inorganic, are limited and describable. However, this again is outside the scope of biology.

c. *Order as an intermediate condition*

There is a further consequence which is also surprising. It becomes obvious as soon as the framework is specified within which order, according to the present definition, has meaning. Order or determinacy content is defined as the arithmetical product of law content times the number of instances ($D = L \cdot a$, cf. equation 18, Section I B4b). Within the realm of organisms, both law content and the number of instances of events have very high values. However, the ratio of the two will differ; indeed they show an inverse tendency. Taking account of the phylogenetic position of systems with very different numerical values for L and a, it becomes obvious that the one group includes very primitive types and the other very highly evolved ones.

1. *Lower and higher forms of order.* We can therefore distinguish between lower and higher forms of order. In addition to the extent of order ($L \cdot a$) there is, so to speak, the

quality of order (Q). This can be specified as the quotient: law divided by number of instances, i.e. $(L \cdot a^{-1})$

$$Q = L/a \qquad\qquad\qquad\qquad (33)$$

In seeking for a general meaning for L we can calculate it for low qualities of order by using the information that is needed to define the positions of atoms rather than by using homologues.[90] We can measure it as usual in $bits_L$ and a will again be the number of identical realizations. Estimated in this way, the quality of order thus varies by 20 to 30 orders of magnitude.

With a cytosine base, for example, there are 34 atoms and the decisions required reach at most 24.5 $bits$ per atom;[91] L is therefore $\leqslant 10^3$ $bits_L$. The number of instances, however, is 5×10^8 identical molecules per average genome, times 10^8 cells on average per organism,[92] times 2×10^6 species, times on average 10^9 individuals; consequently $a = 10^{32}$. The quality of order is therefore very low with $Q = L/a = 10^3 \times 10^{-32} = 10^{-29}$.

On the other hand, the system *Homo* has at least 5×10^5 homologous non-redundant individual structures[93] and thus certainly more than 10^6 $bits_L$. A few years ago the number of instances was still $a \leqslant 10^9$ identical examples in the world population. The quality of order was therefore relatively high with $L/a = 10^6 \times 10^{-9} = 10^3$. It would be at least 19 orders of magnitude above that of a pyrimidine base of the genetic code.

These are coarse approximations. They take no account of the identical representation in other organisms of the homologues of the human ground plan nor the order-producing achievements of the genus *Homo*. It would probably be premature to include the Renaissance and the Enlightenment in the numerator of the equation or the agreement with other ground plans in the denominator. For the present, the expectation of a scale of the quality of order of 20 to 30 orders of magnitude will be enough.

It is interesting how far these quotients of quality of order correspond to the common notion of value. For we assume that the value of an individual decreases with its mass occurrence but increases with rarity. Consider, for example, the difference between the pay of a soldier, on the one hand, and the price that a zoo would be prepared to offer for the extinct or almost extinct Steller's sea cow, on the other hand, if one could be caught. The different qualities of order are as real as the correspondence with our notions of order is obvious.

2. *Evolution in quality of order*. Equally, there is an undeniable evolution in the quality of order of organic structures. Oblique to the time axis it can be proved in all the patterns of order. This is interesting. Nature strives towards higher order in quality as well as quantity.

Mass standard parts break down by differentiation or individualization. Along the axis of evolution, symmetries reduce from spherical through the form with two planes of symmetry at right angles (comb jellies) to only a single plane of symmetry. Even bilateral symmetry often breaks down, as shown in the internal structure of man. We have recognized the mass hierarchy as the primitive form which passes through the dichotomous form to the box-in-box hierarchy of old age. Correspondingly, mass dependences can be recognized which by evolution become individualized to give single connections.

It is therefore a strange thing that order, as understood here, acts to break itself down. *Order appears as a field of transition between primitive and higher conditions of determinacy*. The harmony of living organisms is a transitory condition. It makes a great impression on us, however, because the human apparatus of perception selects entirely according to its pattern. The correspondence with natural harmony can be experienced with never a contradiction.

3. *The evolution of stable conditions*. Even outside living organisms, however, evolution tends to produce stable conditions. This again seems to be a field of transition.

Cosmological theories,[94] both conservative ones and the newer ones of Jordan and Dirac, agree in assuming that the Universe arose as a core which expanded to its present dimensions in 5 to 10×10^9 years and is still expanding. The elements are thought to have arisen from this core during the process of expansion, and hence still to be arising. The elements are very unequal in abundance,[95] for 99.8 per cent of the atoms of the Universe are of the two simplest kinds (83.9 per cent H, 15.9 per cent He). There is therefore a steep decrease in abundance with the quality of order. As nuclear physics advances it has increasingly developed the view that the more complicated elements have developed from the simplest forms, absorbing quanta step-wise. Indeed it is believed that elements are still arising in the super novae and evolve further in the stars and new planets. This view would imply an evolution from simple to more complex, like that of life.

The lowest, most primitive or most original quality of order conceivable would have protons as the numerator and Jordan's Number (10^{40}) as the denominator. (The latter is the total mass of the Universe with $\sim(10^{40})^2$ elementary or proton masses.) This lowest quality of order represents a limiting assumption.

The correspondence with organic evolution is even greater in the so-called 'epigenesis' of regularity. To physicists, biologists, and natural philosophers it is clear that: 'The special causal laws appeared successively in the course of cosmic evolution.'[96] The laws of the lever would not yet have appeared within a cloud of gas, nor the laws of metabolism within a solid body. 'Acting counter to stabilization and levelling-out, the process finally leads to a stage of complication which creates the material conditions for new evolution of order.'[97] Finally the so-called 'chemical evolution'[98] is evidence for total continuity with the morphological laws described in this book.

Indeed all three basic phenomena and their genesis seem to be foreshadowed in the inorganic world, i.e. pattern, breakdown of pattern, and canalization. Standard parts (or units) and symmetries are known, as well as interdependence of parts which is expressed in equations like the hierarchical positions of a bracketed mathematical formula. Symmetries, mass standard units, and hierarchy are known to be differentiated from quanta, on the one hand, to giant molecules on the other, from the ball of gas to the formation of mountains. We also know that, by increasing differentiation, stable conditions arise as a restricted set out of all the conditions that are possible.

Among elementary particles only a few are long-lasting. Among the 1000 different atomic nuclei that these particles *could* constitute, only a third are stable, belonging to 100 elements.[99] Physical laws are embraced in new regularities by chemical structure. And organic structure, as we have seen, exceeds almost all other in accidentalness.

What, however, is the general cause of such epigenesis which adheres to the basic patterns and, in building determinacy up, does not stop at primitive mass order (as in a crystal, for example) but differentiates the repeated events and strives towards the pure regularity, as in a non-periodic solid, of a self-limiting number of conditions? In organic Nature, primitive order seems to be the gain in stability which can be reached with least expense in discovering regularity, with least evolutionary, trial-and-error risk. This primitive order is a cheap order. It has arisen from the mutual action and canalization of events and decisions. It is a preliminary stage which is slowly replaced by the still greater prospects of stability of higher forms. Organic order seems to be only the complex terminal state, being itself a consequence of the order of matter.

4. *The highest states of order.* It therefore seems that the highest orderly state which evolution can strive towards must be that in which there is not only a maximum of determinacy (i.e. the greatest possible predictability, knowledge, and certainty) but also a

minimum of repetition and a maximum of conformity to law (i.e. the greatest possible differentiation, complexity, and individualization).

I have shown that the evolution of organisms has ascended 20 to 30 orders of magnitude in this scale of order-quality. Nevertheless, the number of identical representations far exceeds the law content. Only culture may perhaps reach higher values. The great works of art seem to correspond to these conditions. So also do great ideas which, alone in their class as the scheme predicts, persist longer than their parent generation and by their laws captivate whole centuries.

All this has become unquantifiable so I will give a further consequence. We should expect that the highest order attainable in the universe would be pure conformity to law. In it a would equal 1, and D and Q would equal L. In it nothing would be repeated. However, we know that order (as law and determinacy) can only be recognized by repetition. If this highest type of order existed, we should not be able to comprehend it. The prophets, however, have said this already. Evidently, at this point I must stop.

However, in the general problem of evolutionary theory (Section VIII B7*f*) it will be necessary to come back to this once more. Before this, however, I must discuss our own position in the system of order. I refer to the trans-structural aspect of man, the *sapiens* in the structure of the genus *Homo* — the mechanism of thought and its product.

d. Patterns of thought

The consequences given here again come from comparative anatomy but are used to throw light on the mechanism of human thought. This may seem remarkable, or perhaps even a reversal of causality.

In analysing each of the four patterns of organic order I had to begin, in actual fact, with the methodologically disquieting statement,[100] that without them thought was impossible. Later we found, in detail, that all the four patterns of order are indubitable realities. What is to be made of this? The conjunction of thought patterns and natural patterns is beyond the possibility of accident. They are too similar and too rich in features for the agreement to be accidental. Either they must have an identical cause, or else they cause each other mutually. A common cause cannot be disclosed by the tools of an anatomist. Nevertheless it may exist. In their reciprocal conditioning, however, the natural order, being older, must be the cause, and the order of thought must be the result. I have shown this four separate times[101] and supported it with the observation that the cause of the natural order had itself become clear, without requiring an identical order in thought.

1. *Order in thought as a simulation of natural order*. I have therefore shown, in each case, that thought-order must simulate natural order. In fact this view goes back to Plotinus. Goethe applied it in his morphology. More recently it has been criticized by some (as discussed already in Sections II C4 and VIII B3*b*) but strongly supported by others.

Thus Dessauer says: 'All attempts to prove that man drags order into Nature only by his own thought fail at the fact that his thought is so long mistaken — often centuries long — until, by experience, he adapts himself to a given fact which has become evident.'[102] Philosophy and physics agree that all the patterns of order can be derived from *number*. Thus Strombach says: 'Is it not therefore obvious to suppose that a factual connection exists between natural order and the order of mathematics, which latter is certainly an expression of our logical order of thought? If Nature opens itself to the grasp of mathematics, then its principle of being must somehow correspond to that grasp.'[103] Finally Weizsäcker says: 'Nature is not mental subjectively; it does not think mathematically. However it is mental objectively; it can be thought about mathematically.'[104]

The precondition for this simulation of Nature by thought is identicality of the basic laws. Goethe said this already: 'If the eye were not sun-like, it could not recognize the sun.' This identicality is fully proven. The quanta of the molecules of visual purple, which reacts to the quanta of light, are sun-like. So are the quanta of the hydrogen atoms of the sun, which by fusing to form helium, send the light quanta out.

2. *The cause of the simulation.* This must lie in selection. As Bacon said *'Natura parendo vincitur.'*[105] It is easy to produce models to show that this simulation must have crucial selective advantages. For example, if the similarity data of predators and of their prey can be stored and locked up using just such hierarchical patterns as occur in Nature, this will bring great benefits. 'During the phylogeny of higher organisms up to present-day man, thought has necessarily adapted itself to the logical regularity of the world.' This consequence, drawn by Rensch,[106] has already been defended by several biologists and is confirmed by all the material in this book.[107]

Three methods will furnish the proof. First, there is the analysis by animal psychology of preconceptual thought[108] going back to the primitive amniotes and even primitive tetrapods. Second, we can study the logic of our most recent phylogenetic and ontogenetic predecessors, which is still accessible in primitive peoples and in children. Of course, previous researches were not directed towards the present problem, since the patterns concerned are here defined for the first time. Nevertheless, there is a surprising agreement between the 'systematics' of primitive peoples and scientific systematics.[109] Equally surprising is the ability of children to homologize.[110] This is true although neither is in a position to explain the principles of their comparisons. Lorenz speaks of ratiomorphous achievements rather than rational ones. This is the essential point. Indeed we do not need to go back to children and savages to convince ourselves of the preconceptual abilities of our thinking apparatus.

Some universities encourage systematics, on account of ecology, but suppress comparative anatomical instruction and no longer even know of morphology. I have 'experimented' on students at such universities. By agreement with the student I have prevented access to all the basic concepts of morphology and asked him to undertake a piece of systematic research. The new descriptions resulting have been published in the specialized literature and are in no way worse than many other scientific papers.

The mistrust felt by the superempiricists (cf. Section VIII B3c) is subjectively easy to understand. But objectively their error is equally evident. In large measure, however, it is preconceptual principles of reasoning which produced the whole of natural classification by affinity, or indeed by descent. This was done by intuitive anticipation of the laws of form.

This touches on the universal phenomenon of 'aprioristic' reason, whose existence is so energetically denied by many empiricists. I suppose that when this old controversy is settled, it will prove to be a matter of preconceptual experience, which seems *a priori* to each single individual but is an *a posteriori* experience for its line of its ancestors.

3. *The cause of the conservation of human thought patterns.* This may lie partly in the selective advantage of the simulation of natural patterns, but not entirely, if only because this simulation often leads us astray.

Consider all the exaggerations due to the expectation of pattern. The standard-part pattern leads to overstandardization. Hierarchy leads to throwing things together. We see patterns where none exists. There are tricks of interdependence leading to nonsensical names like 'jelly fish' and many others. And there are tricks of traditive inheritance which fill the language with etymological misunderstandings.

The canalizing cause for conservation must again be the principle of economy. This was already recognized by Simon (1965). The features of all the people, or of all the trees, which each of us has seen would exceed the capacity of the human brain. Even if we lacked merely the storage pattern of mass hierarchy, for example, we should never be able to orient ourselves. For by it we store only the general aspects of 'tree' (excluding bushes and telegraph poles) and the type 'person' (in the first place excluding all individuality). There is an enormous number of redundant events which thereby do not burden the store, as we have already seen in the realm of physics.

The system of thought evolves, of course, both analytically and synthetically. A small child sees every stretch of water, small or large, as a 'puggle'. But later we distinguish between puddles and oceans.[111] On the other hand we synthesize things that we first saw as separate. Heaven and earth become solar systems and galaxies. Space and time even become the system of relativity. The failure caused by every excess in thought-economy, however, leads us back to the four patterns. Without doubt the capacity of our brains is limited – two litres of regularity, well packed though it is – against the volume of the universe. Moreover there seem to be no other patterns for economy. The miracle, therefore, is that our imaginations, full of such inexpensive patterns of order, are on the way to reflecting reality totally, though they are doing it slowly and in canalized fashion with errors and repeats of errors.[112]

e. *The patterns of civilization*

It is likewise a miracle to find our civilization full of primitive patterns. This has already been expounded for each of the four patterns or order.[113] Here, therefore, only a summary is needed. The pattern building of civilized evolution seems to be like that of organic evolution. There is an increase in features, burdens, and dependences so as to produce a maximum of regularity with a minimum of text. For the text involves costs, while regularity brings reliability. In the structural field I expressed this as reaching a maximal determinacy content (D) in the events, with a minimal prospect of accident, and therefore with the least content of law or decision (L). Translated into the equivalent concepts of civilization this means achieving a maximum of orderliness (D, certainty, predictability, recommended action, stipulation, and tranquillity) by means of minimal investment (R, capital, costs, effort, learning or differentiation) and thus with least 'information' (L, wisdom, knowledge, insight, and experience).

The result is the three collective patterns of standard parts or norms, hierarchy, and interdependence (beside one another) and of traditive inheritance (in time). The latter could be called an adoptive order. Modern sociology knows this already.[114] Anatomical experience leads me to add that we seem to be dealing here with the consequences of natural laws rooted deeply in the molecular realm – consequences which we probably accept for this reason. And indeed this is probably why we take their simultaneous action for granted as being a necessity. For example, no standard part can exist without hierarchy, nor conversely. Despite this the contrary assertion repeatedly gives rise to those mass convulsions, which, deplorably enough, we call the turning-points of world history. It is not plurality that deserves our growing criticism. What should be criticized is the primitive condition of those patterns of order together with the consequences of their fixation.

The civilizatory success of these primitive conditions of order depends on a reduction in the quality of order (cf. equation 33, Section VIII B7c). By this, decisions become redundant, as a result of synchronizing the demands upon them, and they are therefore dismantled (L decreases). Events, on the other hand, become easier and easier to repeat identically (a increases). They conform to standards, become ranked hierarchically, become interdependent and are traditively inherited. In the short term this means a decrease in the difficulty of ensuring order and of making new decisions (to an extent measured by the power -*R*). The obvious results are mass standards and mass hierarchies, and thus of mass dependences, being the opposite of self- and individual ordering. These mass standards and hierarchies bring an increase in conformity, along with norms, rank, specializations, ideologies, taboos, and tacit social assumptions. An increase that affects products, producers, and institutions. This increase in conformity is a disappearance of individualization, i.e. of universality, independence, and differentiation. It is mainly the consequence of the order in technical civilization. Its evolution to the higher forms of order brings the gradual dismantling of these mass features and produces the visible phenomenon of culture. As Schrödinger[115] says: 'This will not tend to make production cheaper, but those who are engaged in it happier.'

Paying back yesterday's advantages of civilization will always begin today whenever the environmental conditions change direction — remember the implacable relationship between accident and necessity. What was once credit (with the power of -*R*) must be reckoned as loss (with the power of *R*). High systemization of decisions can in the end make transformation impossible. And if the demands change, the system may break down in chaos. This recalls Spengler's conclusions. One would like to hope, however, that knowing the mechanism, we shall in future be able to put reason on the credit side of the balance. This brings us back to the general problem of evolution, where we left it in Section VIII B7*c*.

f. Two gloomy theories of evolution

The scientific conception of the essence of the Universe is determined by the two great theorems of evolution, which have turned out to be both universal and correct. These are the law of entropy in physics and the law of descent in biology. Their correctness cannot be doubted, but their completeness certainly *can* be.

Schrödinger referred to the consequences of the mechanism of descent as 'gloomy' and as a physicist he sought a way out of such hopelessness.[116] Nevertheless, it seems that every creature's hope is subject to the planlessness (or accidental decision) of a blind designer (mutation) and the pointlessness of a myope (selection) who is called upon continuously to decide between life and death using the measuring rod of momentary opportunism. To a biologist the consequences of the law of entropy are in no way more hopeful. For this teaches us that all order depends on leaving a still greater quantity of chaos behind it. And all the harmony of Creation, the highest values of human cultures, are squandered finally in warming the cold of universal space and in increasing the unspecifiablity and confusion of the whirling atoms.

It is remarkable that we can read of this evolutionary process in the Apocalypse. There heat, death, and chaos symbolize Hell while hopeless, unending struggle for mere survival represents Purgatory (as most recently in Sartre). Certainly there is no doubt about these laws. But where in science have the laws of *genesis* been formulated? Where are we told how order and law arise, whose products are so obvious — worlds and seas, people and their rights. They are much easier to see than extermination and rising chaos. Is it not remarkable that, if we follow Goethe, it was Mephistopheles who first took pen in hand? Is 'the Spirit that always denies' so closely related to the scientific critic?

Obviously no progress can be made in that direction. Those who believe in sensibility will be repelled by such triviality. The critical thinker will despise such fantasizing prophesy. However we know that, preceding all rational knowledge, there is unconscious knowledge. And since this cannot rationalize, the pronouncements made by its prophets cannot be due to reason. In this situation it is helpful to suppose that both theories of evolution are correct, but also incomplete. This is an aim of research, as also a hope. I have already used up more than 276 pages in proving that both these statements are true for living organisms. The evolution of organisms is far from being planless. It is driven by the energy pump and decrease in entropy, by the prospects of accomplishment and conservation. These lead not only to differentiation and diversification, to an increase in accidental improbability, but beyond that to a self-stabilizing harmony of verifiable conformity to law. The laws of transpecific evolution expand the scope of 'scientific meaning', or the goal of living organisms from the 'meaning of survival' to the 'meaning of self-creative order'. This latter meaning has characteristics for which I propose the concepts of self-action, goal-building, and self-ordering.

1. *Self-action*. This goes beyond the apparent passivity of living organisms. It lies between active and passive happening and for it there is no proper name — for which reason I must be excused this new and still empty term. It includes self-steering, feedback, and self-regulation with the steering parts being steered and the steered parts steering. It lies between activeness according to Lamarckism and passiveness according to the genetic dogma — although, as to the mechanism, Lamarckism is unproven while the genetic dogma remains in principle uncontradicted. The intermediate position of this mechanism is easy to see, while that of human existence is more difficult to appreciate. The functional system of events results in systemization of decisions, while the pattern of decisions produces constancy in the functional systems of events. Each part is therefore just as much cause as effect. Indeed, the limits between events and decisions disappear on analysis, as we saw. What remains is systems of decisions which include cause and effect in common. Actively acquired characteristics do not become heritable as the result of individual tendencies, but because of the total of relevent tendencies. Lamarckism overvalued the role of final events, while the genetic dogma overrates that of first decisions. We do not make evolution, nor are we made by it. We ourselves *are* it.

2. *Goal building*. This goes beyond the supposed pointlessness. I now hold it to be certain that the paths of evolution run towards particular conditions, i.e. particular patterns, combinations of events and decisions, morphotypes, and epigenotypes. In other words, they exclude enormous numbers of other possible combinations. Thus evolution is full of goals. But it was not expectable that any primitive fish should become a tetrapod, nor any primitive tetrapod a man. These were only potential possibilities. Everything seems to be predictable necessity except the encounter of the possibility with the requirement. The goal originates only when it is set. It can neither be foreseen, nor given up afterwards. We neither have the potentiality of controlling what the goals of evolution shall be, nor are we given goals from the beginning. The goal arose with us and is not to be taken away from us.

3. *Self-ordering*. This goes beyond the randomness of the forms of order. Order in this world, whatever may have been thought previously, is not merely the establishment of improbable conditions. And increase in order is not mere multiplication. Instead the order contains regularity which repeats itself according to particular symmetrical patterns, which again can be defined mathematically. These patterns penetrate all creatures, their thought as well as their creation. This is an agreement which we respect by calling it the harmony of the Universe. It is a system of patterns which itself evolves from the symmetries and ranks of the molecular code up to the composer's or architect's symmetries and ranks in symphonies and cathedrals. Among the qualities of order that are known, mankind is at the head. Growth in the quality of order is suited to cause an increase in orientation and insight, reason and wisdom. And the distant goals of this growth accord with what we call our highest values. At the same time it must be admitted that this harmony was not conceived in the first place by us — cathedrals and symphonies cannot be a necessary consequence of mammals, or organisms or of biomolecules. They

are only included in them potentially. Order has not been inserted by man into Nature. It is contained in Nature. The differentiation of order — its 'spirit' — arose with order. But the basic form of order — its meaning — did not first arise in living organisms. The meaning of order is a consequence of matter.

This brings us back to inorganic order. There is, first, the special question of how far the organic laws of order are already contained within the inorganic world. This question has been given very different answers, as by Monod and Teilhard de Chardin.[117] But second, and not far removed, there is a general question: What produced the unfolding of law in matter, or what is the principle of order out of which this law does unfold? This question does not arise from the recklessness of a biologist but, quite on the contrary, from the physics of Heisenberg's principle of order. By this principle Heisenberg extended the mathematical law of matter back to the realm where matter arises. In this realm, mass and energy themselves must correspond to each other according to Einstein's equivalence of mass and energy, and must be traced back to a common principle. This principle consists of mathematical structures. As Heisenberg said: 'For modern science, therefore, there stands at the beginning, not the material object, but form, which is mathematical structure in a mental context. And so we can say, with Goethe's Faust: 'Wenn ich vom Geiste recht erleuchtet bin, geschrieben steht: "im Anfang war der Sinn".'[118]

We have traced the correspondence and unfolding of this meaning in the laws of organic form and evolution up to the degree of differentiation where the meaning encounters its own self within consciousness and knowledge — a system in which these molecules are even in a position to think about molecules. The law of self-unfolding order must penetrate the whole of genesis. So far as living things instruct us, it is a law relating to the systems of determinative decisions. The great theorems of evolution must be right, but they cannot have been complete.

g. Determinacy and destination

The reader will now foresee that we shall soon reach the last page that I can add to this book. For obviously we are getting further and further away from the things about which anything concrete can be said — although, on the other hand, we are meeting things which concern us more directly. Well-foundedness seems to be inversely related to relevance, but after what has been said this is no longer surprising. As to the most general matters, we can only ask whether they should be said at all, and if so by whom. The natural scientist, among others, must be allowed his say, because he has the facts and must be allowed to go as far as these facts, in his convictions, will carry him. For, to be frank, what should we learn from all these pebbles of hard fact if they signified nothing for us ourselves, for our life or for our future prospects.

The theory asserts, in essence, that the course of evolution is, to a much greater degree than hitherto supposed, excluded from accident and subject to regularity. This means, in the first place, a widening of determinable prediction. This may be welcome or unwelcome, but that is not the present theme. It is a consequence of the widening of reproducible knowledge. The question of whether we wish to have this knowledge, or not, will not be discussed here. Greater determinacy in evolution means that its blessings and evils will be more predictable, and this predictability must be considered briefly.

1. *The morality of fixated evils.* This is, so to speak, a consequence for the future and is the most important result. Lorenz[119] discovered, if he and the reader will forgive the epigrammatic summary, that God has nailed the Devil into us. People shudder at this result and some who shudder would try to prove, with any available quantitative means, that things are not exactly so. Whatever other results we come to in this book, however,

there can only be the conclusion that they *are* so. However, if man comes to be persuaded that for this reason his inherited evil can be excused — like aggression, for example — then in future he will lay about him in an even more uncontrolled and self-righteous fashion than before. The same things apply, as we saw, for structures and functions equally — the difference is purely methodological. Von Bertalanffy[120] established, concerning the growing-points of evolution, that multicellularity brought death with it, the nervous system brought pain, and consciousness brought anxiety. For this he was accused of subjectivity. But nevertheless he was right. In each of the four patterns of accident and necessity we have shown that gained advantages must be paid for. And they must be paid in the same currency which was formerly accounted as success (mostly success in exterminating the neighbour).

The laws of this process cannot be swindled and its products are full of catastrophic mistakes. But these together are not quite bad enough to make extinction certain. Stability of conditions is a relative matter. It can only be evaluated on the basis of the differentiation that the conditions contain (i.e. the degree of improbability or of order). Man is a complicated animal and biologically one-sided through the excessive development of a single part of the brain. This makes the balancing difficult. It is, therefore, all the more imperative to have a deep knowledge of his fixated deficiencies. Otherwise we shall have to surrender our place on Earth to the rats or the sulphur bacteria. In the field of anatomical structures we have now calculated man's own prospects. And whoever does not soon calculate them for the field of civilized functions will no longer have any prospects.

Epigenetic systems cannot be altered. Since they need to store yesterday's advantages they must today include drawbacks — disadvantages with respect to an environment which we are not adapted to, and which we have ourselves created. This is both the tragedy and the hope of man's position. If the disadvantage disappears it will only be by adapting our environment to the biology of man and his living space.

This confronts us with the consequence of a theory of the environment. This theory contains the converse of what dialectical materialism expects from the environment. Man, the most complex of all products of somatic evolution, cannot be temperamentally restructured. Certainly he cannot be restructured by success-orientated civilizations which *are* successful (i.e. they can exterminate their neighbour so easily) precisely because they consciously pursue the most primitive, many-million-years-old requirements. That is to say, they pile up energy, or in civilized terms they pile up: 'crops, profits, money, influence, power, capital, and armaments'.[121] Goethe wrote further: 'Es sollte stehen: Im Anfang war die Kraft! Doch auch, indem ich dieses niederschreibe, schon warnt mich was, dass ich dabei nicht bleibe!'[122]

2. *Order, and order of values*. Where does this bring us? We have in no way obtained a complete picture. Genesis is not the work of the Devil. He can only collaborate in it. We know enough of evolution to say that evil will be a result of selectional benefits. And these benefits we found again to lie in the directedness that does not merely involve a planless spreading of variety and specialization, but allows order to increase and the forms of order to be elevated. Directedness lies in the expansion of pure law content.

At the same time I stated that this increase in order is identical to the constituents of consciousness, called orientation, predictability, certainty, and tranquillity. These are concepts closely related to right and peace.[123]

I also stated that evolution increases pure law content at the expense of redundancy of events — it raises the quality of order. In social terms, this means the loss of standardized restrictions and of mass quantities, a dismantling of the standards of class, prejudice, taboo, and tacit social assumptions. It means pluralization, individualization, and differentiation — the recognition and stabilization of individual laws at the expense of mass laws. These concepts closely approach our ideals of the dignity of man and humanity. The human environment must be adapted to our survival. This is yet a third discovery. The highest levels of order, towards which matter is directed by the mechanisms of evolution, seem to lie outside the realms of the corporeal. There is an outflowing, by which the order of a time or individual is projected outwards. This outflowing is the product which we call culture. These highest levels of order, which have a minimum of redundancy and mass identicality, but a maximum of individual incomparable regularity, we are accustomed to call our highest values. Indeed, we value them the more highly the more inclusive and unique they are. Indeed, the highest value of all — pure law — we cannot yet even rationally conceive.

The agreement is too extensive to be an accident. Rather we have found the evolution of consciousness to be a consequence of corporeal evolution. We have neither dragged order into Nature, nor did her laws exist before they first appeared. We and evolution are therefore one and the same. This is our canalized hope — the way of escape from the slavery of redundancy, the superfluity of mankind, the prospect of the natural breakdown of dangerously stored energies, the remaining path from Missing Link to man. 'But what we might have taken for luxuriance of construction, now seems to be the requirement of a natural law — a requirement on which our survival depends. Even did we not submit to inherited values, we may perhaps have to submit to recognized laws.[124]

3. *On law and freedom.* Our theory asserts that the course of transpecific evolution — the path of genesis — is mostly subject to law. The consequence of this is a regular fixation of evils and of their corresponding blessings. Ignoring either of these will have no prospect of success. Evolution is self-specified. Contrary to what Leibniz[125] thought this world is in no way the best conceivable. Nor, contrary to Voltaire's[126] riposte, is it the worst. It merely includes both, and our prospects seem to lie in recognizing ourselves in it. In this most general of connections, the relationship between law and freedom is the last of the consequences that I wish to discuss.

For readily understandable reasons, we usually associate law, prohibition, and their uncompromising enforcement with restriction, suppression, and slavery. And we associate freedom with the opposites of these. We find this understandable because we all belong to groups of individuals who are expected to follow laws decreed, not by themselves over themselves, but by others over them. We ourselves are indeed the law of Nature that ought to be followed. But we decree ourselves into communities which need a greater amount of regulated agreement than could be expected from the collective of their members. It would be wrong, however, to equate freedom with chaos.

When we feel free, or strive towards freedom, it is liberation not so much from individual as from collective regularity — a freeing from identicality, redundancy or superfluity, from the norms and self-evident assumptions of the group. It is an individualization[127] which at first is independent of whether the unusual deed will be rewarded by prison or a medal or both. Freedom thus seems to be the attempt to increase individual law at the cost of redundancy. This again is a rise in the quality of order. Sometimes it will run counter to the general ordering process, will find no application, and end in chaos. Sometimes, however, it may cause much higher forms of order to arise and be reckoned as order of high quality. The highest freedom — and the final consequence of the theory — is the applicability of the greatest possible individual order.

It is a triumph of law over identicality. This is the same consequence drawn by Goethe's *Faust* in his final interpretation of the first words of the St. John's Gospel: 'Mir hilft ein Geist! Auf einmal seh' ich Rath und schreib' getrost: im Anfang war die That!'[128]

The order of this world is one of its basic characteristics. It is the product of the unfolding of natural laws and of the prospects of survival of instances of these laws. Along self-ordering paths it follows the possible dependences implied by stratified systems of determinative decisions, and by the conditions for constancy of internal systems within defined external systems. Order is the condition for understanding the Universe. It forms our thought and practice as a consequence of the patterns in all living structure. And these patterns appear as a consequence of the stability conditions for matter, of their possible mathematical symmetries.

The conditions for order are the converse of perplexity, of arbitrariness, and chaos. The genesis of order is the necessary counterpart of entropy.

NOTES

1   von Lwoff (1968, p.94) and Linschitz (1953, p.261).
2   Lwoff (1968, p.93).
3   Dancoff and Quastler (1953, p.268); Quastler (1964).
4   This attitude is defined most clearly in defences of holism, e.g. Weiss (1970 a, b and 1971), Koestler and Smythies (1970), Cannon (1958), Simon (1965), and Koestler (1968, 1970).
5   'Embryo selection' (1943).
6   'Archetype selection' (1957).
7   'Genotype selection' (1958).
8   'Developmental selection' (1964).
9   Rensche (1954).
10 Mayr (1967).
11 Osche (1966, p.844).
12 Dobzhansky (1951), Kosswig (1959), Baltzer (1955), Waddington (1957) etc.
13 Lamarck (1909).
14 A general idea of this total concept can be got from E. Hartmann (1875), Driesch (1909, 1919), Teilhard de Chardin (1961), Schubert-Soldern (1962, 1970).
15 Goethe (1790).
16 I have no space to deal with this here but see Sokal and Sneath (1963, p.69).
17 Sokal and Sneath (1969, p.70).
18 Goethe (1795).
19 Florkin (1962).
20 Florkin (1966, p.7).
21 Wilmer (1970).
22 This distinction was first made by Zuckerkandl and Pauling (1965).
23 Wald (1963).
24 Quoted from Florkin (1966, p.164).
25 Schneirla (1957).
26 Lorenz (1935), and also Tinbergen (1942), and Baerends (1958).
27 Cf. Dilger (1964).
28 Atz (1970).
29 Florkin (1962).
30 This has all been pointed out already by Hennig (1950), Remane (1971), and Simpson (1961).
31 This is evident from the 'operational homology' of numerical taxonomy (e.g. Sokal and Sneath 1963).
32 See the discussion 'how many characters' in Sneath (1957).
33 In classifying morphotypes I here completely follow Remane (1971, p.119 ff.). I differ somewhat from him in the interpretation of morphotypes and diagnoses (cf. Kühn, 1955 p.73 ff.).
34 For example, Simpson is often quoted in this contrast. By contrast see Warburton (1967).

35 Mayr (1967) gives a general review.
36 Mayr (1969).
37 As in N. Hartmann (1950) and (1964).
38 Hassenstein (1958) and, more particularly (1951).
39 The name seems to come from Greek *phainesthai* = to appear, or from phenomenon or phene. Nevertheless the matter has little to do with phenomenology in the accepted sense of Husserl. Instead it has to do with empiricism.
40 This goes back to Cain (1956), Gilmour (1937), and Michener (1957) and led to the numerical taxonomy already mentioned.
41 Sokal and Sneath (1963).
42 E.g. by Simpson (1961) and Mayr (1969).
43 Mayr (1969, p.221).
44 Remane (1971, p.60); Mayr (1969).
45 I recall here that the human preconceptual epistemological apparatus was described in principle by Konrad Lorenz (1973) who brilliantly confirmed the conclusions here independently reached. I shall discuss the preconscious or ratiomorphous mechanism in Section VIII B7*d* and show it to be a product of selection cf. also Brunswick (1957), Popper (1962), D. Campbell (1966 b).
46 The beginnings of such can be found, for example, in Farris (1969) and in Olson and Müller (1958).
47 See Wendt (1953).
48 Mayr (1969, p.221).
49 N. Hartmann (1950, p.689), Baltzer (e.g. 1955).
50 Dobzhansky (1956, p.346).
51 Kosswig (1959, p.215).
52 Mayr (1963, p.606).
53 Britten and Davidson (1969).
54 Waddington (1939).
55 Baltzer (1955).
56 Kühn (1955).
57 Waddington (1957, p.79).
58 This principle was anticipated by Warburton's 'feedback in development' (1955).
59 Kühn (1965, p.541).
60 Baltzer (1957, p.19).
61 Waddington (1957, p.80).
62 The concepts are reviewed in Lerner (1954), Waddington (1957), Nanney (1958).
63 Baltzer (1952, p.295).
64 For a survey of the relevant literature see Britten and Davidson (1969). As concerns man's future the same thought was developed by Medawar (1960).
65 Britten and Davidson (1969, p.356).
66 Britten and Davidson (1969, p.352).
67 Britten and Davidson (1969, p.356).
68 As surveyed by Remane (1971).
69 Rensch (1961, p.127).
70 Osche (1966, p.889).
71 Mayr (1963, p.614).
72 Baltzer (1955).
73 Waddington (1957, p.80).
74 Kosswig (1959).
75 Rosa (1903).
76 Schindewolf (1950).
77 Huxley (1957).
78 Heberer (1958) and (1959b). Compare also Osche (1966, pp.874-6).
79 The phenomenon is discussed e.g. by De Beer (1958).
80 For literature on this subject see especially Haecker (1925), Simpson (1952), Huxley (1958), Osche (1966), Thenius (1969b).
81 Mayr (1963, p.609).
82 Wiedersheim (1893), Romanes (1892), Plate (1925).
83 Riedl (1963 and 1966, p.514).
84 Margalef (1970).
85 Odum (1971). Cf. also Hass (1970).
86 Commoner (1970), Riedl (1973, a, b).
87 The greatest achievements or minimal required stimuli for sense organs are nevertheless astounding enough. A dog can probably smell a single molecule, man can see a few photons and hear sound oscillations with the amplitude of an atomic diameter.
88 Compare e.g. Bavink (1930), Eder (1963).
89 Morowitz (1970, p.168).

90 Following the procedure of Dancoff and Quastler (1953, p.264 ff.).
91 See also Quastler (1964, p.3).
92 Surveyed by Britten and Davidson (1969, p.352).
93 For reasons of caution I take one homologue as a measure for only a few $bits_L$, although its law content is several orders of magnitude higher.
94 Advanced treatments are given in Heckmann (1942) and Jordan (1952).
95 Surveyed by Klüber (1931).
96 Schrödinger (1961). Quoted from Rensche (1961, p.318).
97 Strombach (1968, p.100).
98 E.g. Ponnamperuma (1972).
99 Surveyed in Eder (1963).
100 Sections *Aa* of Chapters IV to VII.
101 Sections C 2*c*4 of Chapters IV and VII, and V C3*d*3 and VI C2*c*2.
102 Dessauer (1958). Something very similar was already said by Kraft (1947).
103 Strombach (1968, pp.59 and 66).
104 Weizsäcker (1958).
105 Strombach (1968, pp.59 and 66) says: 'We must obey Nature, if we wish to conquer her.' (Bacon's words literally mean: 'Nature is conquered by the obedient one.')
106 Quoted from Rensch (1961). Cf. Also Mohr (1965, p.526) and Rensch (1968).
107 Since these lines went to press this interpretation has been confirmed in the most convincing manner by Lorenz (1973). What I have here deduced from the anatomy of the patterns of order, was clearly shown by him by comparing modes of behaviour. See also Brunswik (1934, 1957), Popper (1962), Campbell D. (1966a and 1966b), and Lorenz (1967). I have added only the basic patterns of order and now state that, since the structures of living organisms are selected according to 'hypothetical realism', the patterns of thought and their ratiomorphous preliminary stages are the consequences of these basic patterns.
108 Examples from Köhler (1952) to Lorenz (1973).
109 Berlin, Breedloue, and Raven (1966) and also Diamond (1966).
110 Particularly important here are Lorenz (1965a and 1965b, p.282 ff.) and Lorenz (1973). For a survey see Löther (1972).
111 Compare the thought psychology of the Würzburg school e.g. Bühler (1907-8). See also McGeoch (1952).
112 Compare Lorenz (1971) and the lines of Piet Hein's quoted by Lorenz on p.252.
113 In Sections IV C3, V C4, VI C3, and VII C3.
114 Cf. especially Freyer (1955), Berger and Lockmann (1966), Georgescu-Roegen (1971).
115 Schrödinger (1969, p.125).
116 Schrödinger (1969, p.113). We also recall the views of Bergson (1907), Cannon (1958), and above all Koestler (1968). The latter has dramatically shown how incomplete our current views of evolution are.
117 Monod (1971), Teilhard de Chardin (1959).
118 Quoted from Heisenberg (1959, p.148). It will be remembered that we already encountered these words from the Gospel of St. John in the information theory at the beginning of this book. (Section II C4). ('If the Spirit enlightens me truly, it is written: "In the beginning was meaning" ')
119 Lorenz (1963).
120 Bertalanffy (1955).
121 Riedl (1973a, p.10), Riedl (1973b). See these works for the relevant literature.
122 Goethe, *Faust* Part 1 — the third interpretation of the St. John's Gospel. ('It should be: "In the beginning was Power!" But even as I write this down something warns me that I shall not stay by it!')
123 A similar thought can be found in Weaver (1948).
124 Riedl (1973a, p.16).
125 Theodizee (*Essai de Theodicée sur la bonté de Dieu, la liberté de l'homme et l'origine du mal 1710*) German edition 1879.
126 Candide (*Candide ou l'optimisme*. 1759).
127 Lorenz (1971, p.232) drew a corresponding consequence from studies of behaviour.
128 'A spirit helps me! I see the way and confidently write: "In the beginning was the deed!" '

REFERENCES

In some cases, English-language editions have been inserted in this bibliography though not cited in the text.

Adachi, B., 1933: *Anatomie der Japaner.* 3 Vols., Kenkyusha, Kyoto.

Ardrey, R., 1961: *African genesis; a personal investigation into the animal origins and nature of man.* Dell, New York.

Ardrey, R., 1967: *Adam kam aus Afrika.* Molden, Wien.

Atz, J., 1970: The application of the idea of homology to behaviour. In: Aronson, L., Tobach, E., Lehrmann, D. and Rosenblatt, J., eds., *Development and evolution of behaviour,* 53–75. W.H. Freeman, San Francisco.

Auerbach, C., and Robson, J., 1947: The production of mutations by chemical substances. *Proc. Roy. Soc. Edinb.* **62**: 284–91.

Ax, P., 1961: Verwandtschaftsbeziehungen und Phylogenie der Turbellarien. *Ergebn. Biol.* **24**: 1–68.

Ax, P. and Dörjes, J., 1966: *Oligochoerus limnophilus* nov. spec., ein kaspisches Faunenelement. *Int. Revue Hydrobiol.* **51** (1): 15–44.

Baer, K. von, 1828: Uber Entwicklungsgeschichte der Tiere. Borntraeger, Königsberg.

Baer, K. von, 1876: Studien aus dem Gebiet der Naturwissenschaften. Röttger, St. Petersburg.

Baerends, G., 1958: Comparative methods and the concept of homology in the study of behaviour. *Arch. Néerl. Zool.* **13** (suppl.): 401–17.

Baldass, L. von, 1943: *Hieronymus Bosch.* Schroll, Wien.

Balkaschina, E., 1929: Ein Fall der Erbhomöosis (die Genovariation 'Aristopedia') bei *Drosophila melanogaster. Roux-Arch. Entwicklungs.-Mech.* **115** (1/2): 448–63.

Balss, H., Gruner, H., Buddenbrock, W. and Korschelt, E., 1961: Decapoda. In: *Klass, u. Ordn. d. Tierreichs* **5** (1/7).

Baltzer, F., 1950: Entwicklungsphysiologische Betrachtungen über Probleme der Homologie und Evolution. *Rev. Suisse de Zool.* **57** (11): 451–77.

Baltzer, F., 1952: Experimentelle Beiträge zur Frage der Homologie *Experientia* **8**: 285–97.

Baltzer, F., 1955: Finalisme et physicisme. *Actes Soc. Helvètique Sci. Naturelles* **135**: 92–9.

Baltzer, F., 1957: Über Xenoplastik, Homologie und verwandte stammesgeschichtiche Probleme. Mitt. *Naturforsch. Ges. Bern* **15**: 1–23.

Bavink, B., 1930: *Ergebnisse und Probleme der Naturwissenschaften. Eine Einführung in die heutige Naturphilosophie* (4th ed.). Hirzel, Leipzig.

Beauchamp, P., de, 1961: Généralités sur les Plathelminthes. *Traité de Zoologie* IV (1): 23–212.

Berg, L., 1926: *Nomogenesis or evolution determined by law.* Constable, London.

Berger, P. and Luckmann, Th., 1966: *The social construction of reality.* Doubleday, New York.

Bergson, H., 1907: Evolution créatrice. *Alcan Coll. de Bibl. de Philosophie* (3e Ed.), also: Oeuvres, Press Universitaires de France, Paris.

Berlin, B., Raven, P. and Breedloue, D., 1966: Folk taxonomies and biological classification. *Science* **154**: 273–5.

Bertalanffy, L., von, 1948: Das Weltbild der Biologie. In: Moser. S., ed., *Weltbild und Menschenbild.* Tyrolia, Salzburg.

Bertalanffy, L., von, 1949: *Vom Molekül zur Organismenwelt – Grundfragen der modernen Biologie.* Athenaion (2nd ed.), Potsdam.

Bertalanffy, L., von, 1952: *The problem of life.* Harper, New York.

Bertalanffy, L., von, 1955: Die Evolution der Organismen. In: Schlemmer, D., ed., *Schöpfungsglaube und Evolutionstheorie.* 53–66. Kröner, Stuttgart.

Bertalanffy, L., von, 1968: *General system theory. Foundation, development, application.* Braziller, New York.

Bertalanffy, L., von, 1970: Gesetz oder Zufall: Systemtheorie und Selektion. In: Köstler, A. and Smythies, J., eds., *Das neue Menschenbild,* 71–95. Molden, Wien-München-Zürich.

Beurlen, K., 1932: Funktion und Form in der organischen Entwicklung. *Naturwiss.* **20**: 73–80.

Beurlen, K., 1937: *Die stammesgeschichtlichen Grundlagen der Abstammungslehre.* Fischer, Jena.

Bigelow, R., 1959: Similarity, ancestry, and scientific principles, *Syst. Zool.* **8**: 165–168.

Binding, K., 1872–1919: *Die Normen und ihre Übertretung. Untersuchung über die rechtmäßige Handlung und die Arten des Delikts.* 4 Vols. Engelmann, Leipzig.

Bird, A., 1971: *The structure of nematodes.* Academic Press, New York-London.

Birnstiel, M., Chipchase, M. and Speirs, J., 1970: The ribosomal RNA cistrons. *Progr. Nucl. Acid Res.* **11**: 351–89.

Blackwelder, R., 1967: A critique of numerical taxonomy. *Syst. Zool.* **16**: 64–72.

Bohlken, H., 1958: Vergleichende Untersuchungen an Wildrindern (Tribus Bovini Simpson 1945). *Zool. Jb. (allg. Zool. u. Physiol.)* **68**: 113–202.

Braus, H., 1929: *Anatomie des Menschen* (2nd ed.). Springer, Berlin.

Bresch, C. and Hausmann, R., 1970: Klassische und molekulare Genetik (2nd ed.). Springer, Berlin-Heidelberg-New York.

Bridges, C. and Brehme, K., 1944: *The mutants of Drosophila melanogaster.* Carnegie Institutions of Washington Publication 552, Washington, DC.

Bridgman, P., 1941: *The nature of thermodynamics.* Harvard Univ. Press, Cambridge (Mass.).

Brillouin, L., 1956: *Science and information theory.* Academic Press, New York.

Britten, R. and Davidson, E., 1969: Gene regulation for higher cells: a theory. *Science* **165**: 349–56.

Brunswik, E., 1934: *Wahrnehmung und Gegenstandswelt. Psychologie vom Gegenstand her.* Deuticke, Leipzig-Wien.

Brunswik, E., 1957: Scope and aspects of the cognitive problem. In: Bruner, R. et al., eds., *Contemporary approaches to cognition.* Harvard Univ. Press, Cambridge (Massachusetts).

Bühler, K., 1907–08: Tatsachen und Pobleme zu einer Psychologie der Denkvorgänge. *Arch. Ges. Psychol.* **9**: 297–365 und **12**: 1–92.

Burgers, J., 1965: On the emergence of patterns of order. *General Systems* **10**: 77–90.

Bytinski-Salz, H., 1937: Trapianti di 'organizzatore' nelle uova di Lampreda. *Arch. Ital. Anat. Embryol.* **39**: 177–228.

Cahn, P., 1958: Comparative optic development in *Astyanax mexicanus* and two of its blind cave derivates. *Bull. Americ. Mus. Nat. Hist.* **115**: 73–112.

Cain, A., 1956: The genus in evolutionary taxonomy. *Syst. Zool.* **5**: 97–109.

Camin, J. and Sokal, R., 1965: A method of deducing branching sequences in phylogeny. *Evolution* **19**: 311–26.

Campbell, B., 1967: Biological entropy pump. *Nature* **215**: 1308.

Campbell, D., 1966a: Evolutionary epistemology. In: Schlipp, P., *The Philosophy of Karl R. Popper.* Open Court Publishing Co., Lasalle.

Campbell, D., 1966b: *Pattern matching as an essential in distal knowing.* Holt, Rinehart and Winston, New York.

Cannon, H., 1958: *The evolution of living things.* Manchester Univ. Press., Manchester.

Chen, P. and Baltzer, F., 1954: Chimärische Haftfäden nach xenoplastischem Ektodermaustausch zwischen *Triton* und *Bombinator. Roux' Arch. Entw. Mech.* **147**: 214–58.

Chomsky, N., 1960: *Language and mind.* Harcourt, Brace & World, New York.

Chomsky, N., 1970: *Sprache und Geist.* Suhrkamp, Frankfurt.

Clara, M., 1942: *Das Nervensystem des Menschen.* Barten, Leipzig.

Clark, R., 1964: *Dynamics in metazoan evolution: the origin of the coelom and segments.* Clarendon Press, Oxford.

Claus, C., Grobben, K. and Kühn, A., 1932: *Lehrbuch der Zoologie.* Springer, Berlin-Wien.

Colless, D., 1967: The phylogenetic fallacy. *Syst. Zool.* **16**: 289–95.

Commoner, B., 1970: *Science and survival.* Ballentine, New York.

Corning, H., 1925: *Lehrbuch der Entwicklungsgeschichte des Menschen.* Bergmann, München.

Coulombre, A., 1965: The eye. In: DeHaan, R. und Ursprung, H., eds., *Organogenesis*, 219–53. Holt, Rinehart and Winston, New York.

Cracraft, J., 1967: Comments on homology and analogy. *Syst. Zool.* **16**: 355–9.

Cuénot, L., 1951: *L'evolution biologique.* Masson, Paris.

Dacqué, E., 1935: *Organische Morphologie und Paläontologie.* Bornträger, Berlin.

Dahrendorf, R., 1957: *Soziale Klassen und Klassenkonflikte in der industriellen Gesellschaft.* Enke, Stuttgart.

Dancoff, S. and Quastler, H., 1953: The information content and error rate of living things. In: Quastler, H., ed., *Information theory in biology*, 263–73. Univ. Illinois Press, Urbana.

Danesch, O. and Danesch, E., 1969: *Orchideen Europas; Südeuropa.* Hallwag, Bern-Stuttgart.

Danforth, C., 1923: The frequency of mutations and the incidence of hereditary traits in man. Eugenics, genetics, and family, *Sc. Papers 2. internat. Congr. Eugen. New York* 1921, **1**: 120–8.

Darlington, C., 1969: *The evolution of man and society.* Simon and Schuster, New York.

Darwin, C., 1872: *The origin of species* (6th ed.). Murray, London.

Dayhoff, M., 1969: Computer analysis of protein evolution. *Sci. Amer.* **221** (1): 87–95.

De Beer, G., 1958: *Embryos and ancestors* (3rd ed.). Clarendon, Oxford.

De Groot, S. and Mazur, P., 1962: *Non-Equilibrium thermodynamics.* North Holland, Amsterdam.

Dessauer, F., 1958: *Naturwissenschaftliches Erkennen.* Knecht, Frankfurt (Main).

Diamond, J., 1966: Zoological classification system of a primitive people. *Science* **151**: 1102–4.

References

Dilger, W., 1964: The interaction between genetic and experiential influences in the development of species – typical behaviour. *Am. Zool.* **4**: 155–60.
Doblhofer, E., 1957: *Zeichen und Wunder. Die Entzifferung verschollener Schriften und Sprachen.* Neff, Wien-Berlin-Stuttgart.
Dobzhansky, T., 1951: *Genetics and the origin of species* (3rd ed.). Columbia Univ. Press., New York.
Dobzhansky, T., 1956: What is an adaptive trait? *Amer. Naturalist* **90**: 337–47.
Driesch, H., 1909: *Philosophie des Organischen.* (2 Vols). Engelmann, Leipzig.
Driesch, H., 1919: Der Begriff der organischen Form. *Abh. theoret. Biol.* 3. Bornträger, Berlin.
Driesch, H., 1927: Behaviorismus und Vitalismus. *Akad. 1927/28 (Philosoph. hist. Kl.)* **I**: 1–10. Sitzber, Heidelberg.

Eden, M., 1967: Inadequacies of neo-darwinian evolution as a scientific theory. In: Moorehead, P. and Kaplan, M., eds., *Mathematical challenges to the neo-darwinian interpretation of evolution, Symposium monograph* **5**: 5–19. Wislar Inst. Press, Philadelphia.
Eder, G., 1963: *Quanten, Moleküle, Leben. Begriffe und Denkformen der heutigen Naturwissenschaft.* Alber, Freiburg-München.
Eibl-Eibesfeldt, I., 1967: *Grundriß der vergleichenden Verhaltensforschung.* Piper, München;
Eibl-Eibesfeldt, I., 1970: *Liebe und Haß.* Piper, München.
Eigen, M., 1971: Selforganization of matter and the evolution of biological macromolecules. *Naturwissenschaften* **58** (10): 465–523.
Einstein, A. and Born, M., 1969: *Briefwechsel 1916–1955.* Nymphenburger Verlagshandlung. München.

Farris, J., 1966: An estimation of conservatism of characters *Evolution* **70**: 587–91.
Farris, J., 1967: Comment on psychologism. *Syst. Zool.* **16**: 345–7.
Farris, J., 1969: Successive approximations approach to character weighting. *Syst. Zool.* **18**: 374–85.
Fiedler, W., 1956: System der Primaten. I. In: Hofer, H., Schulz, A. und Stark, D., eds., *Primatologia* 1–266. Karger, New York.
Fisher, R., 1942: *The design of experiments.* Oliver and Boyd, Edinburgh-London.
Flechtner, H. J., 1970: *Grundbegriffe der Kybernetik.* Wiss. Verlagsgesellschaft, Stuttgart.
Florkin, M., 1962: Isologie, homologie, analogie et convergence en biochemie comparée. *Bull. Classe Sci. Acad. Roy. Belg.* (S) **48**: 819–24.
Florkin, M., 1966: *A molecular approach to phylogeny.* Elsevier, Amsterdam-New York.
Freyer, H., 1955: *Theorie des gegenwärtigen Zeitalters.* Deutsche Verlags-Anstalt, Stuttgart.

Garstang, W., 1922: The theory of recapitulation: a critical restatement of the biogenetic law. *J. Linnean Soc. London, Zoology* **35**: 81–101.
Gehring, W., 1966: Übertragung und Änderung der Determinationsqualitäten in Antennenscheibenkulturen von *Drosophila melanogaster*. *J. Embryol. Exptl. Morphol.* **15**: 77–111.
Geist, V., 1966: The evolution of hornlike organs. *Behavior* **27**: 175–214.
Georgescu-Roegen, N., 1971: *The entropy law and the economic process.* Harvard Univ. Press, Cambridge (Mass.).
Ghiselin, M., 1966: On psychologism in the logic of taxonomic controversies. *Syst. Zool.* **15**: 207–15.
Ghiselin, M., 1969: *The triumph of the darwinian method.* Univ. Calif. Press, Berkeley.
Gilmour, J., 1937: A taxonomic problem. *Nature* **139**: 1040–2.
Gilmour, J., 1940: Taxonomy and philosophy. In: Huxley, J., ed., *The New Systematics*, 461–74. Clarendon Press, Oxford.
Glansdorff, P. and Prigogine, I., 1971: *Thermodynamic theory of structure, stability, and fluctuations.* Wiley and Sons, London-New York.
Goethe, J.W. von, 1790a: *Versuch über die Gestalt der Thiere, 261–269.* (II. Weimarer Ausgabe). Böhlau, Weimar.
Goethe, J.W. von, 1790b: *Die Metamorphose der Pflanzen, 23–89* (II. Weimarer Ausgabe). Böhlau, Weimar.
Goethe, J.W. von, 1795: *Erster Entwurf einer allgemeinen Einleitung in die vergleichende Anatomie, ausgehend von der Osteologie, 5–78* (II. Weimarer Ausgabe). Böhlau, Weimar.
Goldschmidt, R., 1940: *The material basis of evolution.* Yale Univ. Press, New Haven.
Goldschmidt, R., 1952: Homeotic mutants and evolution. *Acta Biotheoretica* **10**: 87–104.
Goldschmidt, R., 1961: *Theoretische Genetik.* Akademie Verlag, Berlin.
Goodall, D., 1970: 3rd annual conference of numerical taxonomy. *Syst. Zool* **19** (3): 303–6.
Goss, R., 1969: *Principles of regeneration.* Academic Press, New York.
Gregory, W., 1951: *Evolution emerging,* 2 Vols. Macmillan, New York.
Gross, S., 1969: Genetic regulatory mechanism in the fungi. *Ann. Rev. Genet.* **3**: 395–424.
Grüneberg, H., 1952: *The genetics of the mouse* (2nd ed.). Nijhoff, The Hague.
Grzimek, B. and Schultze-Westrum. T., 1970: Paradiesvögel. In: Grzimek, B., ed., *Grzimeks Tierleben,* IX: 471–481. Kindler, Zürich.

Hadorn, E., 1945: Zur Pleiotropie der Genwirkung. *Arch. Jul. Klaus-Stiftung, Erg-Bd. zu Bd.* **20**: 82–95.

Hadorn, E., 1955: Letalfaktoren. Thieme, Stuttgart.

Hadorn, E., 1961: *Developmental genetics and lethal factors.* Wiley and Sons, New York.

Hadorn, E., 1966a: Dynamics of determination. In: Locke, M., ed., *Major problems in developmental biology,* 85–104. Acad. Press, New York-London.

Hadorn, E., 1966b: Konstanz, Wechsel und Typus der Determination in Zellen aus ♂ Genitalanlagen von *Drosophila melanogaster. Develop. Biol.* **13**: 424–509.

Hadorn, E., 1968: Transdetermination in Cells. *Sci. Amer.* **221** (1): 110–120.

Haeckel, E., 1866: *Generelle Morphologie der Organismen,* 2 Vols. Reimer, Berlin.

Haeckel, E., 1899–1902: *Kunstformen in der Natur.* Bibliographisches Institut, Leipzig.

Haecker, V., 1925: *Pluripotenzerscheinungen. Synthetische Beiträge zur Vererbungs- und Abstammungslehre.* Fischer, Jena.

Haldane, J., 1958: Theory of evolution before and after Bateson. *J. Genet.* **56**: 11–28.

Hartmann, E., von, 1875: *Wahrheit und Irrtum des Darwinismus.* Dunker, Berlin.

Hartmann, N., 1950: *Philosophie der Natur.* De Gruyter, Berlin.

Hartmann, N., 1964: *Der Aufbau der realen Welt* (3rd ed.). De Gruyter, Berlin.

Hass, H., 1970: *Energon. Das verborgene Gemeinsame.* Molden, Wien-München-Zürich.

Hassenstein, B., 1951: Goethes Morphologie als selbstkritische Wissenschaft und die heutige Gültigkeit ihrer Ergebnisse. *Neue Folge d. Jahrb. d. Goethe–Gesellschaft* **12**: 333–57.

Hassenstein, B., 1958: Prinzipien der vergleichenden Anatomie bei Geoffroy Saint–Hilaire, Cuvier und Goethe. *Act. Coll. int. Strasbourg. Publ. Fac. lettr.* **137**: 155–68.

Hassenstein, B., 1965: *Biologische Kybernetik. Eine elementare Einführung.* Quelle und Meyer, Heidelberg.

Hassenstein, B., 1966: Was ist Information? *Naturwissenschaft und Medizin* **3** (13): 38–52.

Hatt, P., 1933: L'induction d'une plaque mèdullaire secondaire chez le Triton par implantation d'un morceau de ligne primitive de poulet. *Compt. Rend. Soc. Biol.* |113 (2): 246–8.

Haupt, H., 1953: *Insekten mit rätselhaften Verzierungen.* Neue Brehm Bücherei 104. Ziemsen, Wittenberg-Lutherstadt.

Heberer, G., 1958: Zum Problem der additiven Typogenese. *Uppsala Univ. Arsskrift* **6**: 40–57.

Heberer, G., 1959a: *Die Evolution der Organismen* (2nd enlarged ed.). Fischer, Stuttgart.

Heberer, G., 1959b: Theorie der additiven Typogenese. In: Heberer, G., ed., *Die Evolution der Organismen* (2nd enlarged edition). Fischer, Stuttgart.

Heckmann, O., 1942: *Theorien der Kosmologie.* Springer, Berlin.

Heikertinger, F., 1954: *Das Rätsel der Mimikry und seine Lösung.* Fischer, Jena.

Heisenberg, W., 1959: Die Plancksche Entdeckung und die philosophischen Probleme der Atomphysik. *Univ.* 14, **2**: 135–158.

Hemleben, J., 1964: *Ernst Haeckel in Selbstzeugnissen und Bilddokumenten.* Rowohlt, Hamburg.

Hendelberg, J., 1969: On the development of different types of spermatozoa from spermoids with two flagella in Turbellaria with remarks on the ultrastructure of the flagella. *Zool. Bidr. Uppsala* **38**: 1–50.

Hennig, W., 1944: Organisches Werden, paläontologisch gesehen. *Paläont. Z.* **23**: 281–316.

Hennig, W., 1950: *Grundzüge einer Theorie der phylogenetischen Systematik.* Deutscher Zentralverlag, Berlin.

Hennig, W., 1966: *Phylogenetic systematics.* Univ. of Illinois Press, Urbana.

Herbst, C., 1916: Regeneration von antennenähnlichen Organen anstelle von Augen, VII. *Arch. f. Enw.-Mech.* **42**: 407–89.

Hochstetter, F., 1946: *Toldts Anatomischer Atlas,* 3 Vols (19th ed). Urban and Schwarzenberg. Wien.

Hofstätter, P., 1959: *Psychologie* (3rd ed.). Fischer, Frankfurt.

Höpp, G., 1972: *Evolution der Sprache und Vernunft.* Suhrkamp, Frankfurt.

Husserl, L, 1928: *Logische Untersuchungen,* 2 Vols. Niemeyer, Halle.

Huxley, J., 1942: *Evolution, the modern synthesis.* Harper, New York.

Huxley, J., 1957: The three types of evolutionary process. *Nature,* 180: 454–5.

Huxley, J., 1958: Evolutionary process and taxonomy. In: Hedberg, O., ed., *Systematics of today. Uppsala Univ. Arsskrift.* 6: 21–39.

Hyman, L., 1959: *The invertebrates: smaller coelomate groups vol. 5.* McGraw-Hill, New York-London-Toronto.

Inglis, W., 1970: The purpose and judgement of biological classification. *Syst. Zool.* 19: 240–50.

Jacob, F. and Brenner, S., 1963: Sur la régulation de la synthèse du DNA chez les bactéries: l'hypothèse du réplétion. *C. R. Acad. Sci. (Paris),* **256**: 298–300.

Jaennel, R., 1950: *La marche de l'évolution.* Presses Universitaires de France, Paris.

Jordan, P., 1952: *Schwerkraft und Weltall.* Vieweg, Braunschweig.

Kaiser, H., 1970: Das Abnorme in der Evolution. *Acta Biotheoretica* **17**(suppl.).

Katchalsky, A. and Curran, P., 1965: *Nonequilibrium thermodynamics in biophysics.* Harvard Univ. Press, Cambridge (Mass.).

References

Kaufmann, A., 1954: *Lebendiges und Totes in Bindings Normentheorie, Normlogik und moderne Strafrechtsdogmatik.* Schwartz, Göttingen.

Kedes, L. and Birnstiel, M., 1971: Reiteration and clustering of DNA sequences complementary to histone messenger RNA. *Nature New Biol.* 230: 165–9.

Kiger, J., 1973: The bithorax complex – a model for cell determination in Drosophila. *J. theor. Biol.* 40: 455–67.

Kiriakoff, S., 1959: Phylogenetic systematics versus typology. *Syst. Zool.* 8: 117–8.

Kiriakoff, S., 1965: Criticism of numerical taxonomy. *Syst. Zool.* 14: 61–4.

Klein, M., 1970: *Einführung in die DIN-Normen.* 6th. ed. Deutscher Normenausschuß, Teubner, Stuttgart.

Klotz, J., 1967: *Energy changes in biochemical reactions.* Acad. Press, New York-London.

Klüber, H., von, 1931: *Das Vorkommen der chemischen Elemente im Kosmos.* Barth, Leipzig.

Kluge, A. and Farris, J., 1969: Quantitative phyletics and the evolution of anurans. *Syst. Zool.* 18: 1–32.

Koehler, O., 1952: Vom unbenannten Denken. *Zool. Anz.* 16 *(suppl.):* 202–11.

Koenig, O., 1970: *Kultur und Verhaltensforschung. Einführung in die Kulturethologie.* Deutscher Taschenbuch-Verlag, München.

Köstler, A., 1968: Das Gespenst in der Maschine (Engl. original: *The ghost in the machine*). Molden, Wien-München-Zürich.

Köstler, A., 1970: Jenseits von Atomismus und Holismus – Der Begriff des Holons. In: Köstler, A. und Smythies, J., eds., *Das neue Menschenbild 192–229.* Molden, Wien-München-Zürich.

Koestler, A., and Smythies, J., 1969: *Beyond reductionism. New perspectives in the life of sciences.* Macmillan, London.

Korschelt, E., 1927: *Regeneration und Transplantation: Regeneration Bd. I.* Bornträger, Berlin.

Korschelt, E. and Heider, K., 1936: *Vergleichende Entwicklungsgeschichte der Tiere.* 2 Vols., Fischer, Jena.

Kosswig, C., 1959: Phylogenetische Trends genetisch betrachtet. *Zool. Anz.* 162 (7/8): 208–221.

Kraft, V., 1947: *Mathematik, Logik und Erfahrung.* Springer, Wien.

Kuenne, R., 1963: *The theory of general economic equilibrium.* Univ. Press, Princeton (New Jersey).

Kühn, A., 1955: *Grundriß der allgemeinen Zoologie* (1. ed.). Thieme, Stuttgart.

Kühn, A., 1965: *Vorlesungen über die Entwicklungsphysiologie.* (2. ed.). Springer, Berlin-Heidelberg-New York.

Kühnelt, W., 1953: Ein Beitrag zur Kenntnis tierischer Lebensformen. *Verhandl. Zool. Bot. Ges. Wien* 93: 57–71.

Kurtén, B., 1958: A differentiation index, and a new measure of evolutionary rates. *Evolution* 12: 146–57.

Lalande, A., 1948: *La raison et les normes.* (*Coll. à la recherche de la verité.*). Hachette, Paris.

Lamarck, J., 1909: *Zoologische Philosophie.* (French original: 1809, *Zoologie philosophique*) Kröner, Leipzig.

Langridge, J., 1958: A hypothesis of developmental selection exemplified by lethal and semi-lethal mutants of *Arabidopsis. Aust. J. Biol. Sci.* 11: 58–68.

Lautmann, R., 1969: *Wert und Norm. Begriffanalysen für die Soziologie.* Westdeutscher Verlag, Köln-Opladen.

Leche, W., 1922: *Der Mensch, sein Ursprung und seine Entwicklung* (2. ed.). Fischer, Jena.

Lehmbruch, G., 1967: *Einführung in die Politikwissenschaft.* Kohlhammer, Stuttgart-Berlin-Köln-Mainz.

Lehninger, A., 1965: *Bioenergetics. The molecular basis of biological energy transformation.* Benjamin, New York-Amsterdam.

Leibniz, G., von, 1879: *Die Theodicée.* Dürrsche Buchhandlung, Leipzig..

Lerner, I., 1954: *Genetic homeostasis.* Wiley and Sons, New York.

Lewis, E., 1964: Genetic control and regulation of developmental pathways. In: Locke, M., ed., *The role of chromosomes in development* 231–52. Acad. Press, New York.

Lima de Faria, A., 1952: Chromomere analysis of the chromosome complement of rye. *Chromosoma* 5: 1–68.

Lima de Faria, A., 1962: Selection at the molecular level. *J. Theor. Biol.* 2: 7–15.

Linschitz, H., 1953: The information content of a bacterial cell. In: Quastler, H., Augenstein, R., *et al.* eds., *Essays on the use of information theory in biology.* Univ. Illin. Press, Urbana: 251–62.

Locke, M., ed., 1966: *Major problems in developmental biology* (The 25th Symposium). Acad. Press, New York-London.

Locke, M., 1968: *The emergence of order in developing systems* (The 27th Symposium for developmental biology). Acad. Press, New York-London.

Lorenz, K., 1935: Der Kumpan in der Umwelt des Vogels. *J. Ornith.* 83: 137–213, 289–413.

Lorenz, K., 1963: *Das sogenannte Böse.* Borotha-Schoeler, Wien.

Lorenz, K., 1966, *On aggression.* Harcourt, Brace & World, New York, Methuen, London.

Lorenz, K., 1965a: Uber die Entstehung von Mannigfaltigkeit. In: *Verhandlgn. der Gesellschaft deutscher Naturforscher und Ärzte.* 103: 80–90. Versammlung zu Weimar, Oktober 1964.

Lorenz, K., 1965b: *Uber tierisches und menschliches Verhalten. Aus dem Werdegang der Verhaltens-lehre.* 2 Vols. Piper, München.

Lorenz, K., 1965c: *Darwin hat recht gesehen.* Neske, Pfullingen.

Lorenz, K., 1966: Stammes-und Kulturgeschichtliche Ritenbildung. *Naturwiss. Rundschau* 19: 361–70.

Lorenz, K., 1967: Die instinktiven Grundlagen menschlicher Kultur. *Die Naturwissenschaften* 54: 377–388.

Lorenz, K., 1971: Knowledge, beliefs and freedom. In: Weiss, P., ed., *Hierarchically organized systems in theory and practice* 231–61. Hafner, New York.

Lorenz, K., 1973: *Die Rückseite des Spiegels. Versuch einer Naturgeschichte menschlichen Erkennens.* Piper, München-Zürich.

Lorenz, K., 1977, *Behind the mirror.* Methuen, London.

Löther, R., 1972: *Die Beherrschung der Mannigfaltigkeit. Philosophische Grundlagen der Taxonomie.* Fischer, Jena.

Ludwig, W., 1940: Selektion und Stammesentwicklung. *Naturwiss.* 28: 689–705.

Lus, J., 1947: Einige Gesetzmäßigkeiten der Vermehrung der Populationen von *Adalia bipunctata* L. – Heterozygotie der Populationen für letale Faktoren. *Dokl. Akad. SSSR n.s.* 57: 825–8.

Lwoff, A., 1968: *Biological Order* (2nd ed.). M.I.T. Massachusetts Inst. of Technology Press, Cambridge (Massachusetts).

Margalef, R., 1970: *Perspectives in ecological theory.* (3. Aufl.). Univ. Chicago Press, Chicago-London.

Margulis, L. and Margulis, T., 1969: A note on equivalance of characters. *Syst. Zool.* 17: 477–9.

Marinelli, W. and Strenger, A., 1959: *Vergleichende Anatomie und Morphologie der Wirbeltiere* (3. ed.). Deuticke, Wien.

Mayr, E., 1942: *Systematics and the origin of species.* Columbia Univ. Press, New York.

Mayr, E., 1963: *Animal species and evolution.* Belknap, Havard Univ. Press. Cambridge, Mass.

Mayr, E., 1965: Numerical phenetics and taxonomy theory. *Syst. Zool.* 14: 73–95.

Mayr, E., 1967: *Artbegriff und Evolution.* (English original: *Animal species and evolution.*) Parey, Hamburg-Berlin.

Mayr, E., 1969: *Principles of systematic zoology.* McGraw-Hill, New York.

Mayr, E., 1970: *Populations, species, and evolution.* Belknap, Harvard Univ. Press, Cambridge (Mass.).

McGeoch, J., 1952: *The psychology of human learning.* McKay, New York.

Meckel, J., 1821: *System der vergleichenden Anatomie.* Reuger, Halle.

Medawar, P., 1960: *The future of man.* BBC, London.

Meyer-Abich, A., 1943: Beiträge zur Theorie der Evolution der Organismen. I. Das typologische Grundgesetz. *Acta Biotheoretica* 7: 1–80.

Meyer-Abich, A., 1950: Beiträge zur Theorie der Evolution der Organismen. II. Typensynthese durch Holobiose. *Biblio. Biotheoretica* 5: 206pp, Leiden.

Meyerhof, O., 1924: *Chemical dynamics of life phenomena.* Lippincott, Philadelphia.

Michener, C., 1957: Some bases for higher categories in classification. *Syst. Zool* 6: 160–73.

Mohr, H., 1965: Erkenntnistheoretische und ethische Aspekte der Naturwissenschaften. In: *Mitt.d. Verb. Deutscher Biologen Nr.* 113: 525–35 (*Beilage zu Naturwiss. Rundschau* 10, 1965).

Monod, J. 1971: *Le hasard et la necessité.* Editions du Seuil, Paris.

Monod, J., 1971: *Zufall und Notwendigkeit. Philosophische Fragen der modernen Biologie.* Piper, München.

Monod, J., 1972: *Chance and necessity. An essay on the natural philosophy of modern biology.* Random House, New York.

Monod, J. and Cohn, M., 1952: La biosynthèse induite des encymes (Adaptation encymatique). *Advanc. Enzymol.* 13: 67–116.

Monod, J. and Jacob, F., 1961: *Cellular regulatory mechanisms.* Cold Spring Harbor, Symp. quant. Biol., New York.

Moore, R., ed., 1965: *Treatise on invertebrate paleontology Part H, Brachiopoda vol I.* Geol. Soc. Am. Univ. Kansas Press, New York.

Morell, P., Smith, I., Dubnau, D. and Marmur, J., 1967: Isolation and characterization of low molecular weight ribonucleic acid species from *Bacillus subtilis. Biochemistry.* 6: (1): 258-65.

Morgan, T., 1907: *Regeneration.* (Translated by Moszkowski, M.,). Engelmann, Leipzig.

Morgan, T., 1929: Variability of eyeless. *Publ. Carnegie Inst.* 399: 141–68.

Morowitz, H., 1955: Some disorder–order considerations in living systems. *Bull. Math. Biophys.* 17: 81–7.

Morowitz, H., 1968: *Energy flows in biology.* Acad. Press, New York-London.

Morowitz, H., 1970: *Entropy for biologists.* Acad. Press, New York.

Müller, A., 1963: *Lehrbuch der Paläozoologie.* Vol. 1: *Allgemeine Grundlagen.* Fischer, Jena.

Müller, A., 1966: *Lehrbuch der Paläozoologie.* Vol. III (1): *Vertebraten; Fische und Amphibien.* Fischer, Jena.

Müller, A., 1968: *Lehrbuch der Paläozoologie.* Vol. III (2): *Vertebraten; Reptilien und Vögel.* Fischer, Jena.
Müller, A., 1970: *Lehrbuch der Paläozoologie.* Vol. III (3): *Vertebraten; Mammalia.* Fischer, Jena.
Müller, F., 1864: *Für Darwin.* Engelmann, Leipzig.
Muller, H., 1928: The production of mutations by X-rays. *Proc. nat. Acad. Sci. USA* 14: 714—26.
Muller, H., 1950: Our load of mutations. *Amer. J. Human Genet.* 2: 111—76.
Muller, H., 1954: The manner of dependence of the permissible dose of radiation on the amount of genetic damage. *Acta Radiol.* 41: 5—20.

Naef, A., 1919: Idealistische Morphologie und Phylogenetik. Fischer, Jena.
Nanney, D., 1958: Epigenic control systems. *Proc. Nat. Acad. Sci. Wash.* 44: 712—17.
Needham, J., 1936: *Order and life.* Yale Univ. Press, New Haven.

Odum, H., 1971: *Environment, power and society.* Wiley and Sons, New York-London-Sydney-Toronto.
Olson, E. and Müller, R., 1958: *Morphological integration.* Chicago Univ. Press, Chicago.
Oppenheimer, J., 1936: Structures developed in amphibians by implantation of living fish organizer. *Proc. Soc. Exp. Biol. Med.* 34: 461—3.
Osborn, H., 1934: Aristogenesis, the creative principle in the origin of species. *Amer. Naturalist.* 68: 193-235.
Osche, G., 1966: Grundzüge der allgemeinen Phylogenetik. In: Gessner, F., ed., *Handbuch der Biologie III* (2): 817—906. Athenaion, Konstanz.
Osche, G., 1972: *Evolution. Grundlagen — Erkenntnisse — Entwicklungen der Abstammungslehre.* Herder (Studio visuell), Basel-Wien.
Owen, R., 1848: On the archetype and homologies of the vertebrate skeleton. *Brit. Assoc. Rep.* **1846:** 169—340.

Pattee, H., 1973: *Hierarchy theory. The challenge of complex systems.* Braziller, New York.
Patzelt, V., 1945: *Histologie. Lehrbuch für Mediziner.* Urban und Schwarzenberg, Wien.
Pernkopf, E., 1952: *Topographische Anatomie des Menschen.* Vol. III. *Der Hals.* Urban und Schwarzenberg, Wien.
Pernkopf, E., 1960: *Topographische Anatomie des Menschen.* Vol IV (2). Urban und Schwarzenberg, München-Berlin-Wien.
Peters, J., 1967: *Einführung in die allgemeine Informationstheorie.* Springer, Berlin-Heidelberg-New York.
Pilgrim, G., 1947: The evolution of the buffaloes, oxen, sheep, and goats. *Linn. Soc. Zool.* 41 (279): 272—86.
Plate, L., 1925: *Die Abstammungslehre. Tatsachen, Theorien, Einwände und Folgerungen in kurzer Darstellung.* Fischer, Jena.
Polanyi, M., 1968: Life's irreducible structure. *Science* 160: 1308—12.
Ponnamperuma, C., 1972: *The origins of life.* Thames and Hudson, London.
Popper, K., 1935: *Logik der Forschung.* Springer, Wien.
Popper, K., 1962: *The logic of scientific discovery.* Harper and Row, New York. Hutchinson, London.
Popper, K., 1967: Times arrow and feeding on negentropy. *Nature* 213: 320.
Portmann, A., 1948: *Einführung in die vergleichende Morphologie der Wirbeltiere.* Schwabe, Basel.
Porzig, W., 1971: *Das Wunder der Sprache.* Francke, München-Bern.
Prigogine, I., 1955: *Introduction to thermodynamics of irreversible processes.* Thomas, Springfield.

Quastler, H., 1964: *The emergence of biological organization.* Yale Univ. Press, New Haven-London.

Reisinger, E., 1960: Was ist *Xenoturbella? Z. wiss. Zool.* 164 (1/2): 188—98.
Remane, A., 1936: Wirbelsäule und ihre Abkömmlinge. In: Bolk, L. *et al.,* eds., *Handbuch der vergleichenden Anatomie der Wirbeltiere 1—206.* Urban and Schwarzenberg, Berlin-Wien.
Remane, A., 1939: Der Geltungsbereich der Mutationstheorie. *Zool. Anz.* 12 (suppl.): 206—20.
Remane, A., 1943: Bedeutung der Lebensformentypen für die Okologie. *Biologia generalis* 17:164—82.
Remane, A., 1971: *Die Grundlagen des natürlichen Systems der vergleichenden Anatomie und der Phylogenetik* (2. ed.) (Authorised reprint of 1st Edition, 1952) (1. Aufl. 1952. Geest und Portig, Leipzig). Koeltz, Königstein-Taunus.
Rensch, B., 1954: *Neuere Probleme der Abstammungslehre* (2nd ed.). Enke, Stuttgart.
Rensch, B., 1961: Die Evolutionsgesetze der Organismen in naturphilosophischer Sicht. *Philosophia Naturalis* 6 (3): 288—362.
Rensch, B., 1968: *Biophilosophie auf erkenntnistheoretischer Grundlage.* Fischer, Struttgart.
Riedl, R., 1963: Probleme und Methoden der Erforschung des litoralen Benthos. Verhandl. Deutsch. *Zool. Ges. Wien. Zool. Anz.* 26 (suppl.): 505—67.
Riedl, R., 1966: *Biologie der Meereshöhlen.* Parey, Hamburg-Berlin.

Riedl, R., 1970: *Fauna und Flora der Adria* (2nd ed.). Parey, Hamburg-Berlin.
Riedl, R., 1973a: Die Biosphäre und die heutige Erfolgsgesellschaft. *Universitas* 28 (6): 587–93.
Riedl, R., 1973b: Energie, Information und Negentropie in der Biosphäre. *Naturwiss. Rundschau* 26 (10): 413–20.
Riedl, R., and Forstner, H., 1968: Wasserbewegung im Mikrobereich des Benthos. *Sarsia* 34: 163–88.
Roberts, P., 1964: Mosaics involving aristopedia, a homeotic mutant of *Drosophila melanogaster*. *Genetics* 49: 593–8.
Roggen, D., Raski, D., and Jones, N., 1966: Cilia in nematode sensory organs. *Science* 152: 515–6.
Romanes, G., 1892: *Darwin und nach Darwin. Eine Darstellung der Darwinschen Theorie und Erörterung Darwinistischer Streitfragen.* Translated from English by Vetter, B. and Nöldecke. B. Engelmann, Leipzig.
Romanes, G., 1892–97: *Darwin and after Darwin: An exposition of the Darwinian Theory and a discussion of post-Darwinian questions.* 3 vols, Longmans, London.
Romer, A., 1956: *The vertebrate body*. Saunders, Philadelphia.
Romer, A., 1959: *Vergleichende Anatomie der Wirbeltiere.* (1st German edition). Parey, Hamburg-Berlin.
Romer, A., 1966: *Vertebrate paleontology* (3rd ed.). Univ. Chicago Press, Chicago.
Rosa, D., 1903: *Die progressive Reduktion der Variabilität.* Fischer, Jena.
Rosa, D., 1931: *L'ologénè se.* Librairic Felix Alcan, Paris.
Rowell, A., 1965: Inarticulata. In: Moore, R., Ed., *Treatise on invertebrate paleontology. Part H, Brachiopoda vol. I:* 260–97. Geol. Soc. Am. Univ. Kansas Press, Kansas.
Russell, F., 1962: The diversity of animals. *Biblio. Biotheoretica* 13 (suppl.).

Salisbury, F., 1969: Natural selection and the complexity of the gene. *Nature* 224 (521–4): 342–3.
Scharp, H., 1958: *Wie die Kirche regiert wird. Papst, Kardinäle, Vatikan.* Heider, Freiburg.
Schindewolf, O., 1936: *Paläontologie, Entwicklungslehre und Genetik.* Borntraeger, Berlin.
Schindewolf, O., 1950: *Grundfragen der Paläontologie.* Schweizerbart, Stuttgart.
Schmalhausen, I., 1949: *Factors of evolution: the theory of stabilizing selection.* Blakiston, Philadelphia.
Schneirla, T., 1957: The concept of development in comparative psychology. In: Harris, D., Ed., *The concept of development* 78–108. Univ. Minnesota Press, Minneapolis.
Schrödinger, E., 1944: *What is life? The physical aspect of the living cell* (1st ed.). Univ. Press, Cambridge.
Schrödinger, E., 1951: *Was ist Leben? Die lebende Zelle mit den Augen des Physikers betrachtet* (2nd German ed.). Francke, Berlin.
Schrödinger, E., 1959: Geist und Materie. In: Westphal, W., *Die Wissenschaft – Sammlung von Einzeldarstellungen aus allen Gebieten der Naturwissenschaft* 113. Vieweg, Braunschweig.
Schrödinger, E., 1961: *Meine Weltansicht.* Zsolnay, Hamburg-Wien.
Schrödinger, E., 1969: *What is life? Mind and matter.* Univ. Press, Cambridge.
Schubert-Soldern, R., 1962: *Mechanism and vitalism.* Fothergill, P., ed., Univ. Notre Dame Press, Paris.
Schubert-Soldern, R., 1970: Der Evolutionismus Teilhard de Chardin. Kath. Acad., Wien.
Schuster, P., 1972: Vom Makromolekül zur primitiven Zelle – die Entstehung biologischer Funktion. *Chemie in unserer Zeit.* 6: 1–16.
Schützenberger, M., 1967: Algorithms and the neo-darwinian theory of evolution. In: Moorhead, P. and Kaplan, M., eds., *Mathematical challenges to the neo-darwinian interpretation of evolution* 73–80. Symposium Monograph No. 5. Wislar Inst. Press, Philadelphia.
Seidel, F., 1953: *Entwicklungsphysiologie der Tiere.* Göschen 1/63. De Gruyter, Berlin.
Seidel, F., 1972: *Entwicklungsphysiologie der Tiere.* (2nd revised ed.). Göschen 7/62. De Gruyter, Berlin-New York.
Shannon, C., and Weaver, W., 1949: *The mathematical theory of communication.* Univ. Illinois Press, Urbana.
Shatoury, H., 1956: Developmental interactions in the differentiation of the imaginal muscles of *Drosophila*. *J. Embryol. exp. Morph.* 4: 228–39.
Shepard, T., 1965: The thyroid. In: De Haan, R. und Ursprung, H., eds., *Organogenesis* 493-512. Holt, Rinehart, and Winston, New York.
Sheppard, D., and Engelsberg, E., 1967: Further evidence for positive control of the L-arabinose system by gene araC. *J. Molec. Biol.* 25: 443–54.
Simon, H., 1965: The architecture of complexity. *General Systems* 10: 63–76.
Simpson, G., 1951: *Horses.* Oxford Univ. Press, Oxford.
Simpson, G., 1952: *The meaning of evolution.* Yale Univ. Press, New Haven.
Simpson, G., 1955: *The major features of evolution.* (2nd ed.). Columbia Univ. Press, New York.
Simpson, G., 1961: *Principles of animal taxonomy.* Columbia Univ. Press, New York.
Simpson, G., 1964a: Organisms and molecules in evolution. Studies of evolution at the molecular level lead to greater understanding and a balancing of viewpoints. *Science* 146: 1535–8.

References

Simpson, G., 1964b: Numerical taxonomy and biological classification. *Science* **144**: 712–3.
Sleigh, M., 1962: *The biology of cilia and flagella.* Pergamon Press, New York.
Smith E., and Margoliash, E., 1964: Evolution of cytochrome C. *Federation Proc.* **23**: 1243–57.
Smoluchowski, R., 1914: *Vorträge über die kinetische Theorie der Materie und Elektrizität.* G. B. Teubner, Leipzig.
Sneath, P., 1957: The application of computers to taxonomy. *J. Gen. Microbiol.* **17**: 201–26.
Sneath, P. and Sokal, R., 1973: *Numerical Taxonomy: the principles and practice of numerical classification.* Freeman, San Francisco.
Sokal, R., and Sneath, P., 1963: *Principles of numerical taxonomy.* Freeman, San Francisco.
Sokolov, J., 1954: Versuch einer natürlichen Klassifikation der Horntiere (Bovidae). *Tr. Zool. Inst. Akad. Nauk. USSR* **14**: 1–295.
Sondhi, K., 1961: Developmental barriers in a selection experiment. *Nature* **189**: 249–50.
Spemann, H., 1936: *Experimentelle Beiträge zu einer Theorie der Entwicklung.* Springer, Berlin.
Spurway, H., 1949: Remarks on Vavilov's law of homologous variation. In: *La Ricerca Scientifica* (Suppl. Pallanza Symposium) 18. Cons. Naz. delle Ricerche, Roma.
Spurway, H. and Callen, H., 1960: The vigour and male sterility of hybrids between the species *Triturus vulgaris* and *Triturus helveticus.* *J. Genet.* **57**: 84–117.
Stammer, H., 1959: Trends in der Phylogenie der Tiere; Ektogenese und Autogenese. *Zool. Anz.* **162**: 187–208.
Steinböck, O., 1963: Origin and affinities of the lower metazoa: the aceloid ancestry of the eumetazoa. In: Dougherty, E., ed., *The lower metazoa,* Univ. California Press 40–54, Berkeley.
Stensiö, E., 1958: Les cyclostomes fossiles ou ostracodermes. In: Grassé, P., ed., *Traité de Zoologie* **13** (2): 173–425. Masson, Paris.
Stern, C., 1968: *Genetic mosaics and other essays.* Harvard Univ. Press, Cambridge (Mass.).
Stern, C. and Schaeffer, E., 1943: On wild-type isoalleles in *Drosophila melanogaster.* *Proc. Nat. Acad. Sci. Wash.* **29**: 361–7.
Steyskal, G., 1968: Number and kind of characters needed for significant numerical taxonomy. *Syst. Zool.* **17**: 474–7.
Størmer, L., 1955: Merostomata. In: Moore, R., ed., *Treatise on invertebrate paleontology. Part P, Arthropoda vol. II:* 5–41. Geol. Soc. Am., Univ. Kansas Press, Kansas.
Strombach, W., 1968: *Natur und Ordnung.* Beck, München.
Stümpke, H., 1964: *Bau und Leben der Rhinogradentia.* Fischer, Stuttgart.
Sturtevant, A., 1954: Social implications of the genetics of man. *Science* **120**: 405-7.
Sullivan, D., Palacios, R., Stavnezer, J., Taylor, J., Faras, A., Kiely, M., Summers, N., Bishop, J. and Schimke, R., 1973: Synthesis of a desoxyribonucleic acid sequence complementary to ovalbumin messenger ribonucleic acid and quantification of ovalbumin genes. *J. Biol. Chemistry* **248** (21): 1530–9.
Szilard, L., 1929: Über die Entropieverminderung in einem thermodynamischen System bei Eingriffen intelligenter Wesen. *Z. Physik* **53**: 840–56.

Tasch, P., 1969: Branchiopoda. In: Moore, R., ed., *Treatise on invertebrate paleontology. Part R, Arthropoda 4, vol. I:* 128–91. Geol. Soc. Am. Univ. Kansas Press, Kansas.
Teilhard de Chardin, P., 1959: *Der Mensch im Kosmos.* Beck. München.
Teilhard de Chardin, P., 1959: *The phenomenon of man* Harper & Row, New York.
Teilhard de Chardin, P., 1961: *Die Entstehung des Menschen.* Beck, München.
Teilhard de Chardin, P., 1964: *The future of man.* Collins, London.
Thenius, E., 1965: *Lebende Fossilien. Zeugen vergangener Welten.* Kosmos, Franckh, Stuttgart.
Thenius, E., 1969a: Stammesgeschichte der Säugetiere (einschließlich der Hominiden), In: *Handbuch der Zoologie* 8/12 (1): 1–722. De Gruyter, Berlin.
Thenius, E., 1969b: Über einige Probleme der Stammesgeschichte der Säugetiere. *Z. f. Zool. Systematik u. Evolutionsforschung* **7**: 157–79.
Thenius, E. and Hofer, H., 1960: *Stammesgeschichte der Säugetiere. Eine Übersicht über Tatsachen und Probleme der Évolution der Saugetiere.* Springer, Berlin-Göttingen-Heidelberg.
Thom, R., 1972: *Stabilité structurelle et morphogénèse. Essai d'une theorie général des modèles.* Benjamin, Reading, Massachusetts.
Thompson, D., 1942: *Growth and form.* Cambridge Univ. Press, Cambridge.
Thorpe, W., 1970: Nachwort. In: Koestler, A. and Smythies, J., eds., *Das neue Menschenbild (Engl. original: Beyond reductionism)* 404–9. Molden, Wien-Müchen-Zürich.
Tinbergen, N., 1942: An objective study of the innate behaviour of animals. *Bibl. Biotheoretica* **1**: 37–98.
Torrey, T., 1965: Morphogenesis of the vertebrate kidney. In: De Haan, R. and Ursprung, H., eds., *Organogenesis* 559–81. Holt, Rinehart and Winston, New York.
Troll, W., 1941: *Gestalt und Urbild.* Acad. Verl. Ges., Leipzig.
Troll, W., 1948: *Urbild und Ursache in der Biologie.* Springer, Heidelberg.

Trusheim, F., 1931: Aktuo-paläontologische Beobachtungen an *Triops cancriformis* Schaeffer (Crustacea; Phyllopoda). *Senckenbergiana* 13: 234–43.

Trusheim, F., 1938: Triopiden (Crustacea; Phyllopoda) aus dem Keuper Franckens. *Paläontol. Zeitschr.* 19: 198–216.

Tschulok, S., 1922: *Deszendenzlehre*. Fischer, Jena.

Vandel, A., 1964: *Biospéléologie. La biologie des animaux cavernicoles*. Gauthier-Villars, Paris.

Voltaire, J., 1759: *Candide ou l'optimisme*. Miret, Paris.

Waddington, C., 1939: *An introduction to modern genetics*. Allen and Unwin, London.

Waddington, C., 1956: Genetic assimilation of the bithorax phenotype. *Evolution* 10 (1): 1–13.

Waddington, C., 1957: *The strategy of the genes*. Allen and Unwin, London.

Waddington, C., 1966: Fields and gradients In: Locke, M., ed., *Major problems in developmental biology*. Acad. Press, New York.

Wainwright, S., 1969: Stress and design in bivalved mollusc shell. *Nature* 224 (5221): 777–9.

Wainwright, S., Biggs, W., Currey, J. and Gosline, J., 1976: *Mechanical design in organisms*. Arnold, London.

Wald, G., 1963: Phylogeny and ontogeny at the molecular level. In: Oparin, A., ed., *Evolutional Biochemistry*. Pergamon, London.

Warburton, F., 1955: Feedback in development and its evolutionary significance. *Amer. Natural.* 89: 129–40.

Warburton, F., 1967: The purpose of classifications. *Syst. Zool.* 16: 241–5.

Watson, J., 1970: *Molecular biology of the gene* (2nd ed.). Benjamin, New York.

Weaver, W., 1948: Science and complexity. *American Scientist* 36: 536–44.

Wedekind, R., 1927: Umwelt, Anpassung und Beeinflussung, Systematik und Entwicklung im Lichte erdgeschichtlicher Überlieferung. *Sitzungsber. Ges. Beförd. Naturw. Marburg* 62: 237–45.

Weippert, G., 1930: *Das Prinzip der Hierarchie in der Gesellschaftslehre von Platon bis zur Gegenwart*. Hanseat. Verl. Anstalt, Hamburg.

Weiss, P., 1939: *Principles of development*. Holt, New York.

Weiss, P., 1969: The living system: determinism stratified. *Studium generale* 22: 45–87.

Weiss, P., 1970a: Life, order, and understanding. A theme in three variations. *The Graduate J.* 8 (suppl.). Univ. Texas Press, Austin.

Weiss, P., 1970b: Des lebende System: ein Beispiel für den Schichten-Determinismus. In: Köstler, A., and Smythies, J., eds., *Das neue Menschenbild* (English original, *Beyond reductionism*) 13–70. Molden, Wien-München-Zürich.

Weiss, P., ed., 1971: *Hierarchically organised systems in theory and practice*. Hafner, New York.

Weizsäcker, C., von, 1958: *Die Geschichte der Natur* (4th ed.). Vandenhoeck und Ruprecht, Göttingen-Zürich.

Wendt, H., 1953: *Ich suchte Adam; Roman einer Wissenschaft* (2nd ed.). Grote Hamm (Westfalen).

Westoll, T., 1949: On the evolution of the Dipnoi. In: Jepsen, G., Mayr, E. and Simpson, G., eds., *Genetics, paleontology, and evolution* 121–184. Univ. Press, Princeton.

Whitehead, A., 1933: *Adventures of ideas*. Cambridge Univ. Press, Cambridge.

Whyte, L., 1960: Developmental selection of mutations (answer to Lewontin and Caspary 1960). *Science* 132: 1692–4.

Whyte, L., 1964: Internal factors in evolution. *Acta Biotheoretica* 16: 33–48.

Whyte, L., 1965: *Internal factors in evolution*. Braziller, New York.

Wickert, J., 1972: *Albert Einstein in Selbstzeugnissen und Bilddokumenten*. Rowohlt, Hamburg.

Wickler, W., 1961: Ökologie und Stammesgeschichte von Verhaltensweisen. *Fortschr. Zool.* 13: 303–65.

Wickler, W., 1965: Die Evolution von Mustern der Zeichnung und des Verhaltens. *Naturwissensch.* 52: 335–41.

Wickler, W., 1968: *Mimikry. Nachahmung und Täuschung in der Natur*. Kindler, München.

Wickler, W., 1969: *Sind wir Sünder? Naturgesetze der Ehe*. Droemer-Knaur, München.

Wickler, W., 1970: *Stammesgeschichte und Ritualisierung (zur Entstehung tierischer und menschlicher Verhaltensmuster)*. Piper, München.

Wiedersheim, R., 1893: *Der Bau des Menschen als Zeugnis für seine Vergangenheit* (2nd ed.). Mohr, Freiburg-Leipzig.

Wiener, N., 1948: *Kybernetik. Regelung und Nachrichtenübertragung im Lebewesen und in der Maschine* (2nd ed. 1963). Econ, Düsseldorf-Wien.

Wiener, N., 1952: *Mensch und Menschmaschine*. Metzner, Frankfurt (Main)-Berlin.

Wiener, N., 1958: *Cybernetics*. Hermann, Paris.

Wiener, N., 1961: Über Informationstheorie. *Naturwissenschaften* 48: 174–6.

Wilmer, E., 1970: *Cytology and evolution*. Acad. Press, New York.

Wilson, J., 1968a: Increasing entropy of biological systems. *Nature* 219: 534–5.

Wilson, J., 1968b: Entropy, not negentropy, *Nature* 219: 535–6.

Winter, R., Walsh, K. and Neurath, H., 1968: Homology as applied to proteins. *Science* **162**: 1433.

Woolhouse, H., 1967: Entropy and evolution. *Nature* **216**: 200.

Zarapkin, S., 1943: Die Hand des Menschen und der Menschenaffen. Eine biometrische Divergenz-analyse. *Z. menschl. Vererb.-und Konstitutionslehre* **27**: 390–414.

Zemanek, H., 1959: *Elementare Informationstheorie*. Oldenbourg, Wien-München.

Zeuner, F., 1946: *Dating the past; an introduction to geochronology*. Methuen, London.

Zietschmann, O., Ackerknecht, E. and Grau, H., 1943: *Handbuch der vergleichenden Anatomie der Haustiere* (18th ed.). Springer, Berlin.

Zinner, E., 1931: *Die Geschichte der Sternkunde*. Springer, Berlin-Heidelberg.

Zuckerkandl, E., and Pauling, L., 1965: Molecules as documents of evolutionary history. *J. theoret. Biol.* **8**: 357–66.

AUTHOR INDEX

SUBJECT INDEX

301

LIST OF ALGEBRAIC SYMBOLS

p.

c	Complexity of a system or subsystem, measured by the number of dependent phenes (measured in homologues)	126
cg	Complexity of a genetic change, measured by the number of altered phenes in the resulting mutant structure (measured in homologues)	166
cp	Complexity required so that a mutated structure would be able to function, for comparison with cg (measured in homologues)	166
D	Determinacy content; the order content or negentropy content of a system (measured in $bits_D$). This is the same as its predictability, which increases with increasing knowledge	9 ff
D	Degree of atomic disorder	4
E	Event: a single event in a message	9
e	Exclusion: the number of events which never actually appear in a message or event although they are included in the potential range of the source	14
F	Fixation; the extent to which a system or feature is fixated. This is proportional to its age and inversely proportional to the amount it has changed (measured in age a and constancy s)	132, 140
h	Degree of homology, as percentage of represented homologues	131
I	Information content of a system (in *bits*)	5
I_D	Indeterminacy content; the factual information content of a system, in the technical sense of its accident content. This is the same as its unpredictability (in $bits_I$) which decreases with increasing knowledge and also decreases as the system becomes more determinative	10
I_I	Maximal information content (in *bits*); the total number of decisions in a system, both accidental and determinative.	10
i	The degree of integration (as percentage of dependent homologues)	127
k	The Boltzmann constant (in ergs $/°C$)	4
L	Law content; the number of non-redundant determinative decisions in a system (in $bits_L$)	13
N	Negentropy; negative entropy (in D)	4
P	Probability of an event or of a condition, or that a supposition is correct	5
P_D	Probability of a particular event, given that a determinative event is expected	7
P_e	Probability of a mutation being a success in terms of natural selection (e = *Erfolg* = success).	80
P_I	Probability of a particular event, given that an indeterminate (accidental) event is expected	7
P_L	Probability of a particular event, given that a determinative and redundancy-free event (part of the law content) is expected	7 ff
P_l	The probability with which we are led to expect the reign of law for a particular event	8
P_{la}	The probability that a supposed law will hold for an event, given the number of instances where the law seems to apply	8
P_m	Probability of mutation, per locus, per reproductive event	80

		p.
P_s	The probability of a phene remaining constant ('steady') over an observed period in the absence of a determinative mechanism to conserve it	131
p	Degree of similarity in proportions (measured in %)	131
Q	Quality of order (as the ratio L/a)	270
R	Redundancy content; number of decisions in a system which can be avoided by systemization (measured in $bits_R$)	12
R'	Visible redundancy content; redundancy of decisions such as results in redundancy of events (measured in $bits_R$)	14
R''	Hidden single redundancy content, due to redundant decisions of the same rank as each other (measured in $bits_R$)	14
R'''	Hidden serial redundancy due to series of redundant decisions (in $bits_R$)	15
r	Relative redundancy content; proportion of redundant decisions in the determinacy content D, especially applied to unsystemized messages and systems in which: $r \approx a - 1$ (measured in D and L)	13
r	Degree of representation of homologous systems (as percentage of related species)	131
S	Entropy (in ergs/°C or in P)	4
s	Degree of systemization; the proportion of dismantled redundancy. This is equal to the ratio of unsystemized redundancy over conserved redundancy (i.e. $= a/r \; syst$). (Conserved redundancy is such decision redundancy as remains after the decisions have been systemized.)	26
s	Constancy ('steadiness') of a system or feature within a definite field of objects of comparison (measured in **r**, **h**, and **p**)	131
Y	Millions of years	131